Tropical Pacific Invertebrates

TROPICAL

PACIFIC

INVERTEBRATES

A Field Guide to the Marine Invertebrates

Occurring on Tropical Pacific Coral Reefs,

Seagrass Beds and Mangroves

by

Patrick L. Colin and Charles Arneson

with photographs by the authors

Dedicated to
Lori Jane Bell Colin
and the memory of Charles E. Cutress

First published 1995

Published by:
Coral Reef Press
270 North Canon Drive, Suite 1524
Beverly Hills, California 90210
U.S.A
Tel. 800-200-7432

Printed in the United States of America by:
Mybar Printing Inc.
2772 Main Street
Irvine, CA 92714

ISBN 0-9645625-0-2

Frontispiece- Reef building corals thrive in the clear, shallow waters of the tropical Pacific. This picture was taken in a shallow pass through the barrier reef at Losap Atoll in the Federated States of Micronesia.

The Coral Reef Research Foundation was established in 1991 to promote research and education on coral reefs and other tropical marine environments. This volume is a contribution by CRRF toward that goal. A portion of the profits from this volume will support activities of the Coral Reef Research Foundation.

A publication of The Coral Reef Research Foundation

Table of Contents

Acknowledgements

The idea to produce a pictoral guide to tropical Pacific invertebrates arose out of two interrelated events, the formation of the non-profit Coral Reef Research Foundation and a contract award to that Foundation by the U.S. National Cancer Institute. Our goal in forming the Coral Reef Research Foundation was to re-establish a marine research laboratory to support basic research on atoll ecosystems in the central Pacific. The idea was to fill the void left by the closure of the Mid Pacific Marine Laboratory on Enewetak in 1983. In fulfillment of that goal we now have two new research laboratories, one in Chuuk and the other in Palau, which serve as bases for further study on marine biodiversity by us and scientists from all nations. The National Cancer Institute research project involves careful collection, database development and identification of marine invertebrate species for cancer and AIDS drug development. This book is a result of our past and current studies and is our first attempt at organizing field notes, photographs and identifications into a volume that can be used by everyone.

We have been fortunate over the last fifteen years to work and live in a number of locations in the western Pacific. The opportunity for us to study, identify and photograph the organisms illustrated here is the result of help by fellow scientists, friends, governments and various research organizations. However, without the permission and interest of local peoples throughout the region, who allowed us to visit, dive and collect in their home waters, our work would not have been possible. While we cannot identify all of these individuals separately, we would first like to acknowledge their contributions and help during the studies that led to the completion of this volume. Needless to say, this book would not be possible without their unselfish support and cooperation.

The formation of the Coral Reef Research Foundation and the establishment of the Chuuk Atoll Research Laboratory helped make most of this work possible and we would like to thank those who helped us early on. These friends include: Donald Benjamin, Mardi Bren, James Cameron, Margo Haygood, Marion Henry, Gale Anne Hurd, The Hurd Foundation, Jim Miller, Ralph Lewin and Lana Cheng.

We would also like to acknowledge those governments, institutions, colleagues and coworkers who have had exceptional involvement with this book. These include, but are not limited to, the following:
Former and present Coral Reef Research Foundation staff: Lori J. Bell Colin, Larry Sharron, Terry Frohm, Chester Gustaf, Peter Schupp, Ron Sjoken, Don DeMaria, Hector Manglicmot, Matthew Mesubed, Emilio Basilius and Chuck Bowling.
Federated States of Micronesia: Chuuk State Government, Marion Henry, Kimiuo Aisek, Gradvin Aisek, Lenny Kolczynski, and crew of the Truk Aggressor. **Republic of the Marshall Islands:** The people of Enewetak Atoll; Scott Johnson, Lisa Boucher Clark, Walt Chidsey and the crew of the Liktanur II. **Philippines:** Gabriel S. Casals, Pedro Gonzales, Domingo Madulid, Marivene Manuel, Hilconida Calumpong, Adrian Zecha and staff of Amanpulo. **Papua New Guinea:** University of Papua New Guinea (UPNG) Motupore Island Research Station, Christensen Research Institute (CRI), Madang Dept. of Environment and Conservation, Max Benjamin, Alan Rabe, crew of the M/V Febrina, Kevin Baldwin, crew of the M/V Tiata, Mike Huber, John Rewald, Sae Gwae, Matthew and Serena Jebb, Larry Orsak, Jim and Debbie Prescott. **Republic of Palau:** President Kuniwo Nakamura, Division of Marine Resources, Department of Resources and Development; former President Ngiratkel Etpison, the Governments of Koror, Melekeok, Ngatpang, and Ngchesar States, Shallum and Mandy Etpison, Kevin Davidson and crew of the Palau Aggressor. **Indonesia:** Hanny and Ineke Batuna, Graham Usher, Roman Palete, crew of the M/V Serenade, Peter Jennings. **Hong Kong:** Brian Morton, Yvonne Sadovy, George Mitcheson, staff of the Swire Marine Laboratory. **Bahrain**: Jassim al Qasser, Roger Uwate, staff of the Directorate of Fisheries.

United States: Bruce Carlson, Marjorie Awai, Phil Helfrich, Jack Randall, Chuck Nicklin, Bonnie Pelnar, Ali Baradar, Nick Holland, Bill Hamner, Fritz Schmitz, John Faulkner, Mike Cameron, Lolita De Palma, Bertha Cutress and Steven Cross. **Guam:** Valerie Paul, Gustav Pauley, Charles Birkeland and Terry Donaldson.

Many of the identifications in this book were made from specimens or photographs by specialist taxonomists. Without their efforts, no order or understanding of the organisms in the sea would come from the chaos superficially apparent when so many forms of life co-exist. We thank the following for their help and identifications:

Sponges- Michelle Kelly-Borges, Ph.D. and John Hooper, Ph.D.
Hydroids and Jellyfish- Dale Calder, Ph.D. and Ron Larson, Ph.D.
Soft corals- Katherine Muzik, Ph.D. (Octocorals), Gary Williams, Ph.D. (Sea Pens)
Hard corals- Gustav Pauley, Ph.D.
Sea Anemones- Hajo Schmidt, Ph.D.
Flat Worms- Leslie Newman, Ph.D.
Polychaetes- Leslie Harris, Ph.D., Kirk Fitzhugh, Ph.D.
Molluscs- Lori J.B. Colin, Terry Gosliner, Ph.D. and Scott Johnson, Roger Hanlon, Ph. D.
Arthropods- Lou Eldredge, Ph.D., Marivene Manuel and William Newman, Ph.D.
Lophophorates- William Banta, Ph.D. (Bryozoa)
Echinoderms- Loisette Marsh, Ph.D., Charles Messing, Ph.D., Gustav Pauley Ph.D., Gordon Hendler, Ph.D.
Urochordates- Francoise and Claude Monniot, Ph.D.s

We would like to thank the following for their review of the manuscript: Nick Holland, Bill Hamner, Jim Miller, Gale Anne Hurd, Michelle Kelly-Borges (sponges), Lori J.B. Colin (molluscs) and Charles Messing (crinoids).

We would like to thank Scott Johnson and Mike Severens for providing a number of additional photographs to fill out coverage of certain groups. The following photographs are credited to them:

Scott Johnson- pages: 140, 160, 161a,b, 243c, 244c; pictures: 655, 633, 634, 674, 682, 743, 744, 954, 955, 971, 998, 1006, 1062, 1067, 1071, 1166, 167, 1169, 1171, 1174, 1179-84.

Mike Severens- page 206d; pictures: 1053, 1060, 1061, 1065, 1066, 1068, 1070, 1230.

We would also like to thank Mandy Etpison, who drew the line illustrations for this book.

Bonnie Pelnar showed us how to print this book in the United States using digital images. Without her advice and assistance this project might never have been completed and we owe her our sincere thanks. John Hull, Dan Weber, Arlene Mercado, Clayton Watson and the staff of Mybar Printing, Irvine, California contributed significantly to the planning, quality and completion of this book. Steve Miller of Miller Imaging, Santa Monica, California helped us scan the original photographic slides to Kodak Photo CD. The entire book was printed from digital images using the latest printing techniques. The traditional printing process is notorious for producing waste that is harmful to the environment. Mybar Printing is to be commended for its environmentally safe printing procedures. They utilize a number of techniques to reduce many hazardous by-products of the printing process: soy based inks, alcohol-free printing and aqueous coating. Soy-based oils are renewable, biodegradable and have lower solvent emissions than petroleum inks. Alcohol-free printing allows the water used in the printing process to be recycled. Unlike varnish, aqueous coating is biodegradable, renewable, nonvolatile and non-hazardous.

We welcome notification of errors detected in this book.

ENVIRONMENTALLY PRODUCED™

Introduction

"The only solid piece of scientific truth about which
I feel totally confident is that
we are profoundly ignorant about nature."

Lewis Thomas, from *The Medusae and the Snail,* 1979

Shallow tropical waters are remarkable places. They nurture organisms which span the scale of evolution, and communities which have existed in varying forms for tens to hundreds of millions of years. The coral reefs, seagrass beds, algal flats and mangroves hold a diversity of animal life more concentrated than anywhere else in the world's oceans.

This book covers the tropical Pacific, that vast region running from Hawaii and French Polynesia in the east to the Philippines, Papua New Guinea and Indonesia in the west. It includes, in addition, all of Micronesia (Marshall Islands, Gilbert Islands, Caroline Archipelago, Nauru and the Marianas), the remainder of Melanesia (Fiji, Solomon Islands, Vanuatu and others) and the remainder of tropical Polynesia (Samoa, Tonga and others). This is the area of the highest marine biodiversity on earth and that diversity is greatest in the shallow waters of the region. It is also an area where new scientific discoveries occur almost every day. The last great undefined marine faunas (primarily invertebrates) can be found in the western Pacific and Indian Ocean, what is collectively called the Indo-west Pacific. Species still undescribed to science can occur in snorkeling depths literally right off the dock or beach.

Of the estimated one million described species of animals, 95% are invertebrates. Most of these are terrestrial insects, members of the Class Insecta in the Phylum Arthropoda. In the sea, coral reef invertebrates are comprised of single-celled animals, principally foraminifera (Protozoa), and multicellular species such as sponges, cnidarians (hydroids, jellyfish and corals), various worm-like animals, molluscs (nudibranchs and sea shells), crustacean arthropods (barnacles, shrimp and crabs), echinoderms (starfish, urchins, crinoids and sea cucumbers) and the ascidians (seasquirts). The division of the animal kingdom into vertebrates and invertebrates can be misleading. The invertebrates, aside from lacking a vertebral column ("backbone"), share no other distinguishing characteristics among them, except common biological characteristics which are also found in the vertebrates. Vertebrates evolved from invertebrates, and the division of animals into vertebrates and invertebrates does not imply an equality of diversity, evolution or complexity between the two groups.

The idea that invertebrates are primitive animals is a popular misconception. These species have continually evolved and adapted to changes in the environment

Opposite- This photo of the coral reef off Pescador Island, in the central Philippines, shows a multitude of invertebrates, including the barrel sponge *Xestospongia testudinaria,* which has feather stars and many sea cucumbers on it. There is blue coral in the lower corner and a variety of other soft and hard corals in the background.

and to changes in the species composition of marine communities for periods far longer than any vertebrates. Today it is understood, even a simple-looking sponge may have highly evolved chemistry which enables it to compete effectively with "higher" animals for living space on the reef. Invertebrate life forms were present on earth during the Precambrian period, at least 600 million years ago. Many groups of marine invertebrates living today have genera that can be found in the fossil record dating back 150 million years to the Triassic period. The long course of evolution and speciation in the large, relatively stable tropical marine environment has produced a level of species diversity found elsewhere only in tropical rain forests.

In this book we cover the marine invertebrates which occur in the tropical western Pacific, primarily species living on coral reefs and nearby habitats. The marine tropics, characterized by coral reefs and mangrove forests, are usually determined by the occurrence of water temperatures in excess of 68° F (20° C). This is generally between the Tropic of Capricorn (22.5° S. Latitude) and the Tropic of Cancer (22.5° N. Latitude); however, tropical conditions exist outside these rough boundaries as long as the seawater temperature does not drop below 68° F. for any substantial period of time. In the central portions of the tropical Pacific, water temperatures vary only 5° to 10° F. over the year, with the annual range from the mid 70's to high 80's F. Generally, within the first 300 feet of the water column, the depth range where most of the animals covered in this book occur, water temperature varies little. Only where there are areas of occasional upwelling of deep ocean water would significantly colder water be found in the tropics.

The warm water of the tropical Pacific is maintained by direct overhead sunlight and seasonal wind patterns that drive the main currents of the central Pacific. The northeast tradewinds of the Northern Hemisphere and the southeast tradewinds of the Southern Hemisphere blow surface water toward the equator across the vast reaches of the open Pacific. The water is heated as it moves west from the continental coasts of North and South America. The surface currents of the tropical western Pacific are dominated by these wind driven currents. The North Equatorial Current flows from east to west, north of the equator, passing through northern Micronesia until it reaches Vietnam and the Philippines where it splits into north and south components. The southern flow turns east near the equator and heads back across the Pacific as the Equatorial Counter Current; the north flow becomes part of the Kuroshio Current. The South Equatorial

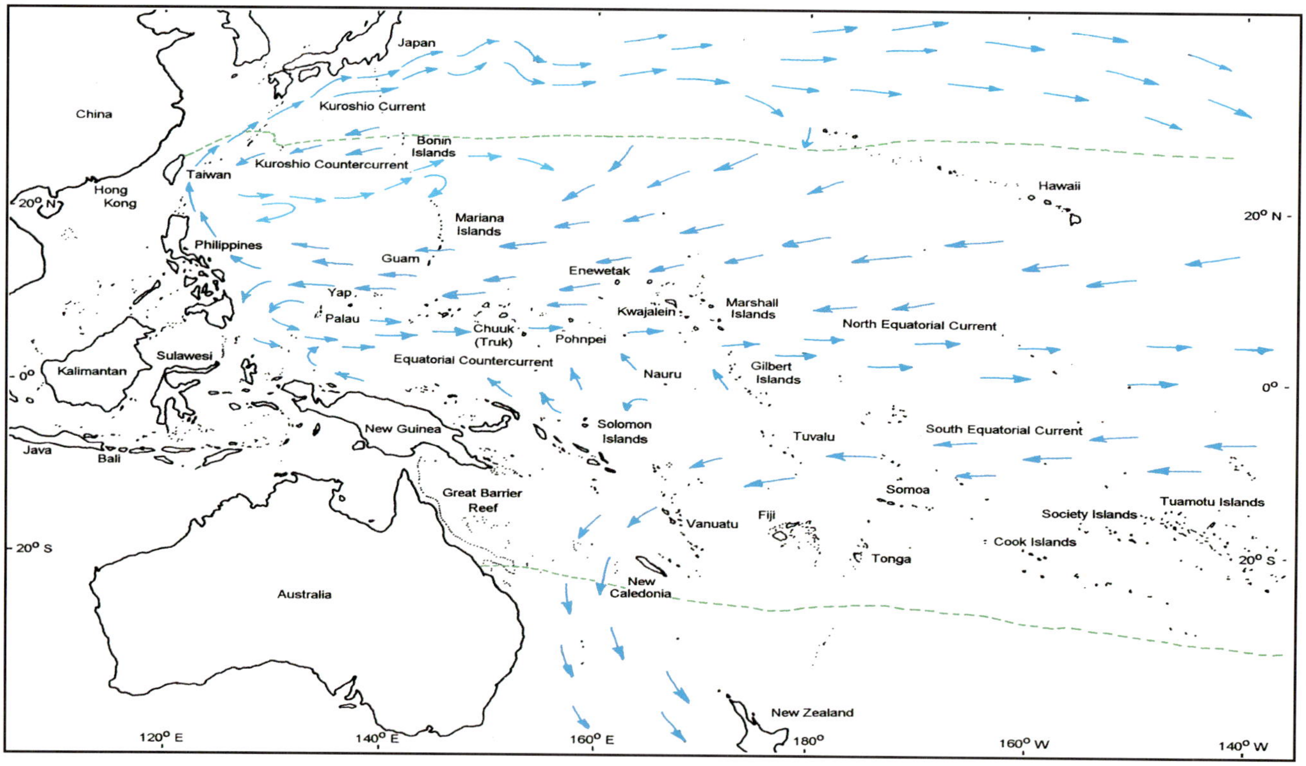

Above- The Pacific Ocean covers nearly one third of the globe as is apparent in this view centered over the Marshall Islands. The approximate limits of the tropics, where water temperature is not below 68°F (20°C) for prolonged periods, are indicated by dashed lines. The directions of major currents are indicated by arrows. Warm currents are shown with red arrows, equatorial currents running east to west and countercurrents running west to east. Colder currents are shown with blue arrows.

Current runs east to west south of the equator, and turns and splits when it reaches the Papua New Guinea-Solomon Islands region. Part of this water is directed back east at the equator to join the Equatorial Counter Current, while part flows south along the east coast of Australia.

To complete this generalized current pattern for the region, many other lesser currents occur on the periphery of the wide Pacific. Water moves east through the Straits of Malacca, between the Malay Peninsula and western Indonesia (Sumatra), turning north between Borneo and Sulawesi. Similarly, water moves eastward through southeastern Indonesia and passes through the Torres Strait between Australia and New Guinea before reaching the Coral Sea. Currents are quite complicated within the northern islands of Papua New Guinea and still poorly understood. In addition to the wind driven currents, tidal fluctuations also play a major role in local current patterns.

Tidal variation is important to all shallow water communities. The area of bottom exposed between high and low tides is known as the intertidal or littoral zone and the area below tidal fluctuation is called the subtidal zone. The distance between high and low tides, the tidal range or amplitude, is not the same in all areas of the tropical Pacific. In the central part of Micronesia, for example, average tidal amplitude is about 3 feet, while in eastern and western Micronesia it is closer to 6 feet. Tides are caused by the gravitational pull of the moon and the sun on the ocean surface and differences in tidal amplitude are usually a function of time of year and of underlying ocean basin and local bottom topography. Tides in some areas are semidiurnal (two highs

Above- This aerial view of Pakin Atoll shows the elements of a coral atoll, the ring of barrier reef with islands surrounding a lagoon and the deep ocean outside. Pakin lacks a deep-water pass.

Gorgonians, feather stars and sponges inhabit this coral outcrop. Strong tidal currents bring planktonic organisms, some of which become food for sedentary reef organisms, to the outer edges of passes and points.

This aerial photo shows a section of the leeward barrier reef at Kuop atoll, just South of Chuuk, in Micronesia. Two passes through the reef are visible. In the background are the high islands of Chuuk Lagoon.

and two lows a day), in others it is diurnal (one high and low a day). A tide table can be consulted for most areas to ascertain the nature of the tides there.

Marine communities are primarily affected by tides in two ways. First, the rise and fall of the tide exposes and submerges shallow areas within the intertidal range. On coral atolls and many fringing reefs, the intertidal zone includes much of the top part of the reef or reef flat. Second, waves breaking on the reef flat when tides are high cause a phenomenon known as "wave-pumping" in which the breaking wave washes up onto the reef flat and transports water across the reef. This produces a directional current on the windward sides of atolls across the reef from ocean to lagoon. At Enewetak Atoll in the Marshall Islands, for example, such reef top currents were reported to be five feet per second or slightly more than one knot. The current speed depends on the height of the tide, becoming less as the tide is lower, until at some point the flow of water across the reef flat is interrupted as the reef flat is exposed by lowering tides. Tides also produce currents in atoll and barrier reef passes which alternate direction, incoming and outgoing, with the tidal cycle. The passages between ocean and lagoon may be fairly shallow or hundreds of feet deep, but all serve as conduits to allow the flow of water back and forth between lagoon and ocean in response to the tides. The strength of the tidal current depends on the stage in the tidal cycle, and there are points in the cycle at which the lagoon and ocean are at approximately equal tidal levels, and the currents stop. Animals that

Above- Onang Island in Chuuk is a typical low coral island on the barrier reef of a coral atoll. Such islands are no more than several feet above sea level and are made of reef rubble and sand thrown up by storms. Soils on such islands are usually poor, but the coconut palm thrives in this environment.

The calcareous algae *Halimeda* produces large amounts of calcium carbonate in the form of plate-like skeletal material. These plates are a major component of reef sediment in the tropics.

This photo of three species of hermatypic corals growing right next to each other illustrates the intense competition for space on the reef. The corals maintain these boundries either through use of chemicals or nematocysts.

are adapted to filter feeding and have excellent means for attachment inhabit these areas of heavy current.

When most divers or snorkelers think of the tropical Pacific, coral reefs immediately come to mind. In ecological terms, coral reefs are referred to as a community. Communities are an assemblage of plant and animal populations occupying a given area. Marine communities of the shallow tropical Pacific can be grouped into several different types, based either on a dominant organism (seagrass bed, coral reef) or some conspicuous non-biological component within that community (rocky shore, sand slope, mud flat). These somewhat arbitrary divisions are a simplistic but useful way to group communities. Still it is important to remember that such labels give no indication of the overlaps, relationships and differences which occur among the organisms identified with these particular communities.

Among shallow tropical communities, coral reefs are the most complex and species diverse. The corals that build reefs (called hermatypic or stony corals) are generally colonies of hundreds or thousands of indvidual coral polyps. Such corals produce the framework of the reef. These corals thrive in relatively shallow, clear water that allows light penetration. The light enables photosynthesis by the symbiotic algae (zooxanthellae) contained within the coral polyps. The algal cells provide essential nutrients to the polyps that help the coral polyps to grow and deposit calcium carbonate in sufficient quantity to build the massive skeletal material that is the reef structure. There are other corals, known as ahermatypic or non-reef building corals, which lack zooxanthellae and do not produce a significant amount of calcium carbonate. Ahermatypic corals are usually found in shaded habitats where they still play an important role in the reef ecosystem. Other organisms such as calcareous algae and the foraminifera also produce large amounts of calcium carbonate on or near reefs. The green calcareous algae *Halimeda* produces huge amounts of flake carbonate material which forms its own habitat, "*Halimeda* beds", in and around reefs. The coralline red algae are generally pink in color and they are important binders of fine reef sediment at all depths. In very shallow water coralline red algae form a ridge of calcareous material which helps to break the force of waves on the reef. Calcium carbonate is produced in lesser amounts by shelled creatures, such as molluscs and echinoderms.

The development of a coral reef is a constant interplay between growth and destruction of the reef structure. Growth rates of reefs are the rate at which the calcium carbonate matrix of the overall reef increases towards the surface. Such rates have been estimated, from a variety of sources, to arrive at a generalized figure on the order of one half of an inch per year. Individual corals can grow at rates much higher than that figure, but these rates do not apply over large areas of reef. The fine branching *Acropora* corals may increase the length of their branches as much as 10 inches (25 cm) a year while a table-like coral *Acropora hyacinthus* grows outward about six inches (15 cm) a year. Reef growth is not just a factor of how fast corals can grow, it is controlled by many factors; for example, death of individual coral polyps, storms, predation by other animals (e.g. crown of thorns starfish) and coral boring organisms which weaken the calcium carbonate matrix of individual coral colonies.

There is fierce competition for space on most coral reefs, both among the corals themselves and with other attached reef organisms. Stony corals fall into three general types; the branching corals, massive corals and plate corals, each with particular advantages when it comes to occupying space on the reef. The branching corals, typified by the genus *Acropora*, are

Above- Idealized cross section of a typical Pacific Ocean barrier reef, near a high island, from shore to the drop off. There are distinct zones along this transect, with mangroves and beach inshore, a lagoon with seagrass coral patches and pinnacles, then the barrier reef with reef flat and fore reef slope.

A shallow water community on a fringing reef, near an island, north of Madang, Papua New Guinea. Soft corals (*Sarcophyton*) and stony corals (species of *Acropora*) are visible in the foreground. The reef grows almost to the shoreline. Many of the very shallow water reef organisms living here will be exposed to air during low tide. Vegetation on land is typical of many off shore reef islands in the tropical Pacific.

the fastest growing, but are relatively fragile, prone to destruction by storm waves and surge. The plate corals, found in many genera, are slower growing, but their flattened or convoluted plate structure allows them to capture light in deeper areas and to grow over nearby corals, cutting off their competitors' light. Some, such as species of *Turbinaria,* are well adapted to low light and high levels of siltation, and the lushness and density of their growth in murky water is often stunning. The massive corals, the brain and star corals, form large heads and clumps reaching ten feet high, but these are the slowest growing. They have the advantage, though, of being the strongest and most resistant to storm damage. In clear tropical water corals dominate other sessile invertebrates and cover large areas of available substrate.

Coral reefs exist in several forms, and these can be grouped into somewhat arbitrary and familiar categories; for example, fringing reefs, barrier reefs, atoll reefs, and patch reefs, among others. The interplay of many factors such as the substratum available for growth, water depth, biological diversity and climate determine the geomorphology and community structure of reefs. Barrier reefs generally occur offshore at continental margins, fringing reefs hug the coastline, and patch reefs are found in shallow water on continental shelves or lagoon bottoms.

Atolls have a number of different reef types. A true coral atoll rises out of deep water and is made up of a ring of reef and low islands, surrounding a lagoon of moderate depth. There are often secondary reefs within the deep water lagoon, either large structures (patch reefs), or high relief small reefs rising to near the surface (pinnacles). Atolls vary in size from less than a mile to over thirty miles across enclosing lagoons of 500 square miles or more in area. There is a positive relationship between the size of the lagoon and its maximum depth; although maximum depths seldom exceed 250-300 feet (75-90 m). The reef may be cleaved by one or more channels connecting the deep ocean with the waters of the lagoon. These "passes" vary in depth from only a little deeper than the top of the reef to a depth reaching almost the maximum depth of the lagoon. Atoll islands are produced from coral blocks, rubble and sand thrown up from the reef primarily during storms. The islands sit atop the barrier reef and are usually no more than about ten feet above sea level. There are many variations on this general plan of a coral atoll, but the basic elements remain in all cases.

Charles Darwin first proposed the widely accepted theory that coral atolls are built upon the remnants of a subsiding volcano and coral growth has kept pace with this subsidence over millions of years to produce the structure which we see today. The reefs around the subsiding volcano go through stages of a fringing reef, barrier reef and finally a true atoll, as the volcano submerges. Actual examples of these stages in atoll formation can all be seen today in the eastern and central Caroline Islands. Kosrae, the youngest of the high islands, has a fringing reef close to shore, lush with coral growth down its flanks, while Pohnpei, further west and older, has reached the barrier reef stage with a rich lagoon a few miles wide between the island and barrier reef. Chuuk, the next high island west, is a classic "almost atoll" with only the remnant peaks of the volcanic basement still protruding above the surface in the central part of the lagoon. Elsewhere in the Carolines true atolls occur, with no trace of their volcanic origin remaining on the surface; here the volcanic rock (basalt) lies a thousand feet or more below sea level.

A different situation exists along a shoreline which is rising, due to tectonic movement, rather than sinking (as occurs in atoll formation). In such a condition, fringing reefs form along the rising shoreline, then are lifted above the sea and become fossil reefs. New reefs grow at the sea surface and are later, lifted again. The best example of such a situation occurs along the Huon Peninsula of eastern New Guinea. There, fossil reefs sit like giant stairsteps on a sloping mountainside and several hundred thousands of years of reef growth and evolution are fossilized. This area has proven a bonanza for geologists studying the growth and evolution of reefs.

Over the area of any reef there are different environmental conditions. The reef can usually be divided into a series of zones running parallel on the long axis of the reef. The zones are based on physical factors such as depth, exposure to waves, turbidity and sediment. Immediate differences are apparent between outer and inner reefs, such as on the ocean and lagoon sides of barrier reefs. On the ocean side of the reef, the water is usually clearer with much greater wave action. Inside the lagoon water is calmer, but murkier.

On the windward (the side facing the prevailing winds) side of atolls, the outer slope of the barrier reef is typically steep, rising out of oceanic depths to the depths where corals can grow (200 feet or so in most cases), then gradually levelling towards the reef top. This zone is called the fore reef, often divided into "deep" and "shallow" portions. The deep fore reef is an area where organisms dependent on light for calcification or photosynthesis have adapted, usually by flattening, to increase their surface area exposed to light. Hermatypic corals become plate-like at depth and a species, such as *Porites rus*, changes from pillars in shallow water to plates at greater depths. Shallow fore reefs are exposed to at least moderate wind and waves and have an area of characteristic geomorphology

Above - The sequence of formation of a coral atoll as visualized by Darwin's theory. In the top view, a gradually subsiding volcanic island has a fringing reef around it. As the volcanic basement sinks further, the fringing reef grows seaward to become a barrier reef and a lagoon is formed between the island and the reef. In the next stage, the "almost atoll", the basement has subsided so that only a few remnants of the original volcanic island remain and the lagoon is broad and deep. In the final stage, the volcanic island is buried under thousands of feet of coral limestone and only the outer rim of reef and islands remains.

Above- Enewetak Atoll. Spur and groove formation is found on the windward shores of atoll barrier reefs. The formation has alternating sand channels (grooves) and fingers of rock (spurs) facing toward the open ocean which is visible in the lower center of the photograph. The pink coralline algal ridge occurs in the area where the surf is breaking, partially obscured by the white water.

known as the "spur and groove" zone, a series of alternating rocky fingers and channels perpendicular to the reef face. The spur and groove zone is actually produced in response to wave action, both through active growth of the reef in response to wave action and by erosion from sand carried by the waves. Spur and groove formation serves to dissipate the energy of incoming ocean waves by breaking up the momentum of the water in the wave. It has been reported, for example, that the spur and groove system on the windward (eastward) side of Bikini Atoll dissipates 95% of wave energy, with the remaining 5% going to wave pumping and maintenance of water level on the reef flat higher than that of the ocean.

Above- This leeward barrier reef rises almost vertically from many hundreds of feet deep to within inches of the surface. The shallow reef is diverse and home to many species of soft and hard corals, some of which are vulnerable to periodic damage by storms when the normal winds change direction.

At the point where the spur and groove meets the reef flat on windward reefs, there is an exceptionally interesting community, the coralline algal ridge. The coralline algal ridge is an assemblage of various coralline red algae. They occur in what is usually the most turbulent water found at coral atolls. The algae grow as a thin crust or as small fan-like forms. They deposit calcium carbonate which produces the slightly elevated algal ridge. A well developed algal ridge starts on the shallow tops of the spurs and continues onto the reef flat, with the various species of corallines having distinct zonation within the ridge area. The healthy coralline algal ridge produces a sponge-like rock structure with channels, cavities and chambers honeycombed into the entire ridge. Animals found here are well adapted for hanging on in the face of extreme wave conditions. Some, such as the spiny lobster *Panulirus penicilatus*, may hide in the reef during the day and then move onto the reef flat at night to feed. Others, such as the sea urchin *Echinometra mathei*, dig deep grooves into the rock on the spurs where they are protected from wave action which would otherwise rip them away from the bottom.

The leeward or downwind sides of coral atolls are quite different. The reef flat is often narrow, without an algal ridge, and then drops precipitously on the ocean side. The leeward dropoff is usually the steepest slope found anywhere at the atoll, and is often undercut with many caves, ledges and crevices in the reef face. The diversity of organisms found in the shelter of the caves and under ledges is high. Most of these organisms occur only in this environment.

Above - This is a section of the west reef of Palau near the area called Blue Holes. The leeward dropoff of most coral atolls is a near vertical escarpment starting only a short distance from the reef flat; it is visible in this photograph where the dark water begins. The large area of back reef sand is typical for coral atolls. This area is home to many burrowing organisms and juvenile fish.

The reef flat on both windward and leeward reefs is often a rocky pavement, flat, with only small grooves and crevices in its surface. A number of organisms occur only on the reef flats. Some remain on the reef flat even at low tides when it is exposed, quite often hiding in crevices. There are other animals which migrate onto the flat to feed only when the tide is high.

Behind the reef flat there is usually a shallow coral area which gives way to a sandy slope as the depth increases. These sediment bottoms can be areas with interesting organisms and extensive biological activity. Many invertebrates remain hidden in the sand during the day. Gastropod molluscs, such as miter shells and cone snails, emerge from the sand to forage at night. A number of sea anemones are found only on sediment bottoms. One of the most interesting is the large *Stichodacyla haddoni,* which usually harbors a small community of commensal crabs, shrimps and anemonefishes; the anemone providing the basis for the presence of the other animals on an otherwise barren bottom.

Sediment bottoms are also home to a diverse burrowing infauna. These organisms live out of sight both day and night, and under normal conditions never emerge from the sediment. Many of these organisms, however, provide evidence of their presence by their activities, which move or disturb sediment. Gastropod molluscs, typified by the auger shells (Terrebridae), leave tracks in the sediment as they plow through the surface layers in search of prey. Many broad sandy areas, both around reefs and in areas

Above- The reef flat at Enewetak Atoll during mid-tide with water flowing over it. Below- A similar reef flat in Papua New Guinea at low tide. Organisms that inhabit this area must be well adapted to strong current and intermittent exposure to air.

This photo of East Fayu Island, northwest of Chuuk in Micronesia shows the large sandy flats that form behind reefs and islands. This sand provides a habitat for many interesting snails, worms and other invertebrates. East Fayu Island is uninhabited and remote. It provides excellent beaches for nesting turtles and relatively undisturbed land for many species of sea birds to mate and reproduce.

Above- This shallow seagrass bed near Tubbataha Reef in the western Philippines is home to many species of juvenile fish and numerous invertebrates. The sand mounds are produced by burrowing polychaetes, shrimp (*Callianassa*) and sea cucumbers. Below- Sandy bottoms and seagrass habitats often merge. This photo shows a common seagrass on a sand bottom.

isolated from hard bottoms, will be found to have conical mounds of sediment packed one next to another over virtually the entire area. These mounds are produced by a group of crustaceans, known as callianassid or ghost shrimp, which form complex systems of mud-lined burrows deep in sediments. The mounds have small depressions at their tops, often with evidence of fresh sand flows down their sides.

Another common back reef habitat is the seagrass bed. The seagrass beds are the meadows of the sea. The organisms which inhabit this area are quite different from those found on nearby reefs. The seagrasses are flowering plants (angiosperms). They usually grow in sediment bottoms with erect, elongate leaves and buried root-like structures (rhizomes). There are several genera of seagrasses, *Cymodocea, Enhalus, Halodule, Halophila, Syringodium, Thalassia*, and *Thalassodendron*. The number of species varies greatly with region. The Great Barrier Reef has 14 reported species of seagrasses. Further north and east, Micronesia has fewer species. Only one, *Thalassia hemprichii,* occurs at some atolls in the Marshall Islands, while Enewetak and Bikini have no seagrasses.

Seagrass beds provide a habitat which is ideal for many animals. There is shelter and cover provided by the often dense blades of the plants, and high production of plant material for food. Many algae also occur within seagrass beds, while the blades of the seagrass provide a surface where other organisms, both plants and animals (epibionts) can grow. These epibionts include macroinvertebrates, such as ascidians, bryozoans and sponges, and plants (diatoms and other algae). These are a food resource which many invertebrates, such as molluscs, holothurians and echinoids, exploit by grazing from the blades.

Seagrass beds are excellent locations for burrowing organisms to live, as the dense mat of rhizomes beneath the sediment surface provides a stabilizing and reinforcing influence on the sediment. Seagrass beds also occasionally have abundant foraminifera or forams living on the sediment bottom. Forams are actually protozoans, single-celled animals which produce a calcium carbonate shell or test, usually less than an inch across. The most often noticed species are like small disks which can be found scattered over the bottom.

Above - This is an aerial view of seagrasss, patch reef and sand bottom habitats near shore in Palau. Mangrove trees are visible along the shore near the upper right corner.

Another tropical marine ecosystem which often co-occurs with nearshore reefs and seagrass beds is the mangrove forest. Generally defined as woody shrubs and trees which grow in salt or brackish water below the high tide level, mangroves can form broad stands or narrow fringes along shores. The larger mangrove swamps can have complex systems of channels, open areas and dense forest. If the tidal range is even moderate, water flows in and out of the mangroves with the tide, producing currents to carry nutrients, which fosters the growth of the epibionts on the mangrove roots. Water in mangrove areas is usually fairly murky, but quite often on a high tide it can be surprisingly clear, with visibility reaching thirty feet or more. This is the best time to observe the often dense growth on the mangrove roots, and is the time when it is easiest to snorkel or dive in mangrove areas.

Above - Seagrasses are angiosperms (flowering plants). Most growth of a seagrass bed takes place through horizontal root like stems. Seagrasses colonize new areas from seeds carried by currents. The seed pod in the center of the photo developed after pollination by male plants of small flowers on female plants. The seed pod holds many seeds.

Land plants are usually not tolerant of high concentrations of salt. The plants which form the mangrove forest assemblage have adapted to this environment. This assemblage varies from place to place and there is a general decrease in the variety and number of species of "mangroves" away from the equator and eastward across the Pacific. Mangrove plants have developed in several plant families, but a single family, the Rhizophoraceae, has a great many of the species considered mangroves. The genus *Rhizophora*, generally known as the red mangrove, is perhaps the most conspicuous and easily identified of the mangroves, with extensive systems of prop and aerial roots. Other important genera include *Avicennia, Bruguiera, Sonneratia* and *Ceriops*.

The environmental conditions under which mangroves grow are extremely difficult. Sediments are usually soft, lack oxygen (anaerobic) and are inundated with salt water. The salinity can vary from near fresh to hypersaline due to rainfall and runoff, or prolonged dry spells. Mangrove plants have adapted to these conditions with elaborate means to excrete large amounts of excess salt and a system of prop roots to support them in the soft sediments. These submerged roots provide a substratum for many other encrusting organisms.

Mangroves are extremely abundant in areas such as Papua New Guinea, Indonesia, the Philippines and other high islands. In Micronesia, mangroves are

Above left- The prop roots of these mangroves near Jellyfish Lake in Palau help stabilize the soft sediment. Above right- Mangrove forests are home to many species of birds and other organisms. They are threatened by development in most parts of the world.

common around the larger high islands, such as Palau, Yap, Chuuk and Pohnpei, but are usually minor communities in coral atolls. East of the Caroline Islands, mangroves are uncommon in the Marshall Islands: Enewetak and Bikini have none, and some atolls in the southern Marshalls have limited stands of only one species considered a mangrove.

It has often been said that the tropical western Pacific is the area of greatest marine biodiversity in the world. While overall there is little doubt that the general statement is correct, there are many examples of biogeographic trends in the region that are not immediately obvious. Present day shallow water marine fauna has in large part evolved from dispersion and subsequent selection of a reef fauna in the Philippine and Indonesian areas that survived the conditions of the last glacial period, which peaked about 17,000 years ago.

Glacial periods produced two significant changes in the marine tropics. First, sea surface temperatures decreased, resulting in a compression North and South of the tropical zones. Second, and perhaps more significant, was the lowering of sea level 300-400 feet by the removal of freshwater from the ocean and its deposition in the polar ice caps (which were much larger than today). This lowering of sea level changed the nature of shallow water communities throughout the region. Broad shallow areas found today were eliminated and replaced by steep slopes (which are today found at 300-400 feet). These slopes provided limited habitat for shallow water species.

Today, the overall center of diversity lies in the Indonesia-southern Philippines region, with steady decreases in the number of genera or species found as one moves away from this area. As species are lost moving away from this center the nature of habitats may also change. For example, the number of seagrass species diminishes moving east from Indonesia so that only a single species remains in the central Caroline Island of Pohnpei and the southern-central Marshallese atolls. That species is lost as one moves further east in the Marshall Islands, only a few hundred miles distant.

Above - In addition to the protection mangroves provide against shoreline erosion, their roots serve as substrate for many benthic invertebrates. Oysters, sponges, ascidians and algae are living on the roots of this mangrove near Madang, Papua New Guinea.

Seagrass habitats show a parallel decline. The fish and invertebrate species also decline, not because they are unable to reach that far east, but due to the lack of the plant species that provide the habitat. Other reef organisms decrease in diversity across this same area, but these patterns are usually defined only in a broad manner. It is surprising how little is really known about the exact limits of most Pacific invertebrate species.

In comparison to the Pacific fauna, the shallow-water fauna of the Western Atlantic and Caribbean Sea is about one third as diverse as the fauna from the Philippine region. Additionally, the Western Atlantic fauna is comprised largely of families and genera that have long-lived planktonic larvae or other means of dispersal; barnacles attached to drifting logs, for example.

The region concerned in this book covers all or portions of four tectonic plates, plates whose movements and relationships have had a considerable effect on the distribution of those invertebrates with limited means of dispersion.

It is difficult to appreciate from the limited number of species we have included in this book the complexity of the marine invertebrate fauna of the tropical Pacific. The reader may wonder why, after over two hundred years of exploration, scientists do not know what organisms occur in the region. The answer lies in the history of exploration, diversity of fauna and remote nature of the tropical Pacific.

Early collections (1700's-1800's) of marine invertebrates came from expeditions sent to map and claim lands of the Pacific. Many of these expeditions had a naturalist on board to collect and observe plants, animals and indigenous peoples. These expeditions were long (up to two years) and arduous. Specimens brought back to Europe and placed in museums were often in bad condition and poorly preserved. Photography was in its infancy, and cumbersome, so specimens were drawn by artists in an attempt to illustrate natural colors. Today these drawings are often the best means of determining the species actually described, as the original specimens may have deteriorated or been unintentionally destroyed. In the 1900's expeditions began producing better and more complete collections of specimens. World War II brought many scientists into contact with the tropical Pacific for the first time. Biological surveys made in association with the nuclear testing programs in the 1950's and 60's, plus the involvement of organizations such as the Smithsonian Institution and universities, resulted in new discoveries and descriptions of Pacific marine biota. The widespread use of SCUBA in the 1960's had tremendous impact on our knowledge of shallow water Pacific invertebrates.

The taxonomy of tropical Pacific marine life is a combination of field work, museum research, literature review and biological detective work. New methods, such as underwater photography, better specimen preservation, electron microscopy and molecular biology aid the taxonomist of today. We are at a point in time today where many, if not most, Pacific invertebrates can be either recognized as undescribed or assigned to a higher level of classification.

Above- The evolutionary tree of invertebrate life. This is an approximate diagram showing possible relationships between the major invertebrate groups. The organisms lowest on the tree appear first in the fossil record. The classification of animals and plants reflects their evolutionary origins. Only through careful study of the structures, life cycles, ecology and biochemistry of marine invertebrates can we learn their correct taxonomic and evolutionary affinities.

About this book.

Tropical Pacific Invertebrates is designed to serve as a general field guide for those marine invertebrates seen or found by divers, snorkelers, naturalists and others in the reefs and shallow water marine environments of the tropical Pacific. We hope the book will prove useful to students of marine biology and ecology, particularly within the region we are covering. We have attempted, within limits, to include most of the common organisms that would be encountered, plus a number of the rarer species which are distinctive and interesting. We have tried to emphasize those groups of invertebrates, such as the sponges and the ascidians, which, because of difficulties in identification, have been treated only superficially in previous popular guides to the Pacific marine fauna. More familiar groups, such as the stony corals and gastropod molluscs, have been the subject of other excellent publications (some of which are referenced at the end of this book) and we have not attempted to include all possible species. For those groups we have chosen to present a selection of typical species which would enable the reader to become familiar with the broadest possible range of tropical marine invertebrates.

Each major division of the animal kingdom, typically a phylum, is included as a separate section and these are presented in approximate phylogenetic order. Each section has two parts: an introduction about the general features of the group, followed by photographs and notes about selected species within that division. The introductory text describes important features of the phylum or division, such as their form (morphology), diversity, reproduction and feeding. Various species are mentioned in these introductory remarks and some photographs are included which illustrate characteristics described. Text remarks are not designed to serve as all inclusive descriptions of each group, but rather to provide a brief introduction to animals concerned, pointing out those things that an observant diver might notice. Readers can look for information at any level they desire in order to gain additional insights into the organisms.

The photograph and notes section illustrates and where relevent, comments on the species selected within that group. Below each photo is a reference number, the scientific name and the country where the photograph was taken. Detailed notes on the taxonomy of organisms and locations of the photographs, indexed by the photo number, are included on the facing page. In many cases the illustrated specimen is backed up by specimens preserved in museum collections and it is these specimens, along with the study of such specimens by a taxonomist, which allow us to assign a scientific name, even if only a genus, to most of the photographs. In some cases, however, it is impossible to determine what species, or even to what genus, the specimen properly belongs and this is noted by the absense of a generic or specific name. Quite a number of the species illustrated are not yet described to science, although their existence is known to taxonomists and descriptions are in the process of being written. The name, taxonomy and locality is followed by information on natural history, geographic ranges, similar species or other interesting notes. Descriptive notes for photographs vary greatly in length, largely related to the amount of interesting relevent knowledge of the organism. Zoogeographic range information is included when it is of interest, but in many cases we can provide no further information regarding the geographic distribution of a species beyond the site records provided by the photographs. We hope the photo notes are an interesting blend of scientific information and personal experience. Each picture has its story, and some of these are included in this section.

In general, most species have been illustrated with a single photograph. Many invertebrates, such as the sponges, show considerable individual and environmental variation and certain examples may look considerably different from the ones we have illustrated. Our photographs attempt to show a typical individual, group or colony in its natural habitat; in a limited number of cases we have provided more than one photo to show radically different forms, or changes during growth, of a single species.

Scientific names are binomials, or two names, written in italics. The first is the genus (the plural is "genera"), always capitalized, while the second is the specific name, which is not capitalized. The genus is a general group into which one or more species fall; a genus may have only a single species within it (monotypic) or many species, but each species within the genus has in common the various characteristics which define the genus.

Some of the species included here are common and well known, their scientific names dating back to Linnaeus and other later authors. Ascribing what we believe to be the correct scientific names to such species is simply a matter of referring to readily available scientific literature which contains authoritative treatment of the particular group in which the species occurs. Other species are much more difficult to identify. Where the letters "cf." (from the Latin *confer*) appear between the generic and specific names, this means the species looks like, but is not necessarily identical with, the particular species named.

Even with the best taxonomy and taxonomists available, there are still many questions regarding the identification of invertebrates pictured. Many of the species illustrated here will prove to be undescribed to science, i.e. they have never been given a scientific name with an adequate description of the species. For others, even a specialist cannot be absolutely certain what a particular specimen is, even with color photographs and a specimen in hand. They may feel it is close to a particular species, but are uncertain whether it is absolutely the same.

Relatively few Pacific marine invertebrates have widely accepted common names. Unless a species has a well known and widely accepted common name, we have avoided using any common names. Many of the common names come from the marine aquarium trade and these are used whenever possible. In a few cases, we have used new common names where they were felt to be particularly appropriate.

How the photographs were taken

Photographs were generally taken using 35 mm single lens reflex cameras in underwater housings with submarine strobes for illumination. In nearly all cases the photographs are of undisturbed animals taken against their natural backgrounds. We have

Scientific Names and Taxonomy

When described (named), each species of animal or plant is given a two part scientific name. This consists of a genus and specific name, such as *Conus geographus*, and is italicized in print. Species are placed into taxonomic categories, each a subunit of the category above it. The major categories break down as follows: Phylum - Class - Order - Family - Genus - Species, and there can be intermediate categories of these, such as subclass or superorder. The idea is to make natural and logical divisions between categories which reflect real biological differences

Taxonomy is the science of identifying, naming and studying the relationships between organisms. Ideally taxonomy creates a system which mirrors the relationships between organisms and how they evolved. Many taxonomists work for museums or universities. Everyone is familiar with the exhibit sections of large museums, such as the Smithsonian Institution, however, most museums also maintain (curate) large research collections. It is these collections and the experience and training of the taxonomists that enables us to determine if an organism is already known, its proper name and to prepare a scientific description of the organism, if it does not already have a scientific name. The descriptions of new organisms are usually published in scientific journals or in books covering a particular group of animals. In addition to their general collections, museums also maintain collections of type specimens, those specimens on which the original description of a species was based.

The scientific names of many reef organisms are occasionally changed. This is not due to the whim of the taxonomist, but because recent study has shown that the species was given a scientific name by an earlier author. Unfortunately, some Pacific invertebrates have been described several times, each subsequent author either not being aware of the previous description or thinking that their specimens represented a new species. In earlier times the type specimens were often not adequately preserved, lost or not sufficiently large to make it easily apparent what the species represented. Workers in earlier centuries did not have the advantages of diving, photography, computers and well maintained collections to document their work and it is easy to understand how confusion occurred with variable species. Where there is more than one scientific name for a species, the oldest name has priority and later names become synonyms and are not used for the species.

tried to choose photographs which are not only a portrait of the species concerned, but which also display some aspect of the biology of the creature. The notes included for each photograph attempt to describe the biological aspects illustrated in the photographs.

Photographs were taken in many different areas, but readers will notice that a number of locality names appear regularly. These are usually localities where we have been able to live or otherwise spend a considerable period of time, often due to the presence of a field marine research facility nearby. These include Madang and Port Moresby in Papua New Guinea, Palau and Chuuk in the Caroline Islands, and Enewetak Atoll in the Marshall Islands. Visits for shorter periods were made to other locations, which because of their species richness were productive areas for photographing marine invertebrates.

One sidelight on the localities is that quite a number of the photographs included here were taken at Enewetak Atoll, the former nuclear test site, in the Marshall Islands. The term "former nuclear test site" has connotations which would imply that the ocean bottom is a nuclear desert, devoid of life. Exactly the opposite is true, since Enewetak has been essentially unfished and its marine life unexploited since the cessation of nuclear testing in 1958. All the organisms that would have been exploited as food by a local Marshallese population have flourished in the absence of human predators, and have approached levels similar to what would probably occur if man did not exist at all.

Collecting Marine Invertebrates

It is an intended purpose of this book to discourage the thoughtless collection of invertebrates, or any other marine life. The coral skeleton broken off today is usually a forgotten piece of trash a year later. Better to leave them where they are able to grow and reproduce, to fulfill their part in the natural scheme of things. While we discourage frivilous collecting, individual specimens are required in the pursuit of scientific studies. Such collections form the basis of much of our knowledge of marine biological science and we owe a debt to the collectors of past centuries who have made much progress towards our goal of understanding the world around us. These efforts are far from finished, as even a brief perusal of this book would indicate, and there is still much left to be learned about the identification of most groups of marine organisms.

About the Coral Reef Research Foundation

The Coral Reef Research Foundation (CRRF) is a non-profit organization founded in 1991 and dedicated to research and education concerning coral reefs and other tropical marine environments. CRRF operates two small marine research laboratories in Chuuk, Federated States of Micronesia and Koror, Republic of Palau. It also conducts basic marine research throughout the western Pacific region. Laboratory projects include elucidation of the marine fauna and flora of the region, and baseline monitoring and mapping of the marine environment.

For further information regarding CRRF and its programs or to order copies of this book please contact::

Coral Reef Research Foundation
P.O. Box 1765
Koror, Palau 96940

Coral Reef Research Foundation
270 N. Canon Dr. Ste. 1524
Beverly Hills, CA 90210
Tel. (800) 200-7432

Sponges*

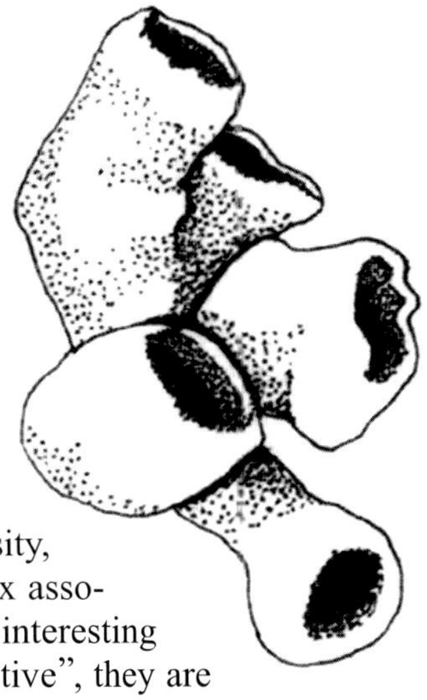

Sponges (Phylum Porifera) are the oldest living group of multicellular organisms (metazoans), first appearing over half a billion years ago. In our region, high species diversity, often spectacular growth forms, exquisite colors, and complex associations with other organisms make sponges exceptionally interesting organisms. While commonly referred to as "simple" or "primitive", they are in fact, very successful and highly evolved organisms which have managed to adapt and survive longer than any other multicellular animal.

The biology of sponges is less well known than that of other organisms. Sponges lack the muscles, nerves and the body organs with which we are all familiar. They have no digestive cavity or mouth. Biological interactions in the sponge take place at the cellular, rather than the level of organs.

Sponges are sedentary filter-feeders. Skeletal support in sponges is provided by a network of hard spicules, flexible fibers, foreign sand or a combination of the three. Spicules are small crystalline structures made of either calcium carbonate (in the mineral forms calcite or aragonite) or silicon dioxide (glass). In addition, collagen and spongin (protein) fibers produce the soft, classically "spongy" skeleton typical of many sponges. The combination of spicule size, type, distribution and their relationship to the fibrous skeleton, is often a primary method used to identify sponges.

Sponges can be thought of as communal associations of cells, loosely arranged to form a network of inhalent and exhalent canals. The inhalent canals originate as small pores (ostia) on the outer surface of the sponge and lead to spherical chambers. These chambers are lined with choanocytes, cells with whip-like flagella that beat in rhythmic waves to pump water through the body in one direction. Water carried into the sponges is filtered for food particles and oxygen and is then expelled through one or several exhalent pores (oscules). Complexity of the canal structure most often increases with sponge size. Sponges vary greatly in growth form and size from thin encrusting sheets a fraction of an inch thick, to large barrels or vases, which may grow to several feet in height and attain a volume of about two cubic yards (about the size of a small cement mixer). From a rather simple body plan, sponges have evolved myriad shapes, sizes, and colors.

*(with contributions by Michelle Kelly-Borges)

Opposite- This blue vase sponge, *Cribrochalina olemda,* is common on reefs throughout much of the region covered in this book. The sponge is frequently found on inshore patch reefs.

These four photographs illustrate some of the variation in growth form and habitat among the sponges. Above left - This encrusting sponge *Aplysilla* sp. shows a beautiful pattern of dendritic excurrent water channels with a single oscule. The incurrent ostia are scattered over the entire surface of the sponge. Below left - This sponge, from Fiji, infiltrates the skeleton of living coral, only the oscules are visible. Above right - This barrel sponge, *Xestospongia,* is one of the largest sponges on coral reefs. It is found in a variety of habitats and colors. Below right - This unidentified sponge grows in small clumps, exposed on the reef. Different internal and external colors and textures are common in many sponge species.

Three classes of sponges are recognized on the basis of their skeletal components. The Class Demospongiae is by far the largest and most diverse group of sponges, it includes the familiar tube, vase, barrel, and fan sponges. Demosponges are characterized by their skeleton, which consists of spongin fibers and siliceous spicules. In some demosponges one or both of these skeletal components may be absent. The Class Calcarea has calcium carbonate spicules in the mineral form calcite. Calcareous sponges are small and delicate with a crunchy texture, due to a lack of spongin and collagen, they occur in limited numbers in all marine environments. In the tropics, calcareous sponges are most often found on vertical reef faces. The Class Hexactinellida, commonly called "glass sponges", has distinct siliceous spicules with six rays. They are seldom found at depths less than several hundred feet and we have not encountered glass sponges within diving depths in the tropical Pacific thus far. A former fourth class, the Sclerospongiae (sclerosponges), has been divided among the first two classes. The species considered to be sclerosponges are known to have had several different origins and do not represent a natural taxonomic grouping. As their name implies sclerosponges are very hard and stony. They typically inhabit caves and shaded ledges on coral reefs. These sponges were believed to be extinct for millions of years until their rediscovery in the 1960's.

Some sponges occur in freshwater habitats, however most sponges are marine. They are found at all latitudes in the marine environment, but reach their greatest diversity on coral reefs in tropical seas. Sponges are also abundant in seagrass beds, mangroves, and other environments.

No one knows exactly how many species of sponges there are, but estimates range from 5,000-9,000 species. As a general rule, each time scientists closely investigate the sponge fauna of a certain region they find it to be more diverse than originally thought. It is certain that in the eastern and central Caroline Islands of Micronesia there are at least 300-400 species of sponges. Palau, in the western Carolines, probably has

600 or more species. In areas such as Papua New Guinea and Indonesia, the total number of species is 1,000 or more. Eventually we will probably find those figures to be conservative underestimates. In addition to the increase in diversity as one moves west across the Pacific, some differences in species numbers can be attributed to available habitats, i.e. the greater the diversity of the habitats, the greater the number of species. Across Micronesia, the lowest diversity in sponges is found in the clear water atolls of the eastern Marshall Islands. These atolls generally lack seagrasses and mangroves and do not have high islands which enhance the diversity of habitats and nutrient enrichment in their lagoons. Areas with high lagoon productivity and many different habitats, such as Pohnpei, Chuuk and Palau, have greater sponge diversity.

The Phylum Porifera provides one of the great challenges to marine taxonomists. The highly diverse western Pacific and Indian Ocean faunas remain the object of study where the basic elucidation and description of species is still far from complete. Consequently, although we have employed the best taxonomic identifications available at the time of publication, a number of the species names will certainly change. The higher taxonomy of sponges (Families, Orders and Classes) is also in a fluid state and many of the assignments to family or order will change in the future. New discoveries by sponge biologists have resulted in the continual revision of groupings to accommodate new information. The reasons for the taxonomic uncertainties are many; the characters used to differentiate many sponge species are few and often variable due to environmental and other factors. Indeed, the concept of what constitutes a species of sponge remains a matter of considerable investigation.

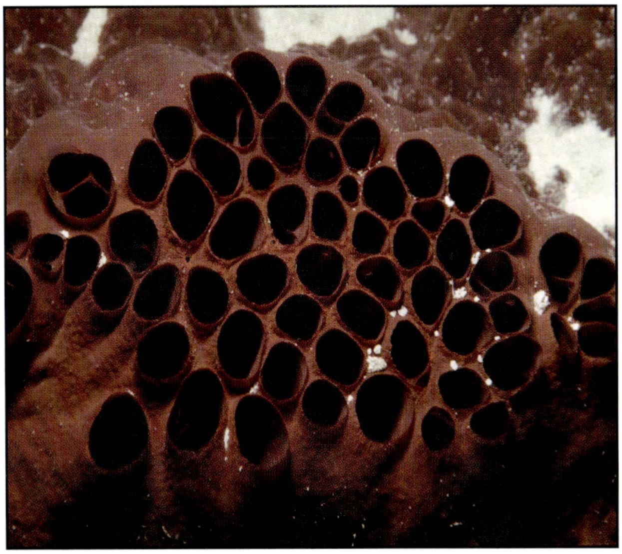

This close up view shows the multiple excurrent oscules of *Spheciospongia vagabunda*, a widely occuring sponge in the tropical Pacific. Water is expelled through these openings after it has been filtered in the body of the sponge.

Sponges reproduce both sexually and asexually. Many individuals are hermaphrodites, producing both eggs and sperm. In sexual reproduction, sperm are released into the water; eggs may be released (oviparous) to undergo fertilization and development in the water or retained and fertilized inside (viviparous) the sponge. In many species there is synchronous release of sperm and eggs triggered by daily and lunar cycles. Fertilized eggs develop into larvae. The larvae swim or creep along the bottom for periods of up to several days, this aids dispersal of the offspring. Asexual reproduction in sponges happens through a variety of methods, including budding, fragmentation and a resting stage known as the gemmule. In many cases, sponges

Above - The external color of many sponges varies with exposure to light. Usually a species that changes color is darker when living in areas exposed to light and pale or white when living in dark areas. One theory for this color change is that the dark color is due to the color of photophilous micro-organisms living on the spong;, in the absence of light the micro-organisms leave the sponge, and the dark color fades. The photograph on the left shows variation in color with exposure to light in *Aaptos* sp. This sponge is living in a partially lighted area, hence its mottled appearance. The photograph on the right shows a normally purple colored sponge living in a cave. It is almost completely white.

Above - Sponges reach their greatest diversity and abundance in tropical seas. In these rich habitats, sponges must compete with other bottom-dwelling invertebrates for attachment sites and hard surfaces. In the picture there are at least eight different species of sponges living on, and competing for, the same two square foot area of wall. Many sponges succeed in obtaining open space through "chemical warfare", producing complex chemical weapons in the course of daily metabolism.

Sponge morphology can be highly variable, both within a species and among different species. The stalked sponge in the photograph, *Podospongia* sp. represents one unusual growth form. This sponge occurs at depths below one hundred feet in Indonesia. Sponges in this genus are uncommon in the tropics, but often found elsewhere in colder water.

can be cut into pieces and each will reorganize itself to survive as a separate individual. This phenomenon forms the basis of the culture of the commercially valuable "bath sponges".

Sponges provide homes for a huge variety of animals including shrimp, crabs, barnacles, worms, brittlestars, holthurians, and other sponges. Perhaps more importantly, but less visibly, a multitude of microbes also lives with sponges inside their canals, between their cells and even inside their cells. We can only guess at the possible trophic (nutritional) relationships between these organisms and the sponge. The external color of many sponges, particularly those with a maroon-brown or greenish surface, is due to the presence of light-loving microbes. In shaded overhangs or caves, the surface color of these sponges disappears due to the absence of the microbes which need light to live.

Sponges must compete with other bottom dwelling invertebrates for attachment sites and living space on hard surfaces. They often succeed through "chemical warfare". Many sponges produce complex protective chemicals as a by-product of their daily metabolism.

Organic chemists have looked at the structure of protective chemicals in some sponges and found the compounds to be diverse and unusual. Some of these chemicals have shown promise as potential sources of new pharmaceutical compounds. Several sponges contain compounds which demonstrate activity against certain tumor cell types. Additionally, other compounds may be effective in treating diseases such as arthritis, heart disease, and AIDS.

While it might seem unlikely, sponges exhibit behavior. *Placospongia* rapidly closes its plate-like surface when touched. *Tethya seychellensis* produces filamentous extensions which it uses to move across the sea bottom. Most sponges are able to vary the rate at which they pump water through their bodies in response to environmental factors. Synchronized spawning behavior in sponges can be a sight to behold. Lunar cycles trigger mass spawning of some species of sponges (and many other invertebrates) on certain days of the year. When triggered, most individuals of a given species begin to spawn at the same time over a large geographic area. Sponges releasing sperm appear to "smoke" while those releasing eggs become sheathed in layers of opaque mucus.

While the majority of sponges will probably not harm humans if handled, there are a number of species which are definitely irritating to human skin. The irritation is the result of chemicals, spicules or both, and individual susceptibility varies greatly. Sponges of the genus *Tedania* have the well earned common name of "fire sponges". In many sponges sharp, spiny spicules easily penetrate skin and cause severe pain, irritation and swelling. We have tried to indicate in the photo notes a few species which we know to be particularly irritating, but this list is far from complete. There are many sponges among those illustrated which will prove to be irritating, so caution is important. In general, it is simply best to leave sponges alone.

Another reason to avoid contact with sponges is that most are fragile and can be easily damaged or dislodged. Large sponges, which may be a hundred years old or more, still have a delicate outer surface. Abrasion of the outer surface can lead to opportunistic infections of the tissue and eventual death.

The fibrous skeletons of species of *Spongia* and *Hippospongia* have been used by humans since antiquity for their water absorbing capacity. Synthetic sponges are unable to match the characteristics of natural "bath" sponges, and these sponges are still in high demand. Bath sponges, of which there are a number of species in our region, lack hard spicules. The best bath sponges have a skeleton of very dense networks of resilient fibers.

No sponges are known to be used as food for humans. Under no circumstances should humans ever attempt to eat sponges. The only instance of ingesting a sponge we are aware of was nearly fatal! Humans aside, sponges are unpalatable to most other marine organisms. However, many nudibranchs, some fishes (angelfishes), and some turtles seem to relish various sponge species; proof of the axiom that (nearly) everything in nature has some kind of predator.

The sponge in the picture can exhibit several different growth forms, even within the same habitat. Three growth forms are visible in the photograph; erect columns, encrusting and small fans. This is one of the reasons many sponges are so diffcult to identify in the field.

The sponge in the picture above is one of many sponges that has other organisms growing on it. There are two other sponges, a hydroid, two species of ascidians and red and green algae living on the sponge in the photograph.

1- *Plakinalopha mirabilis* * Indonesia

2- *Plakinastrella* sp. * Papua New Guinea

3- *Plakortis* sp. * Papua New Guinea

4- *Plakortis* sp. * Papua New Guinea

1- *Plakinalopha mirabilis* * Plakinidae * Homosclerophorida * Indonesia * Manado * fringing reef * 90 ft (27 m). Identified as *Plakinalopha*, this sponge may actually belong to another family, the lithistid sponges. We have kept the old name for convenience until further studies have been completed. This species is kown from Papua New Guinea and Indonesia on reefs from about 30 to 90 feet depth.

2- *Plakinastrella* sp. * Plakinidae * Homosclerophorida * Papua New Guinea * New Britain * offshore reef * 80 ft (25 m). This sponge forms the basis of a tiny biological community we call "turf balls" even though there isn't really any turf involved. The sponge surface is so heavily covered by algae, hydroids, ascidians and other organisms that the sponge itself is no longer visible. These microcommunities are about the size of a tennis ball. The sponge occurs as clumps attached to moderately deep vertical walls of reefs.

3- *Plakortis* sp. * Plakinidae * Homosclerophorida * Papua New Guinea * Port Moresby * Taurama Reef * 60 ft (18 m). Species of this sponge are known as "chicken liver" sponges because of their fleshy texture. These sponges are soft and lack the large spicules typical of other spicule containing sponges.

4- *Plakortis* sp. * Plakinidae * Homosclerophorida * Papua New Guinea * Dyaul Island * 60 ft (18 m). This is another chicken liver sponge. It is typical of the group in that the internal color is virtually identical to the external color.

5- *Plakortis lita* * Plakinidae * Homosclerophorida * Philippines * Pamalican Island * 40 ft (12 m). This chicken liver sponge is common on inshore reefs throughout Micronesia. It often occurs as groups of numerous individual sponges scattered over a small area, rather than as single large individual.

6- *Aciculites ciliata* * Scleritodermidae * Spirophorida * Papua New Guinea * Port Moresby * barrier reef * cave * 60 ft (18 m). This species is an "ear" sponge, a stony flattened sponge which grows in the dark recesses of reef caves. There are a number of poorly-known species in this genus. Their color comes largely from symbiotic organisms, not the sponge itself. Pale sponges are usually found in the darker areas of the caves. At one time these ear sponges would have been considered members of the "stony sponges", the lithistids, but recent work has shown the ear sponges, as broadly considered, to be more correctly separated among other demosponge groups.

7- *Cinachyra schulzei* * Tetillidae * Spirophorida * Papua New Guinea * Madang Ship Channel * 78 ft (24 m). These types of sponges, often called "golf ball sponges", are at first sight so improbable that it is hard to believe they are living animals. Internal color does not generally vary within a species, this one is shocking pink inside. Water is taken in through bright, circular, sieve-like depressions (porocalices) and exits through the large oscules on the top of the sponge. Sediment is trapped between long spicules which protrude from the sponge surface. Internally the sponge is a mass of radiating spicules and fibers, something it shares in common with some other spherical sponges, including the genera *Partetilla* and *Craniella*.

8- *Paratetilla bacca* * Tetillidae * Spirophorida * Palau * marine lake * 3 ft (1 m). This golf ball sponge is always dull yellow internally. It usually grows in caves and beneath ledges on reef areas, but this photograph was taken in a marine lake in Palau. Many organisms, which usually occur somewhat deeper on the reef, are found in some of these shallow lakes.

9- *Craniella abracadabra* * Tetillidae * Spirophorida * Federated States of Micronesia * Chuuk Atoll * Falos patch reef * 40 ft (12 m). The original "punk" sponge, *C. abracadabra* has many soft, flexible spines radiating from its spherical core. The sponge gained its somewhat whimsical specific name when the author of the original description, Prof. Max de Laubenfels, felt this pseudo-magical incantation was an appropriate name because of the bizarre appearance of this sponge. This sponge is found beneath overhangs.

10- *Paratetilla lipotriaenosa* * Tetillidae * Spirophorida * Federated States of Micronesia * Chuuk Atoll * Tonoas Island * 50 ft (15 m). This species of *Paratetilla* seems superficially similar to *Craniella abracadabra* above, but differs in the structure of the spicules which make up the skeleton. Both sponges have radial skeletal morphology. The sponge is growing on a coral of the genus *Porites*.

11- *Paratetilla* sp. * Tetillidae * Spirophorida * Papua New Guinea * West New Britain * 66 ft (20 m). This is another ball-like sponge. Most of these sponges look the same, but their internal color can be used to distinguish between species in the field.

12- *Ancorina acervus* * Ancorinidae * Astrophorida * Federated States of Micronesia * Nama Island * 30 ft (9 m). This sponge does little to reveal its presence beyond the cluster of oscules shown here. The sponge is well camouflaged by algae growing on the surface. The sponge is also yellow to

5- ***Plakortis lita*** * Philippines

6- ***Aciculites ciliata*** * Papua New Guinea

7- ***Cinachyra schulzei*** * Papua New Guinea *

8- ***Paratetilla bacca*** * Palau

9- ***Craniella abracadabra*** * Federated States of Micronesia

10- ***Paratetilla lipotriaenosa*** * Federated States of Micronesia

11- ***Paratetilla* sp.** * Papua New Guinea

12- ***Ancorina acervus*** * Federated States of Micronesia

13- *Myriastra clavosa* * Papua New Guinea

14- *Penares* sp. * Papua New Guinea

15- *Rhabdastrella* sp. * Papua New Guinea

16- *Rhabdastrella* sp. cf. *pleopora* * Federated States of Micronesia

brown internally and is stiff with lots of spicules. A given individual can be quite large and typically this sponge "cements" rocks together, filling in the spaces between large boulder sized pieces of reef rubble. A stony coral of the genus *Pocillopora* is below the sponge. Reddish coralline algae occurs above it.

13- *Myriastra clavosa* * Ancorinidae * Astrophorida * Papua New Guinea * Dyaul Island * flat reef * 23 ft (7 m). These little sponges look remarkably similar to green olives in size and shape. The outer color varies somewhat, from yellow to green to brown.. The sponges attach by their sticky long spicules to many surfaces. The sponges are easily broken loose from the substratum and can roll around, eventually coming to rest and reattaching. Two bright green ascidians, *Didemnum molle*, are visible beneath the sponges.

14- *Penares* sp. * Ancorinidae * Astrophorida * Papua New Guinea * Dyaul Island * reef wall * 33 ft (10 m). This sponge has a hard exterior, and is pale in color due to the low light level where this particular individual was growing.

15- *Rhabdastrella* sp. * Ancorinidae * Astrophorida * Papua New Guinea * Madang * Wongot Island * 66 ft (20 m). The oscules of this sponge are surrounded by a distinctive rubbery membrane with a light colored edge.

16- *Rhabdastrella* sp. cf. *pleopora* * Ancorinidae * Astrophorida * Federated States of Micronesia * Chuuk Atoll * lagoon reef * 60 ft (18 m). The sponge often hosts a community of other encrusting organisms. The ochre internal color can be seen in the lower part of the sponge.

17- *Rhabdastrella* sp. with *Haliclona* sp. encrusting * Ancorinidae * Astrophorida * Federated States of Micronesia * Chuuk Atoll * Pizion Reef * 200 ft (60 m). This is an interesting case of two species of sponges which apparently always occur together. The inner, larger yellow sponge (its color can be seen through the two oscules) is an undescribed species of *Rhabdastrella* (some authorities place it in the genus *Jaspis).* Its outer surface is completely covered (except at the oscules) with a thin layer of a second sponge, a species of *Haliclona.* The *Haliclona* varies in color from a reddish orange, as seen here, to a very pale tan. These two sponges occur on outer reef faces and occasionally on lagoon reefs in Micronesia and the Philippines. The nature of their relationship is unknown.

18- *Dorypleres splendens* * Ancorinidae * Astrophorida * Federated States of Micronesia * Chuuk lagoon * Tonoas Inlet * 33 ft (10 m). This encrusting orange sponge can become massive in calm water. In both cases the sponge surface is distictive, having low, broad, square tubes. This sponge is fairly common in Micronesia where it typically occurs beneath overhangs, often in silty areas.

19- *Melophlus isis* * Ancorinidae * Astrophorida * Guam * Apra Harbor * 66 ft (20 m). The morphology of this sponge is very distinctive, with a globular body topped with a crater-like atrium and "prop legs" attaching it to the bottom. The photo shows a dense group of these sponges in an area of Apra Harbor, Guam, known as the "sponge mound". The surface of the sponge has numerous pits around the incurrent ostia in which live small pinnothirid crabs and brittlestars. The surface of the sponge often has ascidians, algae and other sponges growing on it. This sponge is closely related to *Melophlus (Asteropus) sarasinorum* and could be the same species. *M. isis* is particularly abundant in Micronesia.

20- *Geodia* sp. * Geodiidae * Astrophorida * Papua New Guinea * Dyaul Island * 3 ft (1 m). This *Geodia* sp. was found in a cave in very shallow water. The sponge is stony skin and incompressible. The oscules are grouped in a depression on the upper surface of the sponge.

21- *Thrombus* sp. * Thrombidae * Astrophorida * Papua New Guinea * New Britain * Agu Reef * 93 ft (28 m). This strange sponge looks as if someone stretched a piece of rubber sheet over a rock. The area of reef where this sponge occurred was quite silty. The genus is not particularly well known, but has been recorded previously from Vanuatu, eastern Australia, the northeastern Atlantic Ocean and Caribbean Panama.

22- *Chondrilla* sp. * Chondrillidae * Hadromerida * Papua New Guinea * Dyaul Island * 75 ft (23 m). This sponge adheres tenaciously to the substratum and surface of the sponge is very tough and leathery. It is typically found on the walls of caves and superficially resembles an èncrusting ascidian. This sponge has been seen in Papua New Guinea and Micronesia.

23- *Cliona schmidti* * Clionidae * Hadromerida * Papua New Guinea * Port Moresby * barrier reef * 66 ft (20 m). This is a peculiar sponge which bores into rock on vertical walls or beneath overhangs. The rock into which the sponge bores crumbles fairly easily. The purple/red color is uncommon for this genus and, if touched, the sponge stains hands a blue color.

24- *Cliona* sp. * Clionidae * Hadromerida * Palau * marine lake * 3 ft (1 m). Sponges in the family Clionidae bore into coral heads and reef rock, over time excavating chambers and greatly weakening the structure of the reef.

17- *Rhabdastrella* sp. * Federated States of Micronesia

18- *Dorypleres splendens* * Federated States of Micronesia

19- *Melophlus isis* * Guam

20- *Geodia* sp. * Papua New Guinea

21- *Thrombus* sp. * Papua New Guinea

22- *Chondrilla* sp. * Papua New Guinea

23- *Cliona schmidti* * Papua New Guinea

24- *Cliona* sp. * Palau

25- *Diacarnus* sp. * Federated States of Micronesia

26- *Diacarnus* sp. * Papua New Guinea

27- *Diacarnus* sp. * Papua New Guinea *

28- *Diacarnus spinipoculum* * Fiji

This is particularly true for those coral colonies where clionids bore into the base, reducing the strength of the colony to the point it is easily toppled by storm waves or other physical force. A characteristic of many clionid sponges is the presence of raised sucker-like sieve plates seen on the surface of this sponge.

25- *Diacarnus* sp. * Latrunculiidae * Hadromerida * Federated States of Micronesia * Chuuk Atoll * barrier reef * 40 ft (12 m). This unusual sponge is one of a group of several new species found in the Indo- West Pacific and Red Sea. Important field characteristics in the group are the presence of huge spicule fibers which resemble ligaments, a rubbery texture and large yellow larvae. This species is typically globular, as seen in the photograph, with small mammilate projections over the entire outer surface.

26- *Diacarnus* sp. * Latrunculiidae * Hadromerida * Papua New Guinea * Madang * 75 ft (23 m). This ligament sponge has large branches on which the oscules occur on small siphons near the upper surface of the branches. Individual sponges can be three feet or more across with abundant branches. This species is known only from Papua New Guinea.

27- *Diacarnus* sp. * Latrunculiidae * Hadromerida * Papua New Guinea * Dyaul Island * 60 ft (18 m). This species of *Diacarnus* has many small branches which are oriented vertically and spaced close together. The entire sponge can be large, well over three feet across.

28- *Diacarnus spinipoculum* * Latrunculiidae * Hadromerida * Fiji * Kaimbu Island * 75 ft (23 m). This ligament sponge resembles a large urn or barrel. It is presently known only from Micronesia, Fiji and southeastern Australia.

29- *Placospongia mesobesoides* * Placospongiidae * Hadromerida * Papua New Guinea * Manam Island * 23 ft (7 m). Most species of *Placospongia*, have a horny thick armor that is divided into plates, reminiscent of the hide of an alligator. The ostia and oscules occur in gaps between these paltes. When touched, the sponge contracts, closing the fissures, making it one of the few sponges which "react" rapidly when touched.

30- *Placospongia mesobesioides* * Placospongiidae * Hadromerida * Indonesia * Manado * 80 ft (25 m). While many of the "alligator sponges" are encrusting, firmly attached to the bottom, others form finger-like structures, such as are seen here. The fissures run along the length of the "fingers". The taxonomy of the group is poorly known. There is considerable color variation and the two photographs included here may represent separate species.

31- *Placospongia* sp. * Placospongiidae * Hadromerida * Philippines * Cebu * Santa Rosa * 30 ft (9 m). This rock encrusting sponge is believed to be a species of *Placospongia.* The outer surface is covered with fine white sediment that obscures the ostia. The oscules are large and reveal the inner red orange color of the sponge. The surface is unarmored. This species has been seen in the Philippines and Chuuk.

32- *Atergia* sp. * Polymastiidae * Hadromerida * Palau * Airai Channel * 33 ft (10 m). This interesting little sponge occurs as small lumps of soft sponge encrusting on dead coral and rock in protected inshore areas. Color, within what appears to be the same species, varies from red to near white, perhaps in response to light exposure. This group has somewhat distinctive ostia and oscules which superficially resemble the suckers of an octopus arm. *Atergia* sp. is known from Palau and Chuuk.

33- *Acanthochaetetes wellsi* * Spirastrellidae * Hadromerida * Federated States of Micronesia * Nama Island * reef caves * 40 ft (12 m). This cave-dwelling sponge, commonly called a "sclerosponge", produces a hard skeleton which combines a calcareous basal skeleton with siliceous spicules and an organic matrix of spongin fibers into a rock-like mass denser than that of most corals. Once thought to be extinct, living species of sclerosponges were discovered in the 1960's in reef caves in Jamaica. They were later found in the Pacific and Indian Oceans. Recent work has revealed that the various species, previously grouped together as the Class Sclerospongidea, actually had separate origins. Since the sclerosponges are "polyphyletic" (many lines), the class has been abandoned, although the general term is still used to describe any sponge which produces a calcitic and aragonitic hard skeleton. Sclerosponges are found in at least four orders, including the Hadromerida, Agelasida, Poecilosclerida and Haplosclerida.

34- *Spheciospongia vagabunda* * Spirastrellidae * Hadromerida * Papua New Guinea * Madang * Pig Island * 50 ft (15 m). *S. vagabunda* is the most prominent sponge in this family in our region, previously it was placed in the genus *Spirastrella.* The sponge initially bores into the carbonate substrate of reef rock, but does not excavate as deeply into the rock as members of *Cliona.* Once firmly established in the rock, however, the sponge builds outward, in some areas often forming large masses, more than three feet across. Color varies in *S. vagabunda* from golden brown to near white. This sponge is found in a wide variety of habitats, from shallow inshore areas to deep reefs throughout the tropical Pacific.

29- *Placospongia melobesioides* * Papua New Guinea

30- *Placospongia melobesioides* * Indonesia

31- *Placospongia* sp. * Philippines

32- *Atergia* sp. * Palau

33- *Acanthochaetetes wellsi* * Federated States of Micronesia

34- *Spheciospongia vagabunda* * Papua New Guinea

35- *Spheciospongia* sp. * Palau

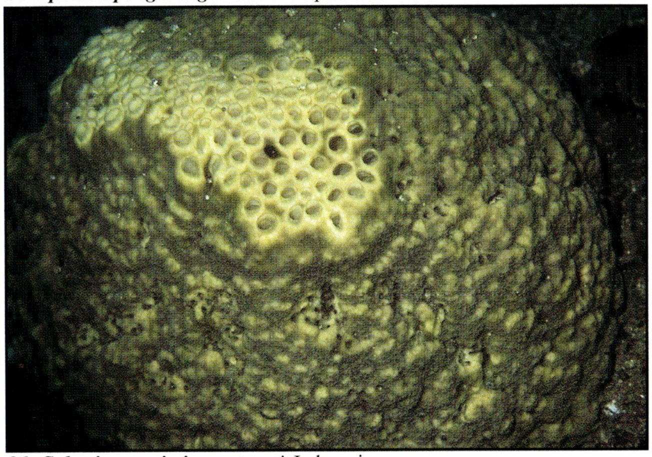

36- *Spheciospongia inconstans* * Indonesia

37- ***Spheciospongia* sp.** * Federated States of Micronesia

38- ***Aaptos* sp.** * Papua New Guinea

39- ***Aaptos* sp.** * Federated States of Micronesia

40- ***Suberites* sp.** * Papua New Guinea

35- *Spheciospongia* sp. * Spirastrellidae * Hadromerida * Palau * marine lake * 3 ft (1 m). This golden brown *Spheciospongia* closely resembles the previous species but is slightly more rubbery.

36- *Spheciospongia inconstans* * Spirastrellidae * Hadromerida * Indonesia * Manado * 66 ft (20 m). This species, photographed in Indonesia, produces large, firm hemispherical growths attached to the bottom. The surface is pitted with keyhole-shaped depressions.

37- *Spheciospongia* sp. * Spirastrellidae * Hadromerida * Federated States of Micronesia * Chuuk Atoll * lagoon reef * 66 ft (20 m). This orange encrusting sponge occurs in thin layers on rock and has large inflated sub-surface channels leading to raised oscules. It is common beneath overhangs on lagoon reefs in Chuuk and probably occurs in much of Micronesia.

38- *Aaptos* sp. * Suberitidae * Hadromerida * Papua New Guinea * Port Moresby * barrier reef * 60 ft (18 m). Species of ***Aaptos*** are always firm with a skeleton that is radial, at least on the outer edges of the sponge. These sponges are often yellow-orange internally, but external color varies depending upon exposure to light. The sponge is yellow-green with darker patches of green in light exposed regions of the sponge. It is found on outer reef slopes, often beneath overhangs, in Papua New Guinea.

39- *Aaptos* sp. * Suberitida * Hadromerida * Federated States of Micronesia * Chuuk Atoll * lagoon reef * 40 ft (12 m). In shallow water this species of *Aaptos* is a deep chestnut brown. On intertidal reefs it is nearly black, in dark caves it is bright yellow, with no difference between the internal and external color. It is common in lagoonal areas of Micronesian reefs.

40- *Suberites* sp. * Suberitidae * Hadromerida * Papua New Guinea * Kavieng * Albatross Channel * 133 ft (40 m).

41- *Terpios granulosa* * Suberitidae * Hadromerida * Papua New Guinea * Madang * Planet Rock * 66 ft (20 m). This blue sponge, *T. granulosa*, is encrusting a small bright red gorgonian. All the gorgonians down the entire side of the reef where the photo was taken were encrusted with this sponge. The site, called "Planet Rock", is a tiny isolated reef rising towards the surface out of water several hundred feet deep about 10 miles offshore of Madang, Papua New Guinea.

42- *Tethya* sp. * Tethyidae * Hadromerida * Federated States of Micronesia * Chuuk Atoll * Satawan Atoll * 86 ft (26 m). The genus *Tethya* is another spherical "golf ball" sponge, superficially resembling *Cinachyra* and *Craniella*. The photographed sponge has several areas of excurrent oscula unlike most species in the genus. White sediment on the surface of the sponge surrounds the incurrent ostia depressions.

43- *Theonella cylindrica* * Theonellidae * Astrophorida * Indonesia * Manado * 170 ft (55 m). This sponge is a "true" lithistid sponge. Lithistids have a special spicule skeleton which interlocks to form a solid rock-like body. Because of the red color, it might be confused with some of the boring clionid sponges, but it is easily recognized that the sponge is growing out from the reef, not boring into it.

44- *Theonella* sp. * Theonellidae * Astrophorida * Papua New Guinea * Port Moresby * Pt. Osbourne * 30 ft (9 m). Many species of *Theonella* are bright red, like this one. These tubes are not soft and flexible as in many pipe sponges, in *Theonella,* they are very hard. The genus *Theonella* is common in the Indo-West Pacific. These sponges have been the source of novel chemicals which have shown desirable bioactivity against a number of human disease agents.

45- *Theonella* sp. cf. *invaginata* * Theonellidae * Lithistida * Indonesia * Manado * 80 ft (25 m). This *Theonella* is a cluster of nearly separate tubes. It has a dusky colored surface. The arms of a brittlestar protrude from one tube while some crinoid arms are also visible extending from the sponge.

46- *Theonella swinhoei* * Theonellidae * Lithistida * Philippines * Cebu * Santa Rosa * 66 ft (20 m). This sponge is common in the Philippines where it occurs on open reef bottom. It can reach a size of more than three feet across, with dozens of individual tubes. The tubes are not as hard as in some other species of the genus.

47- *Theonella* sp. * Theonellidae * Lithistida * Philippines * Cebu * Mactan Island * cave * 80 ft (25 m). This cave-dwelling species is blue in color. The individual photographed has a small flatworm or nudibranch crawling on it, plus a number of whitish encrusting organisms growing on it. The section of cave wall around the sponge has several species of other sponges, small corals and other creatures, which demonstrates the complex nature of such communities.

48- *Agelas* sp. cf *clathrodes* * Agelasidae * Agelasida * Federated States of Micronesia * Chuuk Atoll * Northeast Pass * 100 ft (30 m). This species is sometimes called *Agelas mauritania*, which is incorrect, *Agelas*

41- ***Terpios granulosa*** * Papua New Guinea

42- ***Tethya* sp.** * Federated States of Micronesia

43- ***Theonella cylindrica*** * Indonesia

44- ***Theonella* sp.** * Papua New Guinea

45- ***Theonella sp.cf. invaginata*** * Indonesia

46- ***Theonella swinhoei*** * Philippines

47- ***Theonella* sp.** * Philippines

48- ***Agelas* sp. cf *clathrodes*** * Federated States of Micronesia

49- ***Astrosclera willeyana*** * Papua New Guinea

50- Genus and species undetermined * Papua New Guinea

51- ***Acanthodoryx fibrosa*** * Philippines

clathrodes is an earlier name. The sponge is undescribed but very closely resembles the Caribbean ***Agelas clathrodes,*** hence the comparison in the species name above. Either way, this massive orange sponge is hard to confuse with anything else. It can be as much six feet or more across. The morphology varies greatly, from the large fan-like structures commonly known as "elephant ear sponges" to small encrusting individuals with no particular form. The color and the surface texture varies little with size. The photographed individual has a white didemnid ascidian growing on it, a common occurrence.

49- *Astrosclera willeyana* * Astroscleridae * Agelasida * Papua New Guinea * West New Britain * cave * 50 ft (15 m). This is another cave-dwelling sclerosponge. The skeleton has distinct bands, like tree-rings. The living tissue forms only a thin veneer on the surface. The surface shows distinct astrorhizae, star-shaped depressions where oscules occur.

50- Genus and species undetermined * Anchinoidae * Poecilosclerida * Papua New Guinea * Port Moresby * Basilisk Passage * 60 ft (18 m). This distinctive encrusting sponge is soft and its identity is not yet known. We have only observed it along the southern coast of New Guinea Island where it encrusts on hard objects in areas of high sediment coral reefs.

51- *Acanthodoryx fibrosa* * Coelosphaeridae * Poeciloslcerida * Philippines * Cebu * Mactan Island * 66 ft (20 m). This vermillion red sponge typically forms fans or plates and is somewhat mucousy if touched. It is common in many areas of the Philippines and Indonesia, but is unknown from northern Australia. Its color is consistent throughout this range.

52- *Coelocarteria singaporense* * Coelosphaeridae * Poecilosclerida * Papua New Guinea * West New Britain * 100 ft (30 m). The dark "arms" of this spherical sponge can be open or closed fistules which bear the oscules and inhalent ostia. This general form is found in a number of additional sponge genera which have species that live in the sediment, including *Oceanapia, Aka* and *Orina.* Often, the body of the sponge is completely buried and only the fistulose tubes and oscules reach the sediment surface.

53- *Zyzzya* sp. * Iophonidae * Poecilosclerida * Federated States of Micronesia * Chuuk Atoll * Northeast Pass * 66 ft (20 m). This black sponge bores into coral with initially only the flexible tube-like oscules visible. In older individuals the rock into which the sponge initially bored is essentially gone, leaving the main body of the sponge exposed. This species is found in many areas of Micronesia. The surface of the rock has small algae growing on it with two *Didemnum molle* ascidians in an upper corner.

54- *Monanchora* sp. * Crambidae * Poecilosclerida * Federated States of Micronesia * Chuuk Atoll * lagoon reef * 40 ft (12 m). This is an encrusting sponge with typical sub-surface channels visible. Species of this genus are widely distributed in the Indo-west Pacific, as far south as New Zealand.

55- *Monanchora ungiculata* * Crambidae * Poecilosclerida * Papua New Guinea * West New Britain * 66 ft (20 m). This encrusting sponge is highly inflated and can deflate to a thin sheet of tissue if disturbed. It varies in color, from almost pink to deep red with "frosting" of white, gold or yellow. It is distributed widely in the region, particularly in silty reef habitats.

56- *Crella* sp. cf. *calypta* * Crellidae * Poecilosclerida * Federated States of Micronesia * Chuuk Atoll * lagoon reef * 75 ft (23 m). This semi-burrowing or encrusting sponge is capable of disolving coral. Superficially, this sponge resembles *Acanthodoryx fibrosa*, but it is not as firm and fibrous.

57- *Biemna* sp. * Desmacellidae * Poecilosclerida * Fiji * Kaimbu Island * 33 ft (10 m). This undescribed encrusting species is sticky and incorpo-

52- ***Coelocarteria singaporense*** * Papua New Guinea

53- ***Zyzzya* sp.** * Federated States of Micronesia

54- ***Monanchora* sp.** * Federated States of Micronesia

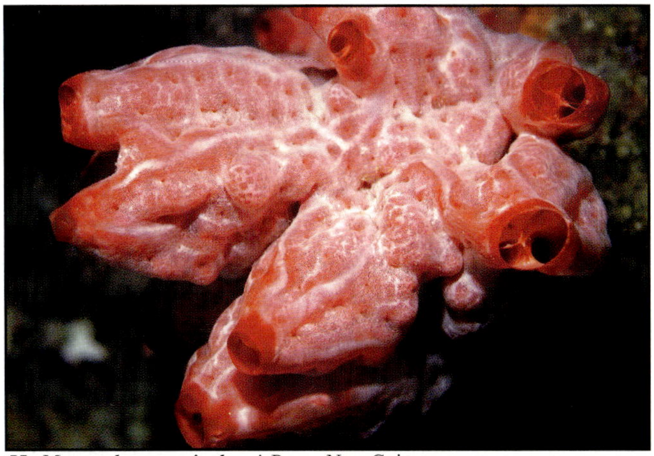

55- ***Monanchora ungiculata*** * Papua New Guinea

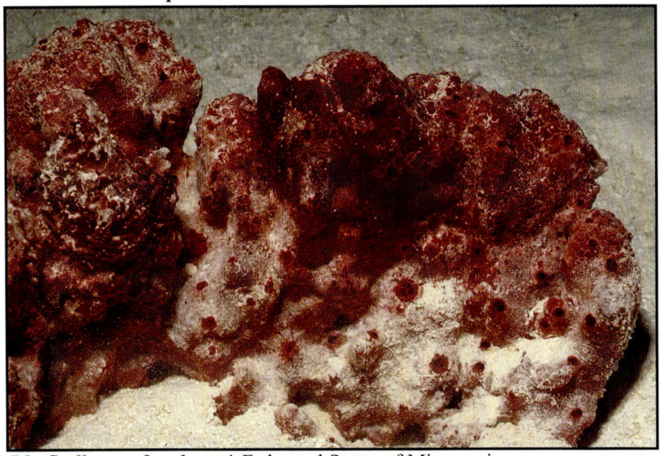

56- ***Crella* sp. cf. *calypta*** * Federated States of Micronesia

57- ***Biemna* sp.** * Fiji

58- ***Batzella* sp.** * Federated States of Micronesia

59- ***Batzella* sp.** * Papua New Guinea

60- ***Iotrochota* sp.** * Federated States of Micronesia

rates sand into its skeleton. Species of this genus from northern Australia and New Caledonia are known to be highly toxic to humans, and it is recommended that these sponge not be handled.

58- *Batzella* sp. * Desmacididae * Poecilosclerida * Federated States of Micronesia * Chuuk Atoll * Tonoas Island * 50 ft (15 m). This sponge is common on lagoon reefs in Chuuk. The dark branches stick out in all directions. The photographed sponges are growing among a variety of stony corals.

59- *Batzella* sp. * Desmacididae * Poeciloslcerida * Papua New Guinea * Eastern Fields * 90 ft (27 m). This species of *Batzella* is yellow. The arms of brittlestars extend out from the edges of the sponge.

60- *Iotrochota* sp. * Desmacididae * Poecilosclerida * Chuuk Atoll * Tonoas Island * algal flat * 150 ft (45 m). Species of *Iotrochota*, like this one, smell strongly of iodine if removed from the water and turn dark in air. Most species are dark purplish-black but this species has a green sheen to the

61- *Desmacella lampra* * Federated States of Micronesia

62- *Liosina paradoxa* * Marshall Islands

63- *Liosina* sp. * Palau

64- *Psammoclemma* sp. * Papua New Guinea

surface. The species occurs in the deep lagoon algal flat community. This community contains an unusual diversity of invertebrate species and algae.

61- *Desmacella lampra* * Desmacellidae * Poecilosclerida * Federated States of Micronesia * Chuuk Atoll * lagoon reef * 23 ft (7 m). This red sponge is growing on branches of *Acropora* coral. It has white pharonids (a lophophorate, see Chapter 7) growing with it. This association is common, but the relationship between the sponge and the pharonid is poorly known.

62- *Liosina paradoxa* * Desmacididae * Poecilosclerida * Marshall Islands * Kwajalein Atoll * 66 ft (20 m). Members of *Liosina* are readily recognized by their surface texture, which has shallow irregular ostial depressions between areas of smooth plate-like surface. Often this sponge is covered with silt, which hides the surface appearance until the silt is fanned away. The present species encrusts on rocks and is a dull orange internally beneath tan skin. There are three species of ***Liosina*** in the Pacific, only one of which is described.

63- *Liosina* sp. * Desmacididae * Poecilosclerida * Palau * Mutremdiu Wall * 66 ft (20 m). This *Liosina* forms tubes which arise from an encrusting basal mass. It has the characteristic texture of the genus, also seen in the previous species. This species is undescribed.

64- *Psammoclemma* sp. * Desmacididae * Poecilosclerida * Papua New Guinea * Eastern Fields * 185 ft (55 m). This sponge occurs along deep reef drop offs where it grows as club-like lobes arising from a basal mass. It is one of the species which is much darker when exposed to light and the photographed individual is mottled indicating it was in an area of moderate light intensity.

65- *Clathria ?plinthina* * Microcionidae * Poecilosclerida * Federated States of Micronesia * Mortlocks * Satawan Atoll * lagoon * 78 ft (24 m). This is the epitome of an encrusting sponge; the vibrant red sponge forms a thin film of tissue over the rock subtrate it grows upon. This species used to be in the genus *Microciona*, but recent work has indicated *Clathria,* being revised by Dr. John Hooper, is a more appropriate genus.

66- *Clathria reinwardti* * Microcionidae * Poecilosclerida * Papua New Guinea * Port Moresby * Lion Island * 50 ft (15 m). This species was once considered to be *Rhapadopholus reinwardti*. It is quite variable in growth, but most often as gnarly branches similar to those shown in the photograph. This is a very common sponge throughout Micronesia and is the most common *Clathria* in the Indo-West Pacific.

67- *Clathria* sp. * Microcionidae * Poecilosclerida * Papua New Guinea * Bagabag Island * 80 ft (25 m). This *Clathria* is an encrusting species. Delicate ostia, which bring in water, lie close to dendritic channels which collect this flow after food particles have been filtered from the water. The dendritic channels increase in size until they reach the large circular oscules, four of which are visible in the photograph.

68- *Clathria* sp. * Microcionidae * Poecilosclerida * Federated States of Micronesia * Chuuk Atoll * lagoon bottom * 100 ft (30 m). This rusty red sponge is found on sandy bottom in Chuuk lagoon. At depth, without a light it appears black to a diver

69- *Clathria basilana* * Myxillidae * Poecilosclerida * Federated States of Micronesia * Chuuk Atoll * lagoon reef * 60 ft (18 m). This beautiful frosted red or orange sponge occurs widely in Micronesia and elsewhere in the region. It is hard to confuse with any other sponge, although the color can vary from orange to red. Most often the sponge is seen in the tubular form shown in the photograph, but at times the tubes can be much shorter relative to their diameter.

70- *Clathria mima* * Microcionidae * Poecilosclerida * Federated States of Micronesia * Chuuk Atoll * Nematon Bay * 40 ft (12 m). This beautiful sponge was previously recognised as *Ophlitaspongia mima.* It encrusts reef substrates and appears to be distributed throughout much of the region. Sometimes the white of the excurrent veins on the surface is not as bright as that shown in the photograph, but there is always a distinct color difference between the lighter channels and the surrounding sponge.

71- *Clathria vulpina* * Microcionidae * Poecilosclerida * Federated States of Micronesia * Chuuk Atoll * Tonoas * algal flat * 160 ft (48 m). *Clathria vulpina* or *C. frondifera* as it was previously known is hollow inside with a fibrous surface. Despite its rugged surface, it is not spiny. It is widely distributed, being known from Australia, the South Pacific, Southeast Asia and Micronesia. In Chuuk it is found on the lagoon bottom among other sponges and algae.

72- *Echinochalina intermedia* * Microcionidae * Poecilosclerida * Indonesia * Manado * 80 ft (25 m). This sponge has distinctive colors, with translucent bluish water channels and areas of coppery flecks on the sponge surface. While occasionally found alone, we most often see it growing in association with the calcareous tubes of an unidentified worm. The species is

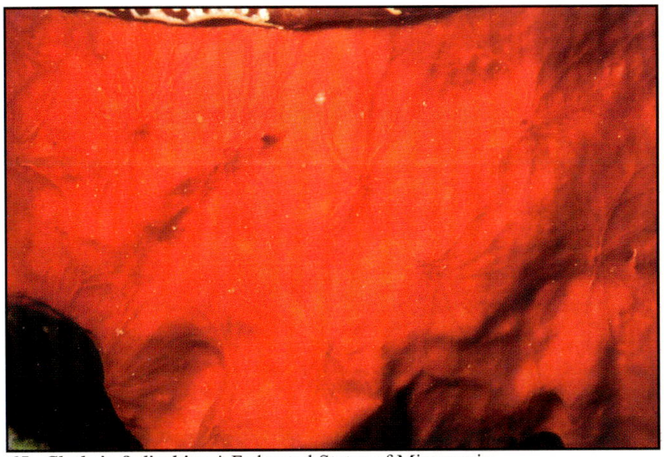

65- *Clathria ?plinthina* * Federated States of Micronesia

66- *Clathria reinwardti* * Papua New Guinea

67- *Clathria* sp. * Papua New Guinea

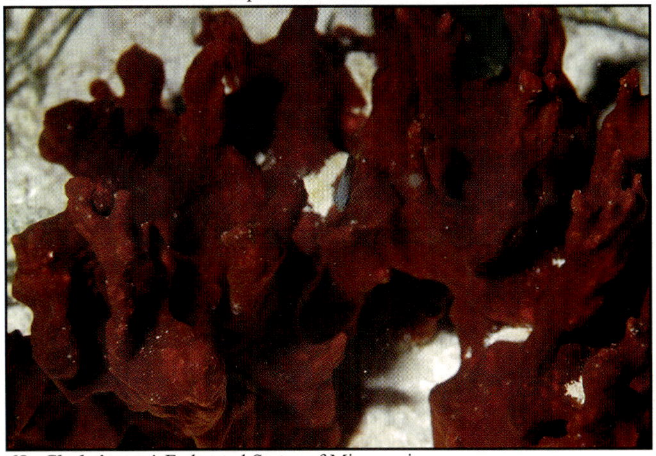

68- *Clathria* sp. * Federated States of Micronesia

69- *Clathria basilana* * Federated States of Micronesia

70- *Clathria mima* * Federated States of Micronesia

71- *Clathria vulpina* * Federated States of Micronesia

72- *Echinochalina intermedia* * Indonesia

known from northeast Australia, New Caledonia, Indonesia and Micronesia.

73- *Carmia ?lampra* * Mycalidae * Poecilosclerida * Palau * Turtle Island Basin * 40 ft (12 m). This bright orange sponge is a common encrusting species in Micronesia.

74- *Mycale* sp. * Mycalidae * Poecilosclerida * Indonesia * Manado * 90 ft (27 m). This sponge looks like the species of *Placospongia* covered earlier, however, it lacks the horny "alligator skin". Rather, it is soft and fragile with the tissue of the sponge hardly more than a flimsy structure full of water. A second sponge protrudes from one side and a crinoid is in the lower corner of the photograph.

75- *Stylinos* sp. cf. *ruetzleri* * Halichondriidae * Halichondrida * Marshall Islands * Kwajalein Atoll * patch reef * 50 ft (15 m). This common orange encrusting sponge is found in lagoonal and protected areas throughout the region. It is soft with elements of the fibrous skeleton protruding out (conulose) and the diffuse sponge tissue stretched net-like between. The photographed individual occurs among other sponges, corals and coralline algae. This species used to be in the genus *Ulosa*, but that genus has been synonymized with *Stylinos*. It is very similar to the Caribbean sponge, *Stylinos ruetzleri.*

76- *Stylinos* sp. * Halichondriidae * Halichondrida * Federated States of Micronesia * Etal Atoll * lagoon * 33 ft (10 m). This soft branching sponge occurs among dense algae on the slope in the lagoon of Etal Atoll. The lagoon of Etal was quite different in terms of species composition, from that found on most Caroline Island atolls, perhaps due to the lack of a deep water passage into the lagoon..

77- *Ectyodoryx* sp. * Myxillidae * Poecilosclerida * Indonesia * Manado * 15 ft (5 m). This species of *Ectyodoryx* has lots of sand incorporated into the sponge, producing a hard structure which does not really have a great deal of strength. The sponge tissue uses the sand grains instead of spongin fiber to hold the sponge together. However sand provides little capacity to inter-lock the skeleton and there is really nothing to keep the sponge from breaking apart once a piece is loosened.

78- Undescribed Poecilosclerid * Poecilosclerida * Papua New Guinea * Kavieng * Albatross Channel * 60 ft (18 m). This is an extremely attractive and distinctive Poecilosclerid encrusting sponge found in Papua New Guinea, but its affinities are not presently known. Many of the encrusting sponges which occur beneath ledges, such as this one, have characteristic appearances and with careful observation a diver can begin to recognize various species, bringing a bit of order to the apparent chaos of the reef ledge and cave community.

79- *Echinodictyum asperum* * Raspailiidae * Poecilosclerida * Papua New Guinea * Madang * Rasch Passage * 33 ft (10 m). This is a readily identifiable sponge which always takes the form of an upright hollow black fibrous mass attached to the bottom by a relatively thin stalk. It occurs from the Arabian Gulf to Micronesia and is sometimes overlooked as it is often covered with sediment. The skeleton is very fibrous and the outer texture appears quite rough.

80- *Echinodictyum* sp. * Raspailiidae * Poeciloslcerida * Papua New Guinea * Duke of York Islands * Makada Reef * 100 ft (30 m).

81- *Raspailia nuda* * Raspailiidae * Poecilosclerida * Indonesia * Manado * 60 ft (18 m). This sponge forms a mass of red-orange fingers, often in the shape of a cup or bowl. It is presently known from northwest Australia, the Philippines and Indonesia.

82- *Raspailia* sp. * Raspailiidae * Poeciloslcerida * Papua New Guinea * Eastern Fields * cave * 115 ft (35 m). Reminescent of chicken feet, this sponge is found along deep reef dropoffs and caves.

83- *Rhabderemia sorokinae* * Rhabderemiidae * Peocilosclerida * Palau * Mutremdiu wall * 100 ft (30 m). This sponge is a mass of yellow gnarled fuzzy fingers. It is known from the Great Barrier Reef, southern Indonesia, Papua New Guinea and Palau.

84- *Acanthella cavernosa* * Axinellidae * Halichondrida * Federated States of Micronesia * Chuuk Atoll * Schieben Island * 60 ft (18 m). *Acanthella cavernosa* is refered to as a axinellid sponge. These red-orange sponges are attached to the bottom by a central stalk-like axis of compact spicules, and are frequently bush-like in morphology. The family Axinellidae is actually an artificial assemblage, but for our purposes here, the disparate elements (which make them readily identifiable as "axinellids") used to lump together this "family" make it a convenient "assemblage" to recognize. These sponges are notoriously difficult to separate into species because most lack sufficiently variable characters to do so. Some are sufficiently characteristic to be easily identified. This species, *A. cavernosa*, is one of the latter.

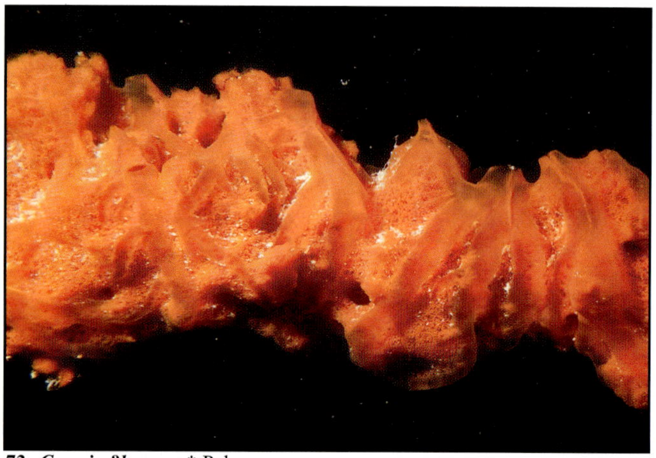

73- ***Carmia ?lampra*** * Palau

74- ***Mycale* sp.** * Indonesia

75- ***Stylinos* cf. *ruetzleri*** * Marshall Islands

76- ***Stylinos* sp.** * Federated States of Micronesia

77- ***Ectyodoryx* sp.** * Indonesia

78- Genus species unknown * Papua New Guinea

79- ***Echinodictyum asperum*** * Papua New Guinea

80- ***Echinodictyum* sp.** * Papua New Guinea

81- ***Raspailia nuda*** * Indonesia

82- ***Raspailia* sp.** * Papua New Guinea

83- ***Rhabderemia sorokinae*** * Palau

84- ***Acanthella cavernosa*** * Federated States of Micronesia

85- *Acanthella* sp. * Papua New Guinea

86- *Acanthella* sp. * Federated States of Micronesia

87- *Axinella* sp. * Federated States of Micronesia

89- *Auletta* sp. * Papua New Guinea

85- *Acanthella* sp. * Axinellidae * Halichondrida * Papua New Guinea * Kavieng * 70 ft (21 m). This is another axinellid sponge, but with a charateristic shape. In this case the sponge has an almost tubular body with one or two oscules at the end, as is shown in the photograph. It is not very common, and is presently known only from Papua New Guinea.

86- *Acanthella* sp. * Axinellidae * Halichondrida * Federated States of Micronesia * Chuuk Atoll * lagoon reef * 40 ft (12 m). The identification of this sponge is uncertain, this species may actually belong in the genus *Phakellia*. This species is common in the Indo-west Pacific. A fairly typical axinellid, it is bushy, usually with multiple conical projections.

87- *Axinella* sp. * Axinellidae * Halichondrida * Federated States of Micronesia * Chuuk Atoll * Lematol Bay * 90 ft (27 m). This red axinellid is found in silty areas at the bases of lagoonal reefs. It isdifficult to recognize as a sponge until the dense covering of sediment is fanned away.

88- *Axinella proliferans* * Axinellidae * Halichondrida * Federated States of Micronesia * Chuuk Atoll * lagoon reef * 40 ft (12 m). This axinellid has a strong stalk with the body of the sponge occurring as a single plane fan.

89- *Auletta* sp. * Axinellidae * Halichondrida * Papua New Guinea * Port Moresby * barrier reef * 60 ft (18 m). This sponge has many fine yellow fingers arising from a single stalk. A number of brittlestars occur among the fingers of this sponge.

90- *Auletta* sp. * Axinellidae * Halichondrida * Papua New Guinea * New Britain * offshore reef * 80 ft (25 m). This axinellid has fine yellowish-tan tubes branching from the single stalk. This sponge is soft and slimy.

91- *Axinosa* sp. * Axinellidae * Halichondrida * Papua New Guinea * Madang Channel * 100 ft (30 m). This sponge is a yellow, slimy fan with wart-like projections and oscules on the surface. It is known only from the bottom of the channel into Madang harbor. A delicate hydroid is growing on the sponge, and beyond it a zoanthid (*Parazoanthus*) occurs.

92- Undescribed Axinellid, * Axinellidae * Halichondrida * Papua New Guinea * New Britain * Kimbe Bay * 50 ft (15 m).

93- *Cymbastella ?marshae* * Axinellidae * Halichondrida * Federated States of Micronesia * Chuuk Atoll * lagoon reef * 78 ft (24 m). This flattened ear sponge is compressible and flexible. genus is known from Australia, New Caledonia and Micronesia.

94- *Pseudaxinella* sp. * Axinellidae * Halichondrida * Papua New Guinea * Dyaul Island * 66 ft (20 m). This is an example of an atypical axinellid. While the sponge is thickly encrusting, individual papillae retain an overall axial structure. This species exudes mucous.

95- *Higginsia* sp. cf Mixta * Desmoxyidae * Halichondrida * Palau * Marine lake * 3 ft (1 m).

88- *Axinella proliferans* * Federated States of Micronesia

90- *Auletta* sp. * Papua New Guinea

91- *Axinosa* sp. * Papua New Guinea

92- Axinellid * Papua New Guinea

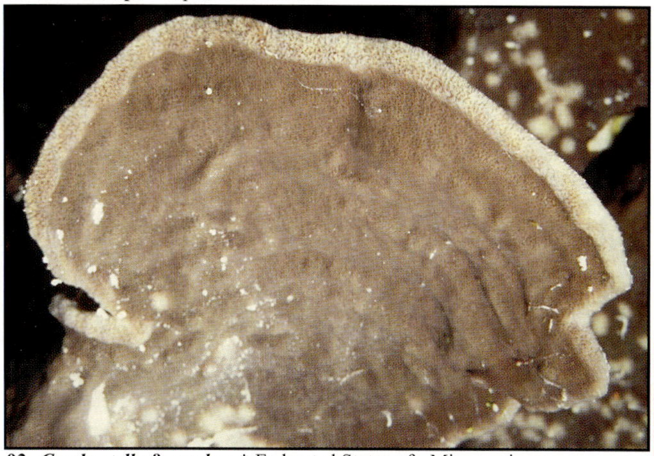
93- *Cymbastella ?marshae* * Federated States of Micronesia

94- *Pseudaxinella* sp. * Papua New Guinea

95- *Higginsia* sp.cf. *mixta* * Palau

96- P*hakellia* sp. * Indonesia

97- *Reniochalina* sp. * Papua New Guinea

98- ***Stylissa flabelliformis*** * Federated States of Micronesia

99- ***Axinyssa*** **sp.** * Papua New Guinea

100- ***Axinyssa*** **sp.** * Federated States of Micronesia

96- *Phakellia* sp. * Axinellidae * order * Indonesia * Manado * 110 ft (33 m). This axinellid sponge is tubular, and thus is superficially similar to members of the genus *Auletta*. However, it has a rough outer surface, while *Auletta* sp. are mostly smooth. The sponge is found on deep reef dropoffs.

97- *Reniochalina* sp. * Axinellidae * Halichondrida * Papua New Guinea * Duke of York Islands * Mioko Channel * 66 ft (20 m). This species is similar to the species of *Axinosa* included previously, but *Reniochalina* is fibrous and has a fuzzy texture without oscules on the surface of the fan. It is common in northern Australia and New Caledonia.

98- *Stylissa flabelliformis* * Axinellidae * Halichondrida * Federated States of Micronesia * Chuuk Atoll * Pizion Reef * 200 ft (60 m). This is another of the stalked axinellids, it has a fan-shaped body. This species occurs along deep reef dropoffs protruding out from the wall.

99- *Axinyssa* sp. * Axinellidae * Halchondrida * Papua New Guinea * Dyaul Island * 80 ft (25 m).

100- *Axinyssa* sp. * Halichondriidae * Halichondrida * Federated States of Micronesia * Chuuk Atoll * 80 ft (25 m). This lagoon sponge is black and exudes abundant mucous. The sponge is lumpy and can form fingers. It has characteristic ostia as shown in the photograph. In Chuuk Atoll it occurs in the lagoon from 50-120 feet deep. A sea cucumber lies next to the sponge.

101- *Hymeniacidon* sp. * Halichondriidae * Halichondrida * Federated States of Micronesia * Chuuk Atoll * lagoon bottom * 165 ft (50 m). This sponge forms small tufts on lagoon algal flats at depths below about 120 feet.

102- *Katiba milnei* * Halichondriidae * Halichondrida * Federated States of Micronesia * Chuuk Atoll * lagoon reef * 50 ft (15 m). This encrusting sponge forms a thin layer on dead coral and other carbonate substrate. It is markedly sticky to the touch and turns a milk chocolate brown when touched.

103- *Myrmekioderma* sp. * Desmoxyidae * Halichondrida * Guam * Apra Harbor * 50 ft (15 m). Species of *Myrmekioderma* are known for producing a lot of mucous when torn. The surface of polygonal plates is divided by narrow grooves that contain the ostia.

104- *Myrmekioderma granulata* * Desmoxyidae * Halichondrida * Federated States of Micronesia * Chuuk Atoll * lagoon algal flat * 113 ft (34 m). The external color of this sponge varies with exposure to light. The dark color is almost certainly due to symbiotic microorganisms. Internally the sponge is orange. The surface has irregular polygonal plates. The sponge is common on deep water sandy areas near reefs and some inshore areas. This species is sometimes confused with *Aaptos* sp., but the irregular polygons of *M. granulata* are a good field character.

105- *Prianos osiros* * Halichondriidae * Halichondrida * Federated States of Micronesia * Chuuk Atoll * Tonoas Inlet * 50 ft (15 m). *P. osiros* generally has a "frosted" appearance on its outer surface and a very flabby texture. Originally described from Chuuk by de Laubenfels, it is found in inshore areas, often on overhangs or beneath ledges.

106- *Prianos* sp. * Halichondriidae * Halichondrida * Indonesia * Manado * 66 ft (20 m). This sponge from Manado in Indonesia appears to be a species of *Prianos* but its identification is still uncertain.

107- *Stylotella aurantium* * Halichondriidae * Halichondrida * Federated States of Micronesia * Chuuk Atoll * Tonoas * 23 ft (7 m). One of the most common and easily recognized sponges on reefs in the

101- ***Hymeniacidon*** **sp.** * Federated States of Micronesia

102- ***Katiba milnei*** * Federated States of Micronesia

103- ***Myrmekioderma* sp.** * Guam

104- ***Myrmekioderma granulata*** * Federated States of Micronesia

105- ***Prianos osiros*** * Federated States of Micronesia

106- ***Prianos* sp.** * Indonesia

107- ***Stylotella aurantium*** * Federated States of Micronesia

108- ***Arenosclera* sp. cf. *psammochera*** * Federated States of Micronesia

109- ***Callyspongia* sp.** * Papua New Guinea

region. Although *S. aurantium* is quite variable in shape and surface texture, it is nearly always the same flourescent orange yellow color shown in the photograph. It is soft and easily compressed, like "soggy bread". Known throughout the region from reefs, seagrass beds and mangroves.

108- *Arenosclera* sp. cf. *psammochera* * Callyspongiidae * Haplosclerida * Federated States of Micronesia * Chuuk Atoll * Northeast Pass * 80 ft (25 m). This sponge is purple and sticky with most being fan-shaped individuals.

109- *Callyspongia* sp. * Callyspongiidae * Haplosclerida * Papua New Guinea * Kavieng * 50 ft (15 m). The genus *Callyspongia* is one of the most common genera, it is found worldwide in the tropics. Quite often *Callyspongia* spp. grow as tubes, but it is also found as sheet or finger sponges. It is often difficult to identify specimens to species, particularly since numerous species are not yet described. The species shown here was growing on a deep reef drop off and although of very distinctive form and color with dense branches, it is probably undescribed.

110- *Callyspongia aerizusa* * Marshall Islands

110- *Callyspongia aerizusa* * Callyspongiidae * Haplosclerida * Marshall Islands * Kwajalein Atoll * 80 ft (25 m). This well known species is a distinctive bluish green color, but can occasionally be light brownish-pink. The species is typically tubular but can also form huge fans. Specimens from deeper algal flats have thinner, more elongate tubes.

111- *Callyspongia* sp. * Papua New Guinea

111- *Callyspongia* sp. * Callyspongiidae * Haplosclerida * Papua New Guinea * Madang * barrier reef * 40 ft (12 m). This undescribed species of *Callyspongia* has a stiff spiny surface texture.

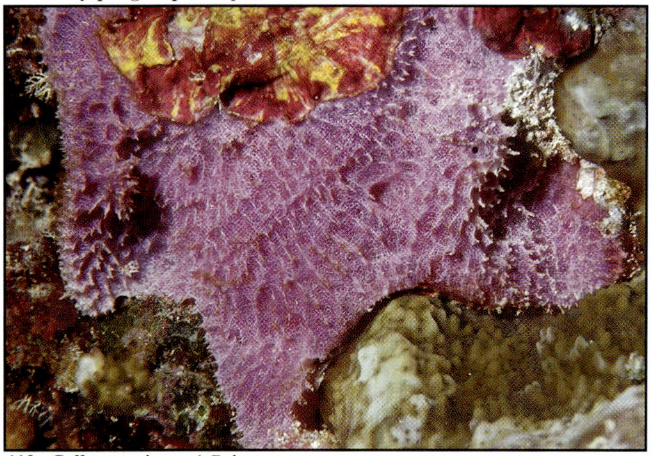
112- *Callyspongia* sp. * Palau

112- *Callyspongia* sp. * Callyspongiidae * Haplosclerida * Palau * Mutremdiu Wall * 50 ft (15 m). This encrusting species of *Callyspongia* has a relatively soft , smooth surface.

113- *Callyspongia* sp. * Callyspongiidae * Haploscerida * Indonesia * Manado * 33 ft (10 m). This tube-forming *Callyspongia* is spongy and springy.

114- *Callyspongia* sp. * Callyspongiidae * Haplosclerida * Papua New Guinea * New Britain * 60 ft (18 m). This species of *Callyspongia* has short robust tubes and is of lovely deep purple color.

115- *Callyspongia* sp. * Callyspongiidae * Haplosclerida * Federated States of Micronesia * Chuuk Atoll * Anaw Channel * 165 ft (50 m). Found on a deep reef dropoff, this *Callyspongia* was a pure white.

116- *Callyspongia* sp. * Callyspongiidae * Haplosclerida * Papua New Guinea * Port Moresby * Bootless Bay * Buna Motu Island * 33 ft (10 m).

117- *Dactylia* sp. * Callyspongiidae * Haplosclerida * Papua New Guinea * Madang * Pig Island * lagoon side * 40 ft (12 m). This lovely tube sponge has thin walls. This sponge lacks spicules and uses clear spongin fibers for support.

118- *Euplacella* sp. * Callyspongiidae * Haplosclerida * Indonesia * Manado * 75 ft (23 m). This genus is closely related to ***Callyspongia,*** but differs in the arrangement of the surface fibers.

119- *Euplacella* sp. * Callyspongiidae * Haplosclerida * Indonesia * Manado * 33 ft (10 m).

120- *Siphonochalina* sp. cf. *fascigera* * Callyspongiidae * Haplosclerida * Papua New Guinea * Bagabag Island * barrier reef pinnacle * 94 ft (28 m). This tube sponge was photographed on the barrier reef of Bagabag

113- *Callyspongia* sp. * Indonesia

114- *Callyspongia* sp. * Papua New Guinea

115- ***Callyspongia* sp.** * Federated States of Micronesia

116- ***Callyspongia* sp.** * Papua New Guinea

117- ***Dactylia* sp.** * Papua New Guinea

118- ***Euplacella* sp.** * Indonesia

119- ***Euplacella* sp.** * Indonesia

120- ***Siphonochalina* sp. cf. *fascigera*** * Papua New Guinea

121- ***Siphonochalina* sp.** * Papua New Guinea

122- ***Adocia* sp.** * Papua New Guinea

123- ***Adocia* sp.** * Federated States of Micronesia

124- ***Adocia turquiosia*** * Palau

125- ***Adocia viola*** * Federated States of Micronesia

Island, an isolated volcanic island. The barrier reef around Bagabag experiences strong currents and sometimes high waves. Despite the surge, the barrier reef at Bagabag is rich with invertebrates which are well adapted to hanging on in the ocean swell.

121- *Siphonochalina* sp. * Callyspongiidae * Haplosclerida * Papua New Guinea * Madang Channel * 100 ft (30 m). This species is very similar to the previous one, but has smaller tubes.

122- *Adocia* sp. * Chalinidae * Haplosclerida * Papua New Guinea * Port Moresby * Basilisk Passage * 100 ft (30 m). This encrusting sponge has the deepest red color we have seen in any reef sponge. It encrusts on deep reef walls. It is smooth with raised, regularly spaced oscules.

123- *Adocia* sp. * Chalinidae * Haplosclerida * Federated States of Micronesia * Chuuk Atoll * lagoon reef * 33 ft (10 m). This soft orange sponge encrusts rocks and overhangs in silty areas in Micronesia.

124- *Adocia turquoisia* * Chalinidae * Haplosclerida * Palau * marine river * 1 ft (0.5 m). This sponge is is soft, easily torn, and as its name implies a dull blue-green color. The species is a common fouling organism in mangrove and seagrass communities throughout the Indo-West Pacific.

125- *Adocia viola* * Chalinidae * Haplosclerida * Federated States of Micronesia * Chuuk Atoll * Anaw Reef * 23 ft (7 m). This soft fragile sponge was described from Guam and Pohnpei by de Laubenfels, the specific name referring to the violet color. In Chuuk we have found it in only one area, a small reef channel near the northern barrier reef.

126- *Sigmadocia amboinensis* * Chalinidae * Haplosclerida * Indonesia * Manado * 60 ft (18 m). This is another sponge which is closely associated with a second species, in this case the blue *S. amboinensis* overlies a yellow species of *Rhabdastrella*, which is not visible in the photograph.

127- *Nara nemetifera* * Chalinidae * Haplosclerida * Federated States of Micronesia * Chuuk Atoll * lagoon reef * 60 ft (18 m). This encrusting sponge has distinctive white threads of collagen within a matrix of purple tissue. This sponge is common and encrusts dead coral surfaces.

128- *Haliclona* cf. *coerulescens* * Chalinidae * Haplosclerida * Federated States of Micronesia * Northeast Pass * 47 ft (14 m). Originally described from the West Indies, de Laubenfels was unable to distinguish the species shown in the photograph from the West Indian specimens of *H. coerulescens.* It is doubtful that the two forms from different oceans are the same.

129- *Haliclona cymaeformis* * Chalinidae * Haplosclerida * Palau * Ngel Channel * seagrass beds * 6 ft (2 m). This curious sponge was previously known as *Sigmadocia symbiotica*. It has an algae closely associated with it, which gives the branches a definite greenish tinge. What is believed to be the same species occurs over a wide range in the western Pacific Ocean. It can vary greatly in appearance depending on locality and local environment.

130- *Haliclona koremella* * Chalinidae * Haplosclerida * Indonesia * Manado * 104 ft (31 m). The sponge looks like a mass of greenish or bluish rubbery spaghetti. Most often it is found occurring on and around seagrasses, but occasionally it will also occur in reef areas. The color comes from an algal symbiont.

131- *Haliclona* sp * Chalinidae * Haploslcerida * Palau * Denges Pass * 50 ft (15 m). This soft, sticky, encrusting sponge is widespread throughout Micronesia and elsewhere. The color may vary from near white to a deep blue-purple, however, the surface appearance as is shown in the photograph remains constant.

126- ***Sigmadocia amboinensis*** * Indonesia

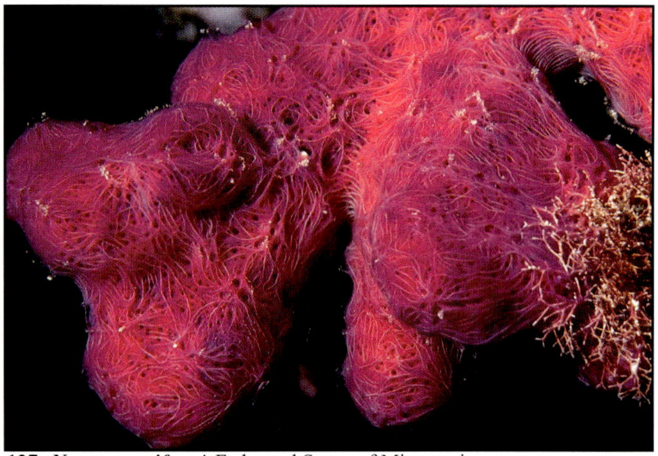

127- ***Nara nematifera*** * Federated States of Micronesia

128- ***Haliclona* cf. *coerulescens*** * Federated States of Micronesia

129- ***Haliclona cymaeformis*** * Palau

130- ***Haliclona koremella*** * Indonesia

131- ***Haliclona* sp** * Palau

132- ***Kallypilidion poseidon*** * Palau

133- **K*allypilidion* sp.** * Indonesia

134- *Kallypilidion* sp. * Indonesia

135- *Kallypilidion* sp. * Indonesia

136- *Reniera chrysa* * Papua New Guinea

137- *Aka coralliphagum* * Federated States of Micronesia

132- *Kallypilidion poseidon* * Chalinidae * Haplosclerida * Palau * marine lake * 3 ft (1 m). The exact taxonomic nature of these sponges has not been determined, but they are close to sponges in the family Chalinidae. This sponge, like others in the genus, forms thin, fragile cups and sheets which appear so delicate it is hard to believe they can exist underwater.

133- *Kallypilidion* sp. * Chalinidae * Haplosclerida * Indonesia * Biak Island * 60 ft (18 m). Superficially similar to the previous species, this sponge forms long tubes.

134- *Kallypilidion* sp. * Chalinidae * Haplosclerida * Indonesia * Manado * 75 ft (23 m). The body of this sponge resembles soggy tissue paper. Coloration can be variable.

135- *Kallypilidion* sp. * Chalinidae * Haplosclerida * Indonesia * Manado * 75 ft (23 m).

136- *Reniera chrysa* * Chalinidae * Haplosclerida * Papua New Guinea * Madang * 66 ft (20 m). This well known thickly encrusting sponge is very soft and delicate, but has a crunchy rather than soggy texture. Its bright yellow-orange coloration makes it easily recognizeable. The sponge is often found beneath overhanging coral heads in Micronesia and is evidently widespread in the Indo-west Pacific.

137- *Aka coralliphagum* * Niphatidae * Haplosclerida * Federated States of Micronesia * Mortlock Islands * Losap Atoll * east channel * 50 ft (15 m). These boring sponges over time excavate large volumes of coral skeleton, They can be one of the major destructive forces on coral reefs. Only the pustule-like ostia and large oscula are exposed, the majority of the sponge is buried in the excavated galleries inside the coral head. This species has a great deal of internal mucous. This genus has generally been known as *Siphonodictyon*, but some authorities believe *Aka* is the correct generic name for these species. This sponge is very similar to the Caribbean sponge *A. coralliphagum*.

138- *Aka* sp. * Niphatidae * Haplosclerida * Papua New Guinea * Taurauma Basin * 33 ft (10 m). Another species of *Aka* which is white, rather than yellow in color. Species of *Aka* are differentiated in the field by their color and the shape of their external siphons. The taxonomy of these sponges needs further study.

139- *Aka* sp. * Niphatidae * Haplosclerida * Federated States of Micronesia * Chuuk Atoll * Northeast Pass * cave * 33 ft (10 m). This is another boring sponge, but it is always dark brown or black in color making a clear contrast with the other illustrated species of *Aka* .

140- *Amphimedon ? viridis* * Niphatidae * Haplosclerida * Federated States of Micronesia * Chuuk Atoll * Tonoas Inlet * 33 ft (10 m). This is a very common sponge in shallow water murky areas of Micronesia. The green color comes from algae living with the sponge. This species closely resembles the Caribbean sponge *A. viridis*.

141- *Amphimedon* sp. * Niphatidae * Haplosclerida * Papua New Guinea * New Britain * 66 ft (20 m).

142- *Amphimedon* sp. * Niphatidae * Haplosclerida * Papua New Guinea * Duke of York Islands * Makada Reef * 50 ft (15 m).

143- *Amphimedon* sp. * Niphatidae * Haplosclerida * Papua New Guinea * Duke of York Islands * Ulu Pinnacle * 60 ft (18 m).

144- *Cribochalina* sp. * Niphatidae * Haplosclerida * Federated States of Micronesia * Chuuk Atoll * lagoon reef * 40 ft (12 m). Species of *Cribochalina* are slightly tougher and less even-surfaced than species of *Amphimedon*, but they are also very sticky and spongey. This is another common sponge of lagoon and outer reefs in Micronesia. It seems to like clear water and the color is darker when the sponge is more exposed to light, implying that the color is produced by microbial symbionts.

145- *Cribrochalina olemda* * Niphatidae * Haplosclerida * Palau * lagoon * 40 ft (12 m). This is one of the most common and easily recognized sponges in Micronesia and the western Pacific. It is found on both inshore and offshore reefs. The tubes are soft and the sponge is sticky to the touch, and the sticky mucous remains on the fingers afterwards. The ends of the tubes curve cleanly inward and the inside of the tube is surrounded by concentric rings.

146- *Cribrochalina* sp. * Niphatidae * Haplosclerida * Indonesia * Manado * 90 ft (27 m).

147- *Cribrochalina* sp. * Niphatidae * Haplosclerida * Papua New Guinea * Eastern Fields * 100 ft (30 m). This white sponge has a convoluted surface and is gelatinous and sticky to the touch. It probably is an undescribed species.

138- *Aka* sp. * Papua New Guinea

139- *Aka* sp. * Federated States of Micronesia

140- *Amphimedon ? viridis* * Federated States of Micronesia

141- *Amphimedon* sp. * Papua New Guinea

142- *Amphimedon* sp. * Papua New Guinea

143- *Amphimedon* sp. * Papua New Guinea

144- *Cribochalina* sp. * Federated States of Micronesia

145- *Cribrochalina olemda?* * Palau

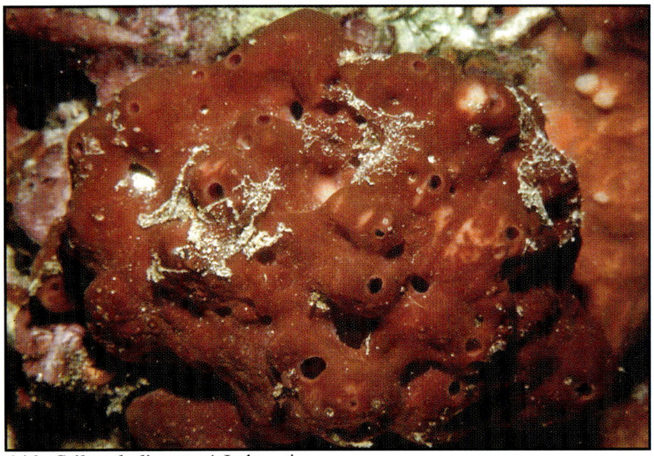

146- ***Cribrochalina* sp.** * Indonesia

147- ***Cribrochalina* sp.** * Papua New Guinea

148- ***Gelloides fibulata*** * Papua New Guinea

149- ***Gelloides?* sp.** * Federated States of Micronesia

148- *Gelloides fibulata* * Niphatidae * Haplosclerida * Papua New Guinea * Port Moresby * Lion Island * 40 ft (12 m). The spicules in this well known species of *Gelloides* fuse together to form what are probably the sharpest, stiffest spines found among tropical Pacific sponges. We like to call these the "cactus" sponge.

149- *Gelloides* sp. * Niphatidae * Haplosclerida * Federated States of Micronesia * Kuop Atoll * outside reef * 110 ft (33 m). The lacey network of this sponge is so delicate, it appears that the sponge is dead and only the underlying fibrous skeleton remains. However, the sponge is indeed alive with fine tissue around and between the fibers. Generally this sponge is found in caves and crevices in Micronesia.

150- *Niphates* sp.cf. *callista* * Niphatidae * Haplosclerida * Indonesia * Manado * 110 ft (33 m). The pinkish-purple color of this sponge can not be missed. The sponge is crunchy, with brittle fibers and exudes mucus. The sponge was originally described as *Gelloides callista* by de Laubenfels, but it very probably represents a new genus.

151- *Petrosia* sp. * Petrosiidae * Petrosida * Papua New Guinea * Madang * Rasch Passage * 80 ft (25 m). Petrosids have a very hard, almost armored texture and they often have the form as shown in this photograph. These sponges may have rounded lobes or fingers and large scooped out oscules which often have a sieve-like grid. The number of undescribed species of *Petrosia* in this section provides some idea of the state of the taxonomy of this diverse and interesting group.

152- *Petrosia capsa* * Petrosiidae * Petrosida * Papua New Guinea * New Ireland * Kalihi Harbor * 66 ft (20 m). The hard green hemispheres of *Petrosia capsa* are distinctive with a single apical osculum and rough outer surface. The green color may come from algae growing within the surface of the sponge.

153- *Petrosia* sp. * Petrosiidae * Petrosida * Indonesia * Manado * 94 ft (28 m). This remarkable sponge has been seen in several locations. It is hard and cup-like in structure. It seems limited to deep reef areas on fairly level or sloping bottoms.

154- *Petrosia* sp. * Petrosiidae * Petrosida * Papua New Guinea * Port Moresby * barrier reef * 40 ft (12 m). This sponge is white when it occurs in caves and beneath overhangs, but is much darker when found in deep water openly exposed to light.

155- *Petrosia* sp. * Petrosiidae * Petrosida * Papua New Guinea * West New Britain * 100 ft (30 m).

156- *Strongylophora* sp. cf. *durissima* * Petrosiidae * Petrosida * Papua New Guinea * Dyaul Island * 66 ft (20 m). This tube-shaped sponge is stony and hard in texture, but in a different way from the lithistids, or stone sponges. Petrosid sponges have internal, concentric rings of thickly packed spicules bound by spongin. Such sponges were long considered to be members of a group called the "lithistids" or stony sponges (due to their stony nature), but recent work has indicated the lithistids were an aritifical assemblage of several different lines of sponge evolution. The lithistids have now been split apart and individual genera or species have been assigned to families and orders which match their other characters.

157- *Strongylophora strongylata* * Petrosiidae * Petrosida * Papua New Guinea * Port Moresby * barrier reef * 60 ft (18 m). This petrosid may have algae growing in its tissues which gives it a green color where exposed to light.

158- *Strongylophora* sp. * Petrosiidae * Petrosida * Papua New Guinea * Kimbe Bay * Agu Reef * 33 ft (10 m).

159- *Xestospongia* sp. * Petrosiidae * Petrosida * Philippines * Cebu * Mactan Island * 50 ft (15 m). Like *Petrosia* this genus is well represented in the region. *Xestospongia*his is a very prominent sponge on Pacific reefs. and it is easy to distinguish in the field due to its "styrofoam-like" texture. The taxonomy of the genus is poorly known and there may be many undescribed species.

160- *Xestospongia* sp. * Petrosiidae * Petrosida * Federated States of Micronesia * Nama Island * cave * 40 ft (12 m).

161- *Xestospongia* sp. * Petrosidae * Petrosiida * Papua New Guinea * Madang * Pig Island * barrier reef * 100 ft (30 m)

162- *Xestospongia* sp. * Petrosiidae * Petrosida * Papua New Guinea * Dyaul Island * 100 ft (30 m). This large yellowish mammilate *Xestospongia* is attractive. Known only from Papua New Guinea.

163- *Xestospongia exigua* * Petrosiidae * Petrosida * Federated States of Micronesia * Chuuk Atoll * Tonoas Island * 13 ft (4 m). This is one of the most common shallow water sponges in the entire region, particularly so

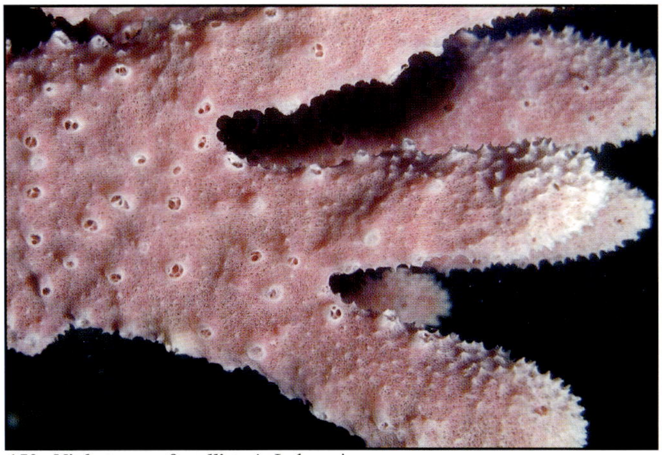

150- *Niphates* sp.cf. *callista* * Indonesia

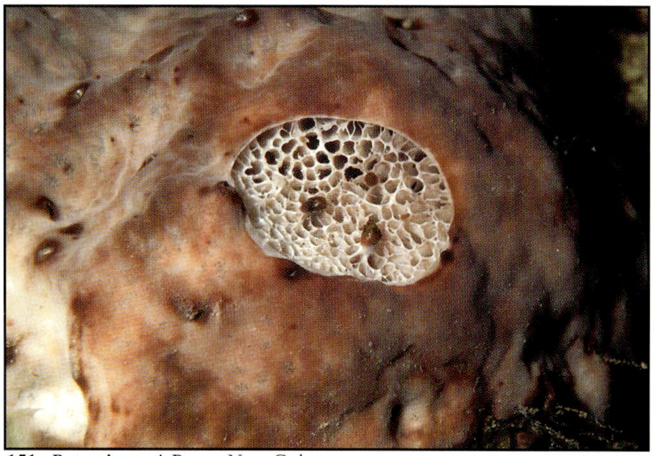

151- *Petrosia* sp. * Papua New Guinea

153- *Petrosia* sp. * Indonesia

152- *Petrosia capsa* * Papua New Guinea

154- *Petrosia* sp. * Papua New Guinea

155- *Petrosia* sp. * Papua New Guinea

156- *Strongylophora* sp.cf. *durissima* * Papua New Guinea

157- *Strongylophora strongylata* * Papua New Guinea

158- *Strongylophora* sp. * Papua New Guinea

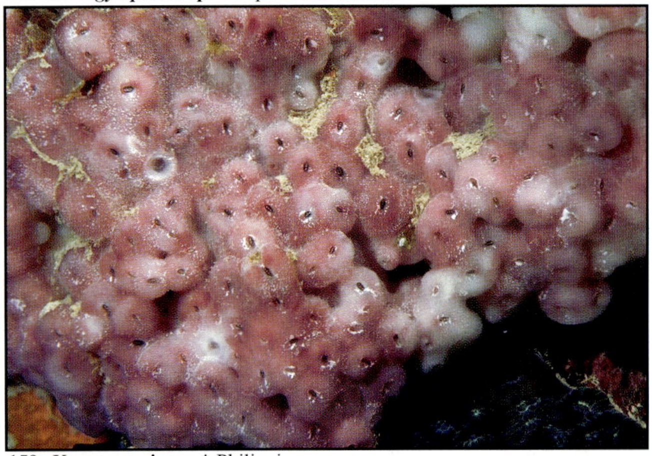

159- *Xestospongia* sp. * Philippines

in lagoonal areas of Micronesia. It takes a wide variety of forms from thin encrusting mats, to encrusting with gnarled projections or as ridges arising from a basal mass. The dark brown color is generally constant, despite the large variations in shape. The small branches are brittle, and when broken reveal a cream colored interior.

164- *Xestospongia testudinaria* * Petrosiidae * Petrosida * Papua New Guinea * West New Britain * 66 ft (20 m). This species is the barrel-sponge of the western Pacific and is remarkably similar in appearance to the well-known *Xestospongia muta* from the Caribbean. In fact the species can not be separated on any morphological basis, such as spicules, growth form, etc. The two sponges, however, have chemical differences which indicate they may be two separate species.

165- *Xestospongia?* sp. * Petrosiidae * Petrosida * Papua New Guinea * Madang * Rasch Passage * 50 ft (15 m). This species of *Xestospongia* is urn-like and somewhat different from most species in the genus.

166- *Vagocia* sp. * Oceanapiidae * Petrosida * Palau * Mutemdiu Wall * 100 ft (30 m). This sponge forms small fans growing along reef drop offs; the sponge is attached to the wall at several points on its undersurface.

167- *Oceanapia* sp. * Oceanapiidae * Petrosida * Indonesia * Manado * 15 ft (5 m). This sponge dwells in the sediment. Only the digitate projections with the ostia and oscules are visible above the bottom. This species is not alone in this type of habitat, virtually every area with stable sediment on reefs, seagrass beds and mangroves will have sponges similarly living embedded into the sediment. This species is widespread in the Indo-west Pacific.

168- *Oceanapia ramseyi* *Oceanapiidae * Petrosida * Indonesia * Manado * 145 ft (43 m). This sponge is common in Northern Australia and in Indonesia. It has a single stalk with a flattened conical head and numerous oscules.

169- *Oceanapia sagittaria* * Oceanapiidae * Petrosida * Palau * Koror * 15 ft (5 m). Known from Gulf of Thailand, Papua New Guinea, Great Barrier Reef, Palau, and Chuuk. A very distinctive sponge, like small purple spires, often with a little purple puffball like capitate structure at the tip.. The puff ball breaks off easily and is believed to serve as both a complex series of exhalant pores and as an asexual dispersal propagule. Most often found in shallow soft bottomed areas, often in turbid water.

170- *Oceanapia* sp. * Oceanapiidae * Petrosida * Papua New Guinea * Manam Island * 100 ft (30 m). This interesting Oceanapid has a series of clustered tubes, each essentially an independent sponge. This specimen was photographed growing on a volcanic sand slope at Manam Island, a very active volcano off the north coast of New Guinea in Madang Province. This slope had a very unusual invertebrate fauna, quite different from that occurring on nearby reefs. We have not seen this sponge elsewhere.

171- *Oceanapia* sp. * Oceanapiidae * Petrosida * Papua New Guinea * Eastern Fields * 133 ft (40 m). The morphology of this species of *Oceanapia* is bizarre, but it is probably adapted for high sediment environments. The main body of the sponge is firm and globular, with apical oscules. A series of blind digitate projections bear the ostia and the entire sponge is firmly attached to a hard substratum. Similar morphology is found in other sponges, particularly those of the genus *Coelocarteria.*

172- *Oceanapia* sp. * Oceanapiidae * Petrosida * Papua New Guinea * Bagabag Island * fringing reef * 66 ft (20 m). This barrel-shaped sponge with a large terminal oscule has tentatively been placed in the genus *Oceanapia*, but it may be more accurately put into *Orina*. If that were the

160- *Xestospongia* sp. * Federated States of Micronesia

161- *Xestospongia* sp. * Papua New Guinea

162- ***Xestospongia* sp.** * Papua New Guinea

163- ***Xestospongia exigua*** * Federated States of Micronesia

164- ***Xestospongia testudinaria*** * Papua New Guinea

165- ***Xestospongia?* sp.** * Papua New Guinea

166- ***Vagocia* sp.** * Palau

167- ***Oceanapia* sp.** * Indonesia

168- ***Oceanapia ramseyi*** * Indonesia

case, it would then belong in another family and order; the Chalinidae in the order Haplosclerida. Such is the state of sponge taxonomy. Its outside surface has small projections which actually give it a fuzzy texture on top of the rather firm barrel part.

173- ***Pellina*** **sp. * Oceanapiidae * Petrosida * Federated States of Micronesia * Nama Island * reef face caves * 40 ft (12 m).** This sponge is a cavern-dweller, living in the twilight zone on the sides of reef caves a short distance inside the entrance. It has a tough skin, but a very soft and mushy interior.

174- ***Carteriospongia flabellifera*** *** Spongiidae * Dictyoceratidae * Papua New Guinea * Port Moresby * barrier reef * 60 ft (18 m).** The fleshy plates of *C. flabellifera* are superficially similar to a number of other sponges in the genera *Phyllospongia* and *Strepsichordaia*. The surface texture of *C. flabellifera* is distinctive. All these sponges have adaptations to capture light for algal symbionts living within the sponge.

175- ***Carteriospongia*** **sp.cf.** ***contorta*** *** Spongiidae * Dictyoceratida * Indonesia * Manado * 66 ft (20 m).** This unusual sponge is easily recognized, although it may be confused with a similar appearing gorgonian, *Soelencaulon* sp.. The blades of the sponge have ribs with the oscules only on one side, the other side is smooth.

169- ***Oceanapia sagittaria*** * Palau

170- ***Oceanapia*** **sp.** * Papua New Guinea

171- ***Oceanapia*** **sp.** * Papua New Guinea

172- ***Oceanapia*** **sp.** * Papua New Guinea

176- ***Coscinoderma*** **sp. * Spongiidae * Dictyoceratida * Papua New Guinea * Kavieng * Albatross Channel * 100 ft (30 m).** This interesting sponge is like a floret, the attachment being a short stalk which expands out to the flattened head of the sponge. We have found this species along northern Papua New Guinea at Kavieng and Madang, between 90 and 150 feet.

177- ***Dactylospongia elegans*** *** Spongiidae * Dictyoceratida * Federated States of Micronesia * Chuuk Atoll * Northeast Pass * 50 ft (15 m).** This is an extremely common sponge in Micronesia. It usually occurs as an encrusting mass, but also may have digitate projections. It has a very rubbery and tough surface texture. It is extremely abundant on the windward slopes of Chuuk Atoll. At depth, because of the decrease in red light, it appears greenish, rather than the yellow orange color seen in the photograph

178- ***Dactylospongia*** **sp. * Spongiidae * Dictyoceratida * Indonesia * Manado * 40 ft (12 m).** This species of *Dactylospongia* has an attractive pattern of excurrent channels leading to the oscules spread among the many lobes which make up the body of the sponge.

179- ***Hippospongia ammata*** *** Spongiidae * Dictyoceratida * Federated States of Micronesia * Chuuk Atoll * barrier reef * 50 ft (15 m).** Members of *Hippospongia* have their fibrous skeleton oriented with most of the fibers in one direction. The sponges have commercial value and when prepared properly they make excellent bath sponges. The skeleton is easily distinguished from that of *Spongia.*

180- ***Leiosella*** **sp. * Spongiidae * Dictyoceratida * Papua New Guinea * Madang * Planet Rock Reef * 100 ft (30 m).**

181- ***Phyllospongia foliascens*** *** Spongiidae * Dictyoceratida * Papua New Guinea * Madang * 90 ft (27 m).** Another of the "fan-sponges", this species is similar in appearance to members of *Carteriospongia* and *Strepsichordaia*. It has algal symbionts.

182- ***Spongia*** **sp.? * Spongiidae * Dictyoceratida * Papua New Guinea * West New Britain * 100 ft (30 m).** This and the next two photographs are all members of the genus *Spongia*, what are commonly known as bath sponges. The taxonomy of this group, unfortunately, is in chaos. Most bath sponges are nominally identified as *Spongia officinalis*, a Mediterranean species, but the chances of the Pacific Ocean bath sponges being the same species are remote.

183- ***Spongia*** **sp. * Spongiidae * Dictyoceratida * Papua New Guinea * New Ireland * fringing reef * 66 ft (20 m).** This close-up photograph shows a small *Spongia* with its smooth surface texture and apical oscule. Sponges of this genus are classical bath sponges as their skeletons are soft and lack spicules. In order to be prepared for use, sponges need to have the tissue removed from the fibrous skeleton by soaking them in water, then wringing the sponge out at regular intervals until all the tissue is gone. Most bath sponges have been called *S. officinalis,* a Mediterranean species, but the chances of the Pacific bath sponges being the same species are remote.

173- *Pellina* sp. * Federated States of Micronesia

174- *Carteriospongia flabellifera* * Papua New Guinea

175- *Carteriospongia sp.cf. contorta* * Indonesia

176- *Coscinoderma* sp. * Papua New Guinea

177- *Dactylospongia ?elegans* * Federated States of Micronesia

178- *Dactylospongia* sp. * Indonesia

179- *Hippospongia amata* * Federated States of Micronesia

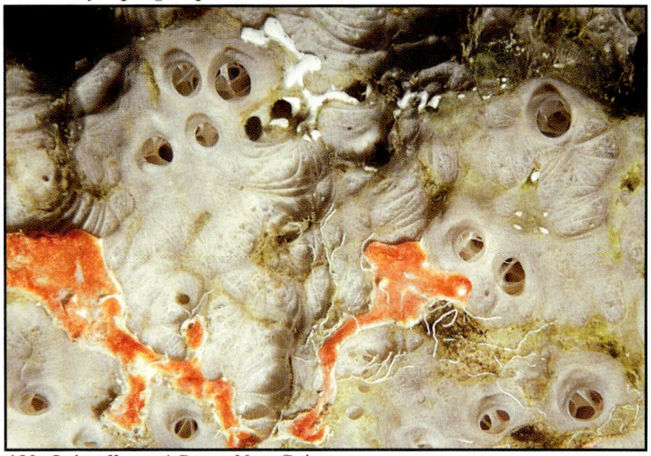

180- *Leiosella* sp. * Papua New Guinea

181- *Phyllospongia foliascens* * Papua New Guinea

182- *Spongia* sp. * Papua New Guinea

183- *Spongia* sp. * Papua New Guinea

184- *Spongia* sp. * Spongiidae * Dictyoceratida * Philippines * Santa Rosa 6 * 165 ft (50 m). This sponge, again nominally *S. officinalis*, appears quite different, being a brilliant white, perhaps in response to the depth where it was photographed. A number of white synaptid holothurians (sea cucumbers) are on the outer surface of the sponge.

185- *Strepsichordaia aliena* * Spongiidae * Dictyoceratida * Philippines * Pamalican Island * 50 ft (15 m). This unusual sponge occurs as flattened or bent plates attached by a single stalk. While flattened like other phototrophic sponges, it is much thicker, like a large pancake.

186- *Coscinoderma* sp. * Spongiidae * Dictyoceratida * Papua New Guinea * Dyaul Island * 100 ft (30 m). Another spongiid which in this case has a velvety surface texture.

187- *Hyrtios erecta* * Thorectidae * Dictyoceratida * Federated States of Micronesia * Chuuk Atoll * Tonoas * 30 ft (9 m). This is one of the most common inshore sponges found in Micronesia. It forms very dark solid finger-like projections with a surface that is regulalary covered in sharp little conules which are lighter than the background color of the sponge. It is found from shallow water to deep reefs. The sponge incorporates sand into its structure, and the more sand it contains, the more brittle are the branches; in such cases the branches break rather than bending.

188- *Hyrtios mela* * Thorectidae * Dictyoceratida * Federated States of Micronesia * Chuuk Atoll * Northeast Pass * 30 ft (9 m). The branches of this very common sponge are superficially similar to *H. erecta*, but are usually larger in diameter and have a reddish or orange tinge which is absent in the gray or black branches of *H. erecta*.

189- *Ircinia ramosa* * Irciniidae * Dictyoceratida * Federated States of Micronesia * Chuuk Atoll * lagoon patch reef * 50 ft (15 m). These rubbery branches form an anastomosing network and are extremely tough and hard to tear, a common feature of many species of *Ircinia*. The surface has small conules and oscules scattered over it. This sponge is common on many lagoon reefs in Micronesia and it, or a closely related species, inseparable at present, occur to depths of about 180 feet where the branches become more slender.

190- *Ircinia* sp. * Irciniidae * Dictyoceratida * Philippines * Cebu * Mactan Island * 100 ft (30 m).

191- *Ircinia* sp. * Irciniidae * Dictyoceratida * Papua New Guinea * New Britain * Garove Island * 33 ft (10 m). This encrusting sponge is rubbery and hard to tear.

192- *Hippospongia metachromia* * Thorectidae * Dictyoceratida * Federated States of Micronesia * Chuuk Atoll * Gosei Maru * 33 ft (10 m). This is a common sponge in Micronesia, and although actually yellow in color it usually appears green at depth. It is abundant on many of the shipwrecks in Chuuk lagoon.

193- *Luffariella* sp. * Thorectidae * Dictyoceratida * Federated States of Micronesia * Chuuk Atoll * barrier reef * 50 ft (15 m). This sponge is very common on reefs in Micronesia. It can also grow in the form of small vases.

194- *Sarcotragus* sp.cf. *arbuscula* * Irciniidae * Dictyoceratida * Papua New Guinea * Dyaul Island * 100 ft (30 m). This genus has a rough texture, with a conulose surface.

184- *Spongia* sp. * Philippines

185- *Strepsichordaia aliena* * Philippines

186- ***Coscinoderma* sp.** * Papua New Guinea

187- ***Hyrtios erecta*** * Federated States of Micronesia

188- ***Hyrtios mela*?** * Federated States of Micronesia

189- ***Ircinia ramosa*** * Federated States of Micronesia

190- ***Iricinia* sp.** * Philippines

191- ***Iricinia* sp.** * Papua New Guinea

192- ***Hippospongia metachromia*** * Federated States of Micronesia

193- ***Luffariella* sp.** * Federated States of Micronesia

194- *Sarcotragus* sp.cf. *arbuscula* * Papua New Guinea

195- *Thorectandra* sp. * Federated States of Micronesia

196- *Aplysilla sulphurea* on *Collocalypta* sp. * Papua New Guinea

197- *Aplysilla* sp. * Federated States of Micronesia

195- *Thorectandra* sp. * Thorectidae * Dictyoceratida * Federated States of Micronesia * Chuuk Atoll * barrier reef * 100 ft (30 m). These more or less spherical sponges are distinctive from the superficially similar to "golf ball" sponges. In *Thorectandra* the surface is hard, formed into ridges, with much sand incorporated into it and with apical oscules. The illustrated species is relatively common in the Caroline Islands on reefs at moderate depths.

196- *Aplysilla sulphurea* on *Collocalypta* sp. * Aplysillidae * Dendroceratida * Papua New Guinea * Hansa Bay * reef * 33 ft (10 m). This is another one of the remarkable cases of two sponges occurring together. The yellow *A. sulphurea* occurs on the outside of white tubes of *Colocalypta* sp. which grows in the crevices formed between plates of coral. These sponges have been seen together in Papua New Guinea, Indonesia and the Federated States of Micronesia. The oscules of the sponges in the photo are protruding among many small hydroids.

197- *Aplysilla* sp. * Aplysillidae * Dendroceratida * Federated States of Micronesia * Chuuk Atoll * Tsis Island * 33 ft (10 m). Members of *Aplysilla* are usually encrusting, with great reduction in their fibrous skeleton. These sponges lack spicules and rely on small tree-like fibers for skeletal support.

198- *Chelonaplysilla* sp. * Aplysillidae * Dendroceratida * Palau * Airai Channel * 10 ft (3 m). While this and the following species of *Chelonaplysilla* can be the size of a fist or larger, there is very little mass to the sponge, most of the volume is made up of water contained within it. The fibrous skeleton is greatly reduced and overall the sponge has the appearance of a fragile tent held up by a limited number of supports (fibers). This species is always the purple brown color of the photograph, both internally and externally.

199- *Chelonaplysilla* sp. * Aplysillidae * Dendroceratida * Federated States of Micronesia * Chuuk Atoll * lagoon reef * 33 ft (10 m). This species of *Chelonaplysilla* is always gray in color, as seen in the photo. Both species occur on reefs and rocks in moderate depths, but little is known of their distribution limits.

200- *Dendrilla* sp. * Aplysillidae * Dendroceratida * Papua New Guinea * Rabaul * Pidgin Islands * 60 ft (18 m). Unlike *Aplysilla,* species of *Dendrilla* arise tree-like from an encrusting base.

201- *Euryspongia* sp. * Dysideiidae * Dendroceratida * Federated States of Micronesia * Chuuk Atoll * Tonoas * algal flat * 160 ft (48 m). This sponge has been found only on a deep algal flat in Chuuk Lagoon.

202- *Dysidea avara* * Dysideidae * Dendroceratida * Federated States of Micronesia * Chuuk Atoll * 66 ft (20 m). The species is pink or lavender in color and smells strongly of garlic out of water. Known from a number of areas in the western Pacific, this sponge quite possibly has a circumtropical distribution. The species was originally described from the Mediterranean Sea, and there is some doubt whether the Pacific populations are really the same species. Interestingly this sponge was the original source of a compound known as averol which, although not directly used as a drug, was the source used in developing the drug AZT used in AIDS therapy.

203- *Dysidea granulosa* * Dysideidae * Dendroceratida * Federated States of Micronesia * Nama Island * 50 ft (15 m). This species of *Dysidea* grows on rocky substrata of coral heads. The photographed specimen is beginning to lap onto the living coral and it is unknown whether or not the sponge would actually kill the coral with which it comes in contact. There are many examples of sponges and ascidians growing over and killing corals. This is a fairly common sponge throughout Micronesia, being found in both lagoon and outer reef areas. While it is most commonly seen in the branch-like structure, it can also occur as small sheets of sponge without visible branches.

204- *Dysidea herbacea* * Dysideidae * Dendroceratida * Palau * Ulong Channel * 40 ft (12 m). The flattened growth form of *D. herbacea* helps to capture maximum light for symbiotic algae living within its tissues.

205- *Dysidea herbacea* * Dysideidae * Dendroceratida * Federated States of Micronesia * Nama Island * 40 ft (12 m). *Dysidea herbacea* takes many different morphologies dependent upon environmental conditions and depth. The sponge growth form of encrusting on the bottom with vertical projections, is often found in *D. herbacea.* The species may also be yellow or green in color as seen in the photograph.

206- *Dysidea* sp. * Dysideidae * Dendroceratida * Indonesia * Sulawesi * 50 ft (15 m).

207- *Dysidea* sp. * Dysideidae * Dendroceratida * Palau * Idim's Corner * 50 ft (15 m).

198- ***Chelonaplysilla* sp.** * Palau

199- ***Chelonaplysilla* sp.** * Federated States of Micronesia

200- ***Dendrilla* sp.** * Papua New Guinea

201- ***Igernella* sp.** * Federated States of Micronesia

202- ***Dysidea avara*** * Federated States of Micronesia

203- ***Dysidea granulosa*** * Federated States of Micronesia

204- ***Dysidea herbacea*** * Palau

205- ***Dysidea* sp.** * Federated States of Micronesia

206- ***Dysidea* sp.** * Indonesia

207- ***Dysidea* sp.** * Palau

208- ***Dysidea* sp.** * Papua New Guinea

209- ***Dysidea* sp.** * Papua New Guinea

208- *Dysidea* sp. * Dysideidae * Dictyoceratida * Papua New Guinea * Kavieng * Dyaul Island * 66 ft (20 m). This species occurred along a deep reef dropoff.

209- *Dysidea* sp. * Dysideidae * Dendroceratida * Papua New Guinea * Port Moresby * barrier reef * cave * 66 ft (20 m). This soft little sponge occurs within reef caves, hidden away from light. An almost fluorescent white color, the tissue of the sponge is lost by even minimal handling, the cells washing away in a milky exudate, until there is nothing left but the skeleton of the sponge.

210- *Dysidea* sp. * Dysideidae * Dendroceratida * Federated States of Micronesia * Chuuk Atoll * Fourup Reef * crevice * 60 ft (18 m). This sponge is usually visible only by the purple tubular oscules, often as a cluster of tubes which project out from crevices on the reef. Most of the sponge is hidden away inside the reef and is actually almost colorless. This sponge occurs in Chuuk, Palau and Papua New Guinea, but is probably much more widely distributed.

211- *Halisarca* sp. * Halisarcidae * Dendroceratida * Palau * Ngerkuul Pass * 33 ft (10 m). Members of *Halisarca* lack a skeleton of spongin fibers. This sponge appears to have compensated by growing over a calcareous alga whose skeleton forms a secondary skeleton of the sponge.

212- *Halisarca* sp. * Halisarcidae * Dendroceratida * Federated States of Micronesia * Chuuk Atoll * east barrier reef * 40 ft (12 m). This encrusting *Halisarca* superficially resembles a verongid sponge, the following group, in having projections of its fibrous skeleton reaching to the surface of the sponge. However, in this case the sponge appears to have overgrown a calcareous algae and somehow incorporated calcareous elements into its structure. This sponge is most common on the windward outer reef of Chuuk, but also occurs occasionally on lagoon reefs.

213- *Aplysina* sp. * Druinellidae * Verongida * Federated States of Micronesia * Chuuk Atoll * Northeast pass * 110 ft (33 m). This encrusting verongid exhibits many of the typical characters found in the order Verongida. Principal among these is a lack of spicules (calcareous or siliceous), and a characteristic dendridric or reticulate fiber skeleton. This sponge is very similar to *Aplysina* from the Caribbean. The sponge turns dark on exposure to air, if handled the sponges will also turn fingers dark (without apparent harm).

214- *Aplysinella rhax* * Druinellidae * Verongida * Papua New Guinea * Madang * Rasch Passage * 50 ft (15 m). This encrusting sponge is very common in the Caroline Islands, being one of the most abundant sponges in Chuuk below about 60 feet. It varies somewhat in appearance over the range of environments it inhabits, but the shape of the oscules and the papillate texture seem to be reasonably consistent. It is usually quite inflated with water and like *Aplysina*, members of the genus *Aplysinella* turn dark on exposure to air. Its distribution is believed to include Micronesia, Indonesia, Papua New Guinea and the Philippines.

215- *Aplysinella strongylata* * Druinellidae * Verongida * Papua New Guinea * Kavieng * 66 ft (20 m). This and the following species resemble a described species named *A. strongylata*, but it is unknown whether either one really represents that species and/or if the two illustrated species represent the same sponge. The sponge photographed here is found on deep reef faces, often to depths of 180 feet or more. Similar or identical sponges are found in Chuuk, Palau and elsewhere. This genus is very close to *Pseudoceratina*, but doesn't turn blue in air like *Pseudoceratina*.

216- *Aplysinella strongylata* * Druinellidae * Verongida * Papua New Guinea * Bagabag Island * cave * 66 ft (20 m). This tightly clustered group of tubes is typical of many verongids; the fibrous skeleton protruding to the surface of the sponge and the lack of spicules which allow the sponge to be easily torn. The apical opening of each tube is similar to that seen in the previous species. Like most verongids it feels slippery, and has no strong smell.

217- *Pseudoceratina* sp. * Druinellidae * Verongida * Federated States of Micronesia * Chuuk Atoll * algal flat * 115 ft (35 m). This sponge is known to occur only on deep algal flats, where it can be very common. We first found this sponge on a "*Halimeda* ridge", a narrow ridge made up of the skeletal material of that algal genus, rising off the deeper lagoon bottom in Chuuk. While many members of *Pseudoceratina* are hard, this species is quite soft.

218- *Pseudoceratina* sp. * Druinellidae * Verongida * Indonesia * Talisei * 3 ft (1 m). The fibrous skeleton of this verongid is clearly visible where it reaches to the surface as small tubercles. This sponge is tough and rubbery, and the photographed individual was found in very shallow water on a reef.

219- *Pseudoceratina* sp. * Druinellidae * Verongida * Papua New Guinea * Rabaul * Pidgin Islands * 86 ft (26 m). This verongid is found on reefs in northern Papua New Guinea. Its knobby surface is a distinguishing field character; however the species is undetermined at this time.

210- *Dysidea* sp. or new genus * Federated States of Micronesia

211- *Halisarca* sp. * Palau

212- *Halisarca* sp. * Federated States of Micronesia

213- *Aplysina* sp. * Federated States of Micronesia

214- *Aplysinella rhax* * Papua New Guinea

215- *Aplysinella* sp.cf. *strongylata* * Papua New Guinea

216- *Aplysinella* sp.cf. *strongylata* * Papua New Guinea

217- *Pseudoceratina* sp. * Federated States of Micronesia

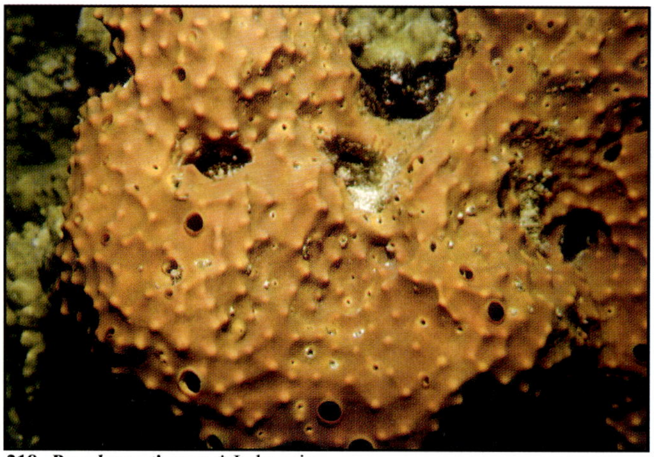
218- ***Pseudoceratina* sp.** * Indonesia

219- ***Pseudoceratina* sp.** * Papua New Guinea

220- ***Pseudoceratina* sp.** * Mortlock Islands

221- ***Pseudoceratina pedunculata*** * Papua New Guinea

220- *Pseudoceratina* sp. * Druinellidae * Verongida * Federated States of Micronesia * Mortlock Islands * Ettal Atoll * 66 ft (20 m). This occurs along overhangs of reefs in moderate depths in the Caroline and Marianas Islands.

221- *Pseudoceratina pedunculata* * Druinellidae * Verongida * Papua New Guinea * Bagabag Island * 90 ft (27 m). This distinctive sponge has a large spherical "head" on a slender stalk which is attached to the reef. Typically, this species is found on offshore reef dropoffs from about 40 to 150 feet deep. It occurs in Papua New Guinea. The photographed individual has brittlestars with extremely long arms associated with it: the disk of the brittlestar is in the osculun of the sponge and the arms are draped over the outer surface of the sponge.

222- *Pseudoceratina verongitea* * Druinellidae * Verongida * Federated States of Micronesia * Chuuk Atoll * lagoon reef * 50 ft (15 m). This encrusting verongid occurs on coral rocks. The oscules are distinctive with their yellow rim.

223- Unidentified Druinellid * Druinellidae * Verongida * Federated States of Micronesia * Chuuk Atoll * east barrier reef * 40 ft (12 m). This yellow green verongid is extremely common on offshore and lagoon reefs in the Caroline Islands, particularly around Chuuk. It is dense, rubbery and turns dark on exposure to air.

224- *Ianthella basta* * Ianthellidae * Verongida * Indonesia * Biak * 50 ft (15 m). The genus *Ianthella*, with fan-like sponges up to three feet or more across, is almost impossible to confuse with any other and the species illustrated here is probably the most common in the region. *Ianthella basta* can be yellow, green, purple, blue or brown in color and can be fan-like, cone shaped or a mixture of the two. When removed from the water, the sponges turn dark due to oxidation of their pigments. *Ianthella basta* is believed to have the widest distribution of any species in the genus, occuring in Indonesia, Papua New Guinea, the Philippines and Australia. It is not known from the Caroline Islands or further east in Micronesia. To further complicate matters there are, however, there are at least three undescribed species within the genus in northern Australia and southern Papua New Guinea.

225- *Ianthella basta* * Ianthellidae * Verongida * Guam * Apra Harbor * 66 ft (20 m). The great fans of *I. basta* sometimes occur in high concentrations in some areas, looking like the shallow water "sea fan" gorgonians of the tropical Atlantic, a matter of convergent evolution in their morphology. The fans of *I. basta* are thin with vertical ribbing on their surfaces, but a similar species, *Ianthella flabelliformis,* is thicker with thick ridges on its surface. The two species can occur in the same area, and despite the variability in color of *I. basta*, it is easily distinguished from *I. flabelliformis*. These sponges occasionally have a small goby, *Pleurosicya elongata*, living on the surface of the sponge.

226- *Ianthella basta* * Ianthellidae * Verongida * Indonesia * Biak * 50 ft (15 m).

Class Calcarea

The remaining species within this chapter are calcareous sponges, members of the Class Calcarea. These sponges have spicules which are made only of calcium carbonate, and only in the mineral form known as calcite. The calcareous sponges are much fewer in number in shallow tropical environments than the Demosponges, but are still important in our region. They are most common in temperate waters, with about 500 species having been described worldwide. This is a group for which a great deal of basic taxonomic work remains to be done, and as can be noted from the following, many tropical species certainly remain undescribed.

227- *Clathrina* sp. * Clathriniidae * Clathrinida * Papua New Guinea * West New Britain * cave * 75 ft (23 m). Calcareous sponges of the genus *Clathrina* are delicate in appearance with a fine web of tubes filled with water. There is really little substance to the actual sponge. They lack large spicules and are soft to the touch. Most often they are found in reef caves. The photographed individuals were found in very dark, extensive crevices and caves incised into a vertical face on a reef far offshore.

228- *Clathrina* sp. * Clathriniidae * Clathrinida * Papua New Guinea * Garove Island * Dudu Rock * cave * 66 ft (20 m). This small species of *Clathrina* is probably a different species than the previous one. The location where it occurred, Dudu Rock, is a small rocky island at the mouth of the harbor at Garove Island, north of New Britain. The harbor at Garove is actually a submerged volcanic crater, open to the ocean along one side and hundreds of feet deep in its center. Gases bubble from the sediments and the area around the bubbling gases is largely devoid of reef life. Dudu Rock, however, had much life growing on it and the sponge was abundant in small caverns in the rock.

222- ***Pseudoceratina verongitea*** * Federated States of Micronesia

223- Druinellid * Federated States of Micronesia

224- ***Ianthella basta*** * Indonesia

225- ***Ianthella basta*** * Guam

226- ***Ianthella basta*** * Indonesia

227- ***Clathrina*** **sp.** * Papua New Guinea

228- ***Clathrina*** **sp.** * Papua New Guinea

229- ***Leucetta primigenia*** * Indonesia

230- ***Leucetta*** **sp.** * Federated States of Micronesia

231- ***Leucetta*** **sp.** * Papua New Guinea

232- ***Pericharax heterorhaphis*** * Federated States of Micronesia

229- *Leucetta primigenia* * Leucettidae * Clathrinida * Indonesia * Biak * 86 ft (26 m). This sponge is one of the largest calcareous sponges in the Indo-west Pacific and can occur in both inshore and offshore reef areas. It also occurs commonly on the wrecks in Chuuk lagoon. The brown exterior and creamy interior is normal for the species. It is very spiculous and can pierce skin with its large sharp spicules.

230- *Leucetta* sp. * Leucettidae * Clathrinida * Federated States of Micronesia * Chuuk Atoll * 100 ft (30 m). This sponge, with spiculous yellow lobes, is found in reef caves and along deep drop offs in many areas. It occurs throughout the region.

231- *Leucetta* sp. * Leucettidae * Clathrinida * Papua New Guinea * Bagabag Island * fringing reef face * 50 ft (15 m). We have seen this distinctive sponge only at Bagabag Island, a volcanic island offshore of Madang on the north coast of New Guinea. It has the spiculous nature of so many calcareous sponges and forms convoluted masses which appear as if fashioned from melted candle wax, often "dripping" down slopes and forming fantastic shapes.

232- *Pericharax heterorhaphis* * Leucettidae * Clathrinida * Federated States of Micronesia * Chuuk Atoll * Fourup Reef * 60 ft (18 m). This sponge is common on reefs in much of Micronesia, but is not seen so often in other areas such as Papua New Guinea. Small individuals, such as the two shown in the photograph, are easily recognized; a lobate or conical shape with an oscular opening and the outer surface mottled brown with green underneath. Large individuals, as much as two feet in length, particularly those in calm water, can assume grotesque shapes with the mass of the sponge appearing to "drip" down a vertical face. The outer surface can also become very dark when exposed to abundant light. The internal color is always a lime green. This sponge was also described as *Leucetta avocado* by deLaubenfels, but since *P. hererorhaphis* is the earlier name (1884 vs. 1954), it has priority.

233- *Pericharax* sp. * Leucettidae * Clathrinida * Philippines * Cebu * Mactan Island * 60 ft (18 m). This small reef-dwelling calcareous sponge is believed to be a member of *Pericharax*, but is almost certainly an undescribed species. It is like a thin walled bag attached to the bottom by a delicate stalk. We have photographed it in the Philippines and Indonesia, but know nothing more of its distribution.

234- *Dendya prolifera* * Leucosoleniidae * Leucosoliniida * Chuuk Atoll * Tonoas * 40 ft (12 m). This distinctive sponge occurs in Micronesia, Papua New Guinea and the Philippines. It is locally abundant in some areas. The sponge has a thin walled main body with a large opening opposite the attachment. Projections with opaque nodules occur along its surface, resulting in a rough appearing sponge. This species is quite fragile and soft, unlike so many of the heavily spiculed Calcarea. It is most often found in small caves and crevices, sometimes on the undersurface of large coral heads, where it often grows hanging down from its attachment.

235- *Leuconia palaoensis*? * Leucosoleniidae * Leuconsoleniida * Papua New Guinea * Bagabag Island * 50 ft (15 m). This species was growing on the vertical face of a fringing reef around Bagabag Island off Madang. The fringing reefs of Bagabag were one the richest, most interesting areas we have ever seen, with beautiful reef caves occurring at reasonable depth on the near vertical face of the fringing reef. The fringing reef wall was coated with a huge variety of marine invertebrates.

236- *Leucosolenia* sp. * Leucosolenidae * Leucosoleniida * Papua New Guinea * Pidgin Islands * 66 ft (20 m). This pale orange calcareous sponge is hardly more than a gelatinous layer, often folded into ruffles, which grows on the undersurfaces of coral heads and rocks. What is apparently the same species occurs in the Carolines Islands and Papua New Guinea.

237- *Sycon* sp. * Sycettidae * Leucosoleniida * Papua New Guinea * West New Britain * 66 ft (20 m). These are really tiny sponges, but of such distinctive form, they are readily identified, even if an exact scientific name can not be placed on them. The photograph shows a group of individual tubes, each one actually a separate sponge. These sponges have among the simplest of water pumping systems.

238- *Sycon* sp. * Sycettidae * Leucosoleniida * Philippines * Pamalican Island * 66 ft (20 m). This closeup view shows the flask-like structure of a species of *Sycon*.

233- ***Pericharax*** **sp.** * Philippines

234- ***Dendya prolifera*** * Federated States of Micronesia

235- ***Leuconia palaoensis*****?** * Papua New Guinea

236- ***Leucosolenia*** **sp.** * Papua New Guinea

237- ***Sycon*** **sp.** * Papua New Guinea

238- ***Sycon*** **sp.** * Philippines

This photograph, taken inside a cave at a depth of 100 feet near Cebu in the central Philippines reveals several species of calcareous sponges in their natural habitat.

Phylum Cnidaria

Hydroids, Jellyfishes, Corals, Sea Anemones and Black Corals

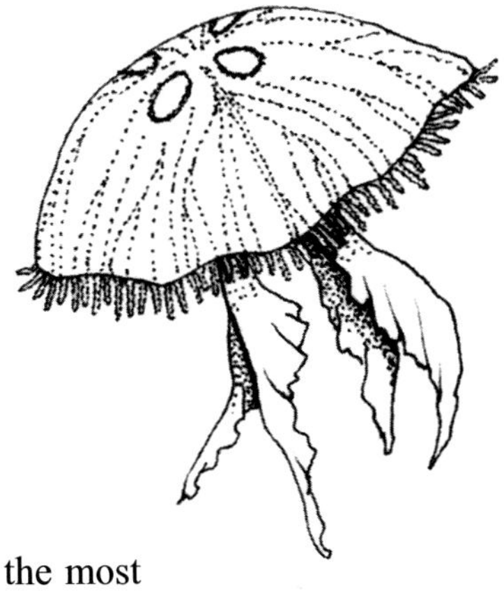

Cnidarians, also known as coelenterates, are arguably the most common and conspicuous invertebrates found in shallow tropical Pacific waters because this phylum includes the many species of corals which build coral reefs. With more than 10,000 species, the cnidarians are predominantly marine and reach a level of diversity and importance in shallow tropical waters unequalled by any other phylum. They form the basis of many tropical reefs ecosystems, but are also abundant in colder water. There are four classes: the Hydrozoa (hydroids), Scyphozoa (jellyfishes), Cubozoa (sea wasps), and Anthozoa (corals, corallimorpharians, sea fans, sea anemones, zoanthids and black corals), distinguished on the basis of life history and morphology. They are united by certain characteristics: radial symmetry, a central mouth surrounded by tentacles, a single opening through which food is ingested and expelled (coelenteron), a jelly-like middle germ layer (the mesoglea), and intracellular stinging structures called nematocysts.

Nematocysts (also called cnidae) are unique to the Cnidarians. Microscopic in size, they are used by the animal to capture food, protect against predators and to attach themselves to substrate. Nematocysts are formed by special cells (cnidoblasts) borne on the tentacles and other parts of the animal. They consist of an internal coiled thread that ends in a dart and a capsule. When the organism receives the proper stimulus, the nematocysts are fired like miniature harpoons, the dart injecting a small amount of painful toxin into the hapless victim while the thread aids in entangling and holding on to the prey.

Two body forms, the polyp and the medusa, are found in cnidarians; some species having only one form while others have both. The polyp lives attached to the substrate, has a fleshy body and an upward directed mouth surrounded by tentacles. The medusa is free swimming; the body is a dome-shaped bell with the mouth underneath and tentacles arranged around the margin. In the Hydrozoa, Scyphozoa and Cubozoa most species have alter-

Left- This vertical wall at Pescador Island near Moalboal in the central Philippines is covered with Cnidarians and other invertebrates. The species diversity in this area is very high. A large soft coral is visible in the center, also visible are numerous orange colonies of ahermatypic coral.

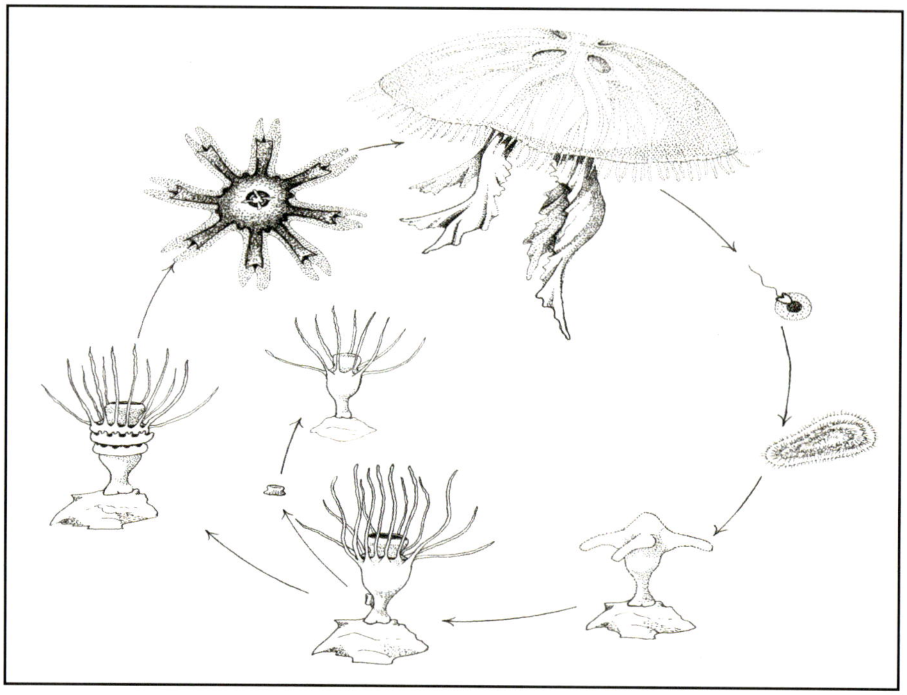

Above left- This photo shows three polyps of the cubomedusae *Carybdea alata* in the process of metamorphosis. The individual at the far right is almost ready to swim away as a small jellyfish. The polyp at the far left has not started to change and is still able to produce asexual buds. Below left- This is a young cubomedusae, *Carybdea alata,* several days after it has been released. As the jellyfish grows, the tentacles will increase in length and the bell will become transparent. Within a year the jellyfish will be sexually mature. Above right- Life history of *Aurelia*. Sperm and eggs produced by mature medusae are released into the water where fertilization takes place. The fertilized eggs develop into a planula larvae which attach to the substrate and develop into polyps. The polyps grow, increase the number of their tentacles and begin to bud off secondary polyps. Many clone-like secondary polyps are produced. After a period of about nine months some of the older polyps begin to produce medusae. Unlike cubomedusae where one polyp converts to one medusa, *Aurelia* polyps produce many juvenile jellyfish (ephyrae). There are numerous variations of this life history within the Cnidaria, in many species, only medusae or polyps develop.

nate polyp and medusa stages. Members of the remaining class, Anthozoa, exist only as polyps, either solitary or forming colonies. Within the Anthozoa the sea anemones corallimorpharians and cerianthids are solitary, the stony corals and zoanthids have species which are either colonial or solitary, while the sea fans, sea pens, softcorals (otcocorals) and black corals (antipatharians) are colonial.

The most abundant Cnidarians on coral reefs, the stony corals (Scleractinian corals), are enormously important in the ecology of tropical seas. In warm, clear water they build massive reefs, which are highly productive "oases" in otherwise relatively barren water masses. The process of calcification, in which stony coral polyps take carbon dioxide and water and with the aid of intracellular symbiotic algae (zooxanthellae) produce calcium carbonate, appears simple, but it is actually quite complex and still incompletely understood. The end result, however, is the enhanced growth of individual coral colonies which coexist and form much of the framework of most coral reefs. Those corals which contribute skeletal material to the overall framework of the reef are called hermatypic corals. Hermatypic corals generally belong to the scleractinia (stony corals), but can also include a few species of octocorals and hydrozoan corals such as *Heliopora* and *Millepora* respectively. The ahermatypic, or non-reef building corals, generally do not contain zooxanthellae; they include many species of solitary hydrozoans and scleractinians and do not contribute a significant amount of calcium carbonate to the reef structure..

Cnidarians capable of stinging humans include free-swimming jellyfishes of the Hydrozoa and Scyphozoa and benthic fire corals, hydroids, sea anemones, corallimorpharians, zoanthids and the polyp stage of some jellyfish. The injury they can cause humans ranges from no discernible effect to the potentially fatal stings produced by the powerful nematocyst

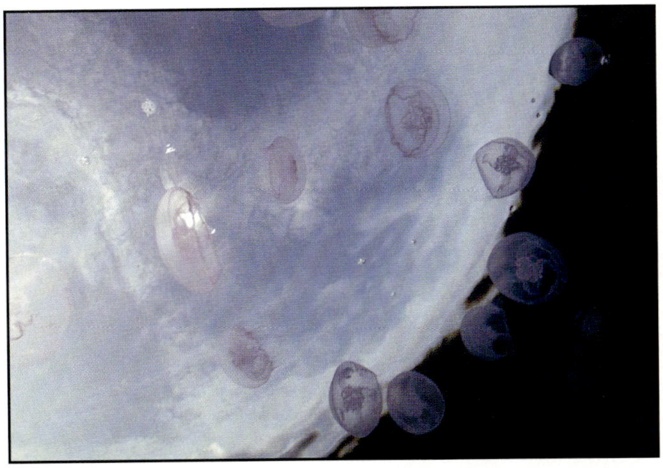

Above left- These Scyphozoan jellyfish, *Aurelia*, swim just beneath the surface of the water on calm days. From a a divers position below them looking up, they resemble flying saucers in outer space. These jellyfish feed on small planktonic animals which have also gathered together near the surface. Below left- Many other animals form symbiotic associations with Cnidarians. These juvenile fish (jacks) seek protection within the bell of the jellyfish. They are not imune to the sting of the medusa and will quickly become prey if they become too tangled in the tentacles. Above right- These mushroom corals (*Fungia* sp.) and related genera, are solitary unattached coral polyps as adults. Mushroom corals are able to roll over if turned upside down and they can move about the reef. They are often found in aggregations on patch reefs. The color of these corals is variable and due to the presence of symbiotic algae. Below right- This photo shows a section of reef in the Philippines, covered with soft and hard corals and other invertebrates.

toxins of some sea wasps and the Portuguese man-of-war. Hydroids, such as *Millepora, Aglaeophenia cupressina* and *Lytocarpus philippinus* and certain corallimorpharians can be very abundant in areas frequented by divers and swimmers. The sting and subsequent rash that results from contact with these Cnidarians is painful and may last for several days.

The nematocysts of octocorals usually will not sting humans, however many octocorals possess spicules, small calcareous structures which can scratch and penetrate human skin. Additionally, sea fans are home for certain species of small brittlestars, with abundant needle-like spines. Touching a sea fan with brittlestars, even with a gloved hand, may result in a very unpleasant burning sensation. The best treatment for these stings is 95% alcohol.

Some of the zoanthids are also potentially dangerous to humans. Members of *Palythoa* and *Zoanthus* have toxins, generally known as palytoxins, which are present in mucous and the gonads. Palytoxins are among the most toxic substances found in nature; fortunately they are not highly concentrated in the surface tissues of the zoanthid. While casual contact with zoanthids generally will not cause harm to humans, contact with *Palythoa* or *Zoanthus* through an open wound can be very painful and dangerous. The toxicity of *Palythoa* was well known to the ancient Hawaiians who coated spear tips with it to make them more deadly.

Among commonly observed hydrozoans on coral reefs are the hydroids, which are colonial, polyp-like animals. Large, bushy or plume-shaped, hydroids at first glance look more like algae than animals. Hydroids, less than an inch in height, are quite common although usually overlooked due to their small size and often cryptic habits. Some hydroids, such as the *Stylaster* and *Millepora* spp. (fire coral), have calcareous skeletons and resemble scleractinian corals. Hydroids and hydromedusae are carnivorous and feed on small planktonic animals.

Hydrozoans occur as either polyps or medusae or both. Hydromedusae are small jellyfish, generally less than an inch or so in diameter, and are usually transparent. They can often be observed near the surface on calm days in clear water. Some of these small jellyfish, and larger jellyfish-like hydrozoans such as the Portuguese man-of-war, are capable of producing painful stings when they come in contact with a swimmer. Use caution whenever dealing with hydroids. There are other animals which put these powerful nematocysts to their own uses. Hydroids are the prey of various nudibranchs who store unfired, ingested nematocysts in specialized pockets of their digestive tract . The nudibranch is then able to sting and use the nematocysts for its own defense.

Most hydroids have male and female individuals, but their life cycle is highly variable, sometimes complex and poorly understood. Generally, attached colonial hydroids develop male or female medusae which bud off, then swim away, develop gonads and reproduce. The fertilized egg divides and develops into a free swimming larvae that subsequently attaches to the bottom and grows into a new hydroid. Hydroids may live for as little as a few weeks or, as in the case of *Millepora,* many years. This alternation between medusae and attached polyp has led to much confusion in naming hydroids; often there are separate generic names for polyp and medusae of the same species.

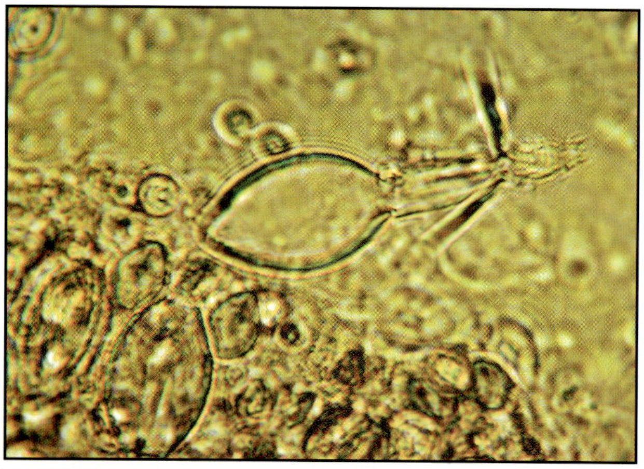

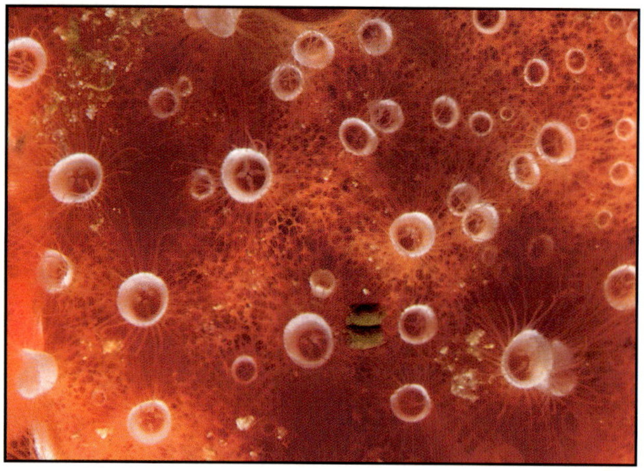

Above left- Photomicrograph of nematocysts from the Portugese man of war. One large exploded nematocyst is visible in the center, two large unfired nematocysts are just behind it, many small round nematocysts are also visible. Below left- This photo shows a small unidentified hydroid which lives with a sponge. Hydroids are often found growing on other organisms. Right- A large head of the fire coral *Millepora* sp. with a crinoid on top is in the center of the photo.

239- *Solanderia* sp. * Solanderiidae * Hydroida Papua New Guinea * Madang * barrier reef * 20 ft (6 m). The family and genus is found worldwide in the tropics. While a hydroid, this genus resembles a small sea fan or gorgonian with its branches usually in one plane, perpendicular to the current or wave action. There are several species, including some which live as deep as 300 feet or more. *S. misakinensis* is known from Japan and Hawaii. *S. minima* is known from Zanzibar and possibly Hawaii. *S. secunda* is known from the central Pacific. The most common members are always found in exposed areas on wave swept shallow outer reefs.

240- *Aglaophenia cupressina* * Plumaridae * Hydroida * Papua New Guinea * Madang * Cape Croiselles * 30 ft (9 m). This large hydroid often covers large areas of reef and, unfortunately, stings humans badly. The sting it produces is sharp and painful, almost like an electric shock, rather than the burning sensation produced by contact with the fire corals of the genus *Millepora*. The rash which results from even a small sting may last several days. Since *Aglaophenia* is large and conspicuous, however, it is possible to avoid contact. Special care should be exercised when diving around large concentrations of *A. cupressina*, as an inadvertent swell or wave might carry a diver into contact with it.

241- Unidentified hydroid * Plumaridae * Hydroida * Palau * Mutremdiu Wall * 66 ft (20 m). This and the following species are closely related to *Aglaophenia*. Both species are widely distributed in our region. The sting which results from contact with these species is as painful as that of *Aglaophenia*.

242- Unidentified hydroid * Plumaridae * Hydroida * Marshall Islands * Enewetak Atoll * 90 ft (27 m).

243- *Lytocarpus phoenicea* * Plumaridae * Hydroida * Federated States of Micronesia * Chuuk * lagoon reef * 10 ft (3 m). Once you have touched this hydroid, you will not soon forget it. It stings badly and instantly on contact. Fortunately it is distinctive and once you know what it looks like it is easy to avoid. The species is common in Micronesia, Papua New Guinea and the Philippines.

244- *Lytocarpus philippinus* * Plumaridae * Hydroida * Hong Kong * Breaker Reef * 15 ft (5 m).
This species is closely related to *L. phoenicea*, and like the former species this hydroid is also a bad stinger. Both species are filter feeders and usually inhabit coral outcrops and ledges along drop offs.

245- *Cnidoscyphus* sp. * Plumaridae * Hydroida * Marshall Islands * Enewetak Atoll * cement ship * 20 ft (6 m). This small hydroid overgrows dead patches of substrate. It may form dense bush-like clumps that resemble red algae or form sparse linear colonies of just a few upright stalks. The polyps occur on small branches off a central stalk. The stalk may attain lengths of about four inches.

246- *Plumularia* sp.* Plumaridae * Hydroida * Palau * Lighthouse Reef * 10 ft (3 m). This genus has a central stalk with alternating side branches that bear polyps. Most species of *Plumularia* are white or grey, this one is a beautiful orange color. These hydroids are usually found in current swept areas and clean water. This species is also capable of stinging.

247- Unidentified hydroid * Plumaridae * Hydroida * Marshall Islands * Enewetak * Medren patch reef * 30 ft (9 m). There are many species of small hydroids on Pacific coral reefs, most are poorly known. This species is usually found growing near sponges.

239- *Solanderia* sp. * Papua New Guinea

240- *Aglaophenia cupressina* * Papua New Guinea

241- Unidentified hydroid * Palau

242- Unidentified hydroid * Marshall Islands

243- ***Lytocarpus phoenicea*** * Federated States of Micronesia

244- ***Lytocarpus philippinus*** * Hong Kong

245- ***Cnidoscyphus* sp.** * Marshall Islands

246- ***Plumularia* sp.** * Palau

248- Unidentified hydroid * Plumaridae * Hydroida * Marshall Islands * Kwajalein Atoll * Roi Namur Island * 40 ft (12 m).

249-Unidentified hydroid * Plumaridae * Hydroida * Indonesia * Biak* 30 ft (10 m). This hydroid is common on shallow reef throughout much of the region. It can cause mild skin irritation if handled.

250- *Antennellopsis integerrima* * Plumaridae * Hydroida * Papua New Guinea * Duke of York Islands * Ulu Pinnacle * 66 ft (20 m). This unusual little hydroid looks like tufts of segmented yellow filaments. We have seen it only occasionally and the location where the photograph was taken is the only area where we found it to be common.

251- *Myronema* sp. * Plumaridae * Hydroida * Papua New Guinea * Madang * sea grass bed * 3 ft (1 m). This species is common in very shallow water where it overgrows rocks and dead tree branches. From a swimmer's point of view, the colonies appear as an algal mat, upon closer inspection the tentacles are visible. The polyps are about one inch tall. The brown color is due to zooxanthellae.

252- *Rhizogeton* sp? * Clavidae * Hydroida * Federated States of Micronesia * Chuuk * Pis Moen Channel * 20 ft (6 m). This species lines the furrows in coral heads which are produced by alpheid shrimp.

253- *Tubularia* sp. * Anthomedusae * Hydroida * Philippines * Cebu * Mactan Island * reef wall * 50 ft (15 m). *Tubularia* is often found living on sponges, soft corals and other reef invertebrates. Unlike most other Hydrozoa the family Tubulariidae is characterized by the lack of a free-swimming medusa stage in its life cycle. The spherical pink structures below the tentacles are reproductive organs (gonophores) which produce eggs that develop into larva and later, new polyps. *Tubularia*, like many other hydroids, can reproduce by asexual division and is often found growing in dense patches. Members of this family have some of the largest polyps among the Hydrozoa; this one is about 1/4 inch in length.

254- *Halicordyle disticha* * Halichordylidae * Hydroida * Palau * Mutremdiu Wall * 30 ft (9 m). This hydroid is about six inches tall. It grows on coral rock on near-shore reefs. This species is also a member of the "fouling community", those organisms which attach and grow on pier pilings, ship hulls, etc. This species buds off small medusae.

255- *Millepora* sp. * Milleporidae * Milleporina * Federated States of Micronesia * Chuuk * Northeast Pass * 15 ft (5 m). The fire corals in the genus *Millepora* are common on all reefs of the tropical western Pacific, but not from Hawaii. The taxonomy of the genus is not very well known, although there are a number of described species. All the species in the genus cause a stinging or burning sensation when touched, but to many people the pain is not as instant and intense as stings from some hydroids.

256- *Millepora* sp. * Milleporidae * Milleporina * Federated States of Micronesia * Chuuk * Northeast Pass * 40 ft (12 m). There are quite a few described species of *Millepora*, but no one really knows how many of those are valid. For a number of specimens, even with collected material in hand, we cannot ascribe them to a particular species. We have opted instead to illustrate some forms we believe are sufficiently distinct and common to be included here but only refer to them as *Millepora* sp.

247- Unidentified hydroid * Marshall Islands

248- Unidentified hydroid * Marshall Islands

249- Unidentified hydroid * Indonesia

250- *Antennellopsis integerrima* * Papua New Guinea

251- *Myronoema* sp. * Papua New Guinea

252- *Rhizogeton* sp.?* Federated States of Micronesia

253- *Tubularia* sp. * Philippines

254- *Halicordyle disticha* * Palau

255- *Millepora* sp. * Federated States of Micronesia

256- *Millepora* sp. * Federated States of Micronesia

257- *Millepora platyphylla* * Federated States of Micronesia

258- *Millepora* sp. * Marshall Islands

257- *Millepora platyphylla* * Milleporidae * Milleporina * Federated States of Micronesia * Chuuk * South Pass Pinnacle * 6 ft (2 m). This species of *Millepora* occurs in shallow, often wave-swept waters. It consists of vertical plates which are often variously joined and fused, producing a stronger structure than if the plates all stood alone. This species is hard to confuse with other *Millepora,* except when it is small and the vertical plate structure is not well developed.

258- *Millepora* sp. * Milleporidae * Milleporina * Marshall Islands * Enewetak Atoll * Medren * patch reef * 10 ft (3 m).

259- *Millepora tuberosa* * Milleporidae * Milleporina * Federated States of Micronesia * Chuuk Atoll * Weno * west fringing reef * 40 ft (12 m).

260- *Stylaster* sp. * Stylasteridae * Hydroida * Papua New Guinea * Pidgin Islands * 40 ft (12 m). The small ahermatypic colonies of the genus *Stylaster* are some of the most delicate and beautiful cnidarians found on coral reefs. The genus occurs world-wide, even into polar seas, but it has its greatest flowering in the reefs of the Indo-west Pacific. Like so many groups of Cnidarians, the taxonomy of *Stylaster* and the related *Distichopora* are poorly known. There are at least 48 described species of *Stylaster*, but it is uncertain how many of those are really valid.

261- *Stylaster sanguineus* * Stylasteridae * Hydroida * Palau * Ngerkuul Pass * 100 ft (30 m). While most species of *Stylaster* are found beneath overhangs and in reef caves, *S. sanguinensis* is also very common in some deeper reef areas which are alternately swept by strong currents and slack water with the tides. The location where the photograph was taken, Ngerkuul Pass in Palau, is such an area and has an interesting assortment of creatures at 30-100 foot depths.

262- *Stylaster* sp. * Stylasteridae * Hydroida * Indonesia * Taliseo Island * 50 ft (15 m).

263- *Stylaster* sp. * Stylasteridae * Hydroida * Marshall Islands * Enewetak Atoll * leeward barrier reef * 66 ft (20 m).

264- *Distichopora borealis* * Stylasteridae * Hydroida * Federated States of Micronesia * Mortlock Islands * 50 ft (15 m). This is a typical branched colony of this genus. This species is common on reefs, under ledges, in Micronesia.

265- *Distichopora irregularis* * Stylasteridae * Hydroida * Indonesia * Manado * 20 ft (6 m). There are at least 34 described species of *Distichopora*, but the actual number of true species is not known. As with many groups of cnidarians, determining the taxonomic affiliation of a particular specimen is often very difficult due to lack of adequate collections, variation within individuals of a single species and a lack of understanding as to what exactly constitutes a species of many cnidarians. This problem has held back studies of these organisms and still presents a major challenge to contemporary taxonomists.

266- *Distichopora violacea* * Stylasteridae * Hydroida * Federated States of Micronesia * Chuuk * west reef channel * 40 ft (12 m). Colonies of *Distichopora* are most often found in caves and overhangs on patch reefs and outer dropoffs. These beautiful corals are fragile and easily damaged. They appear to grow very slowly.

259- *Millepora tuberosa* * Federated States of Micronesia

260- *Stylaster* sp. * Papua New Guinea

261- *Stylaster sanguineus* * Palau

262- *Stylaster* sp. * Indonesia

263- *Stylaster* sp. * Marshall Islands

264- *Distichopora borealis* * Federated States of Micronesia

265- *Distichopora irregularis* * Indonesia

266- *Distichopora violacea* * Federated States of Micronesia

267- ***Distichopora* sp.** * Marshall Islands

268- ***Distichopora* sp.** * Marshall Islands

269- ***Distichopora* sp.** * Marshall Islands

270- ***Distichopora* sp.** * Federated States of Micronesia

267- *Distichopora* sp. * Stylasteridae * Hydroida * Marshall Islands * Enewetak Atoll * leeward barrier reef * 33 ft (10 m). Some species of *Distichopora*, such as this one, are often abundant on the open reef face on the outer dropoffs. There is a species of violet ascidian growing at the base of the coral colony.

268- *Distichopora* sp. * Stylasteridae * Hydroidea * Marshall Islands * Enewetak Atoll * leeward barrier reef * 40 ft (12 m).

269- *Distichopora* sp. * Stylasteridae * Hydroida * Marshall Islands * Kwajalein Atoll * west barrier reef * 66 ft (20 m).

270- *Distichopora* sp. * Stylasteridae * Hydroida * Federated States of Micronesia * Chuuk lagoon * patch reef * 40 ft (12 m).

271- *Porpita pacifica* * Chondrophora * Hydrozoa * Indonesia * Bali * open water. This highly modified hydrozoan is more properly part of the open ocean community. It drifts ashore on reefs and beaches when blown in by wind and current. Often it is found in association with a small nudibranch, *Glaucus glaucus*, which feeds on the soft parts of the jelly.

272- *Physalia physalis* * Siphonophora *Hydrozoa * Marshall Islands * Enewetak Atoll * open water. The Portuguese man-of-war is one of the most painfully stinging and dangerous Cnidarians. Like all siphonophores it is a highly modified colony of small medusa-like individuals and specialized polyps. *Physalia* has no means of propulsion and relies on wind to move it across the surface. The tentacles may extend more than 20 feet underwater and care should be taken when swimming in the water near this species.

273- *Physophora hydrostatica* * Siphonophora * Hydrozoa * Marshall Islands * Enewetak Atoll * open water. This little siphonophore is about four inches long. The gas float at the top and the clear structures beneath it are modified medusae; they aid in flotation and propulsion. The tentacle-like structures are modified polyps. This species is only rarely seen near reefs.

274- *Olindias* sp. * Olindiadidae * Limnomedusae * Indonesia * Manado * open water. This little jellyfish has a bell about two inches across, yet may extend its tentacles more than ten feet below itself. This species is found near shore, often in murky water where it feeds on small fish and zooplankton.

275- *Aequorea australis* * Aequoreidae * Leptomedusae * Federated States of Micronesia * Chuuk * lagoon * 6 ft (2 m). *Aequorea* is common seasonally throughout the world's oceans. There are several species, and this one varies in bell diameter from two to ten inches, depending upon age. It is usually seen with many tentacles streaming out below it. The bell is almost without pigment and transparent.

276- *Timoides agassizi* * Timoididae * Leptomedusae * Marshall Islands * Enewetak Atoll * open water. This small medusa is known from both the Indian and Pacific Oceans where it is not often observed, but does appear in large numbers on occasion. The orange parts hanging below the bell are parts of the reproductive organ. The bell is about one inch across.

271- ***Porpita pacifica*** * Indonesia

272- ***Physalia physalis*** * Marshall Islands

273- ***Physophora hydrostatica*** * Marshall Islands

274- ***Olindias* sp.** * Indonesia

275- ***Aequorea australis*** * Federated States of Micronesia

276- ***Timoides agassizi*** * Marshall Islands

Class Scyphozoa - Jellyfish

The Scyphozoans are the animals we normally think of when someone mentions the word jellyfish. These relatively large and often brightly-colored medusae are some of the most beautiful animals in the sea. They often occur seasonally in large aggregations and are easily observed as they swim slowly near the surface. Scyphozoans, unlike their Hydrozoan relatives, not only tend to be larger, but also have different types of nematocysts. Most species are several inches to a foot in diameter, however, some attain bell diameters of several feet, which makes them the largest Cnidarians, except for some colonial scleractinan corals. Scyphozoans are entirely marine with their general habitats near shore. They sometimes are blown or carried into brackish water estuaries.

The life cycle of most Scyphozoans resembles the alternation between polyp and medusa stage found in the Hydrozoa, except that the polyp stage is much reduced. Based on the early development of the newly-settled larvae and a few morphological characters, the Scyphozoa comprise four groups; the Stauromedusae, Coronate medusae, Seameostomae, and Rhizostomae. The Cubomedusae (sea wasps), formerly included in the Scyphozoa, are now thought, by some taxonomists, to represent a separate class of Cnidaria, the Cubozoa, based on their early life history, however; for convenience, we have included them with the Scyphozoa .

Stauromedusae are jellyfish that do not swim. Usually less than an inch long, they have a stalk arising from the top of their bell which they use to adhere to blades of sea grass. Coronate medusae live primarily in very deep water but are represented in shallow tropical waters by several species. The Coronate medusae have larger polyps than most jellyfish and some of these polyps have been classified in genera different from the medusae due to previous lack of knowledge concerning their life history. The Seameostomae and Rhizostomae include the medusae we usually think of as jellyfish; *Aurelia* (moon jelly) and *Cassiopea* (upside-down jelly) respectively. They are common near shore, often colorful and at times can be observed with commensal fishes, nudibranchs or shrimp.

The Cubomedusae are nearly transparent and have four tentacles or bundles of tentacles that originate at the four corners of the squarish bell. They are represented by at least five species in the tropical Pacific, all of which can inflict a painful sting.

Little is known about the longevity of jellyfish; however, it appears most mature within a year's time, reproduce and then die. The following year new medusae bud off the surviving polyp stage and enter the water column to repeat the life cycle.

Jellyfish appear to have few predators. Several species of butterfly fish have been observed eating pieces of large jellyfish at Chuuk in Micronesia and there are reports of large turtles eating jellyfish. Certain large species are harvested commercially for food, however the bell must be prepared properly in order to remove all nematocysts before being consumed.

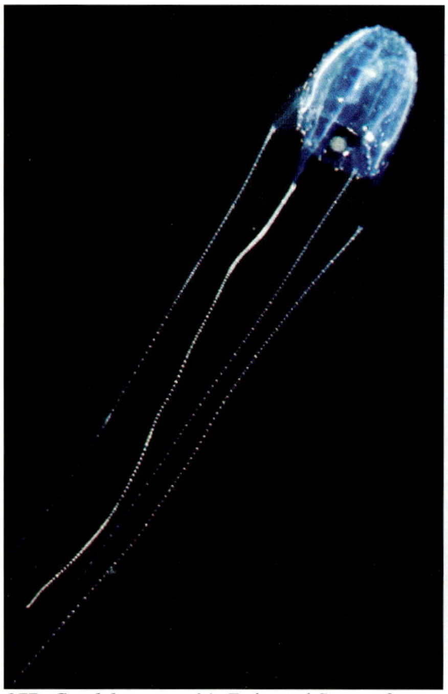

277- *Carybdea rastoni* * Federated States of Micronesia

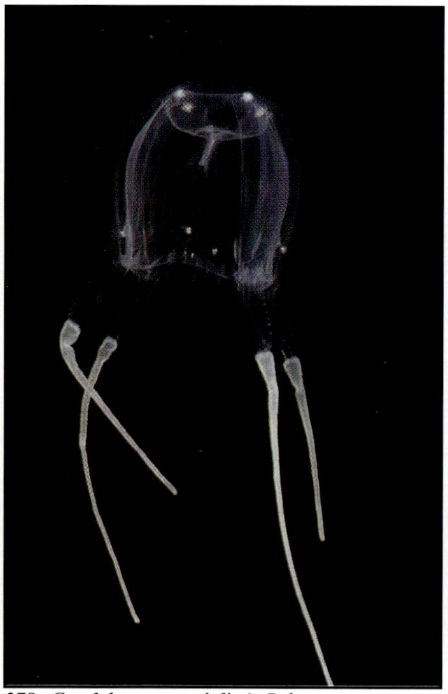

278- *Carybdea marsupialis* * Palau

279- *Chiropsalmus* sp. * Philippines

277- *Carybdea rastoni* * Carybdeidae * Cubomedusae * Cubozoa * Federated States of Micronesia * Chuuk Atoll * open water. This is the juvenile medusae of *Carybdea alata* several days after it has metamorphosed from polyp to medusa. There are two species of cubomedusae (sea wasps) known from Hawaii, *C. alata* and *C. rastoni.* Both occur in the tropical western Pacific as well. In other areas of the tropical Pacific the species occurring are not well known.

278- *Carybdea marsupialis* * Carybdeidae * Cubomedusae * Cubozoa * Palau * open water. This sea wasp is found worldwide in tropical waters. The square bell has four long tentacles, one from each corner. The bell is longer than it is wide. Although this species produces only a mild sting, other cubomedusae have extremely painful toxins in their nematocysts, hence the name sea wasp is sometimes applied to the group. The sting from these and other Cnidarians can be treated with hot seawater or alcohol which prove to be effective in denaturing the proteinaceous toxin. Cubomedusae are most often seen at night when they rise to the surface to feed. They are attracted to lighted areas such as wharves and boats in their search for food.

279- *Chiropsalmus* sp. * Chirodropidae * Cubomedusae * Cubozoa * Philippines * Palawan * open water. This is closely related to *Chironex fleckeri*, the deadly box jellyfish of Australia. *C. fleckeri* occurs in the Austro-Malayan region, and is typically found in turbid coastal waters, occasionally around mangroves. *Chiropsalmus* sp. has a severe sting, stronger than that of the Portuguese man-of-war (*Physalia physalis*).

280- *Stephanoscyphus* sp. * Nausithoeidae * Scyphozoa * Indonesia * Manado * 10 ft (3 m). These rather inconspicuous polyps are something to watch out for! They are the polyp stage of a sea thimble jellyfish similar to *Nausithoe punctata*, and are in the genus *Stephanoscyphus.* There is considerable confusion regarding the species of *Nausithoe*, so whether or not the polyp-form shown here is a described species can not be determined at present. They have been called "stinging algae", although they are not plants, because of their resemblance to algae. Irrespective of their identification, contact with these polyps rapidly produces intense pain and wounds which ulcerate and can last for weeks. While the polyps of *Stephanoscyphus* are cryptic and small, if you are suspicious they are lurking, there is any easy way to reveal their presence. We gently flick or touch the suspected organism with a finger nail (if you are brave) or an inanimate object (like a pencil or dive slate). If it is *Stephanoscyphus* it will flash a white color as the polyps contract and the white outer sheath of the polyp becomes visible.

281- *Stephanoscyphus* sp. * Nausithoeidae * Scyphozoa * Palau * barrier reef * 40 ft (12 m). *Stephanoscyphus* is found mixed in with algae in shallow water mangrove and seagrass areas and can be quite abundant in limited areas. On the reef, as is shown in the photograph, it may occur as a small bush-like colony which could be mistaken for a sponge or soft coral. In the photo the central part of the colony has the polyps partially retracted showing the white collars of the polyps. This colony was found among *Dendronephthya* soft corals. In this case an unidentified sponge has overgrown a *Stephanoscyphus* colony making it difficult to detect the stinging cnidarian. Notice how the polyps contract by pulling in from four corners.

282- *Linuche* sp. * Linuchidae * Coronatae * Schyphozoa * Marshall Islands * Enewetak Atoll * open water. This small thimble jelly seasonally swarms on the surface. The brown color is due to symbiotic algae (zooxanthellae) within the jellyfish.

283- *Aurelia aurita* * Ulmariidae * Semaeostomeae * Scyphozoa * Federated States of Micronesia * Chuuk * lagoon * open water. The moon jelly is one of the most widely distributed and frequently encountered jellyfish; it sometimes occurs in dense aggregations. The bell, which can reach a diameter of two feet, is fringed with numerous small stinging tentacles.

280- *Stephanoscyphus* sp. * Indonesia

281- *Stephanoscyphus* sp. * Palau

282- *Linuche* sp. * Marshall Islands

283- *Aurelia aurita* * Federated States of Micronesia

284- *Sanderia malayensis* * Bahrain

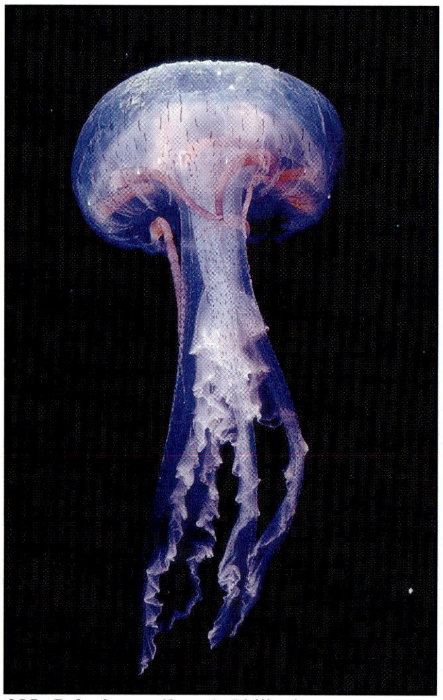

285- *Pelagia noctiluca* * Philippines

286- *Phyllorhiza punctata* * Hawaii

287- *Mastigias papua* * Papua New Guinea

284- *Sanderia malayensis* * Pelagiidae * Semaeostomeae * Scyphozoa * Bahrain * open water. Although sometimes found at sea, this species more commonly occurs near shore in bays and estuaries. The body attains lengths of nearly ten feet and the tentacles can extend twenty feet or more. *Sanderia* is a strong swimmer, but most often is carried with currents. Like most jellyfish, part of the life of *Sanderia* is spent as an attached (benthic) form. Seasonally, the polyps metamorphose to produce free-swimming jellyfish.

285- *Pelagia noctiluca* * Pelagiidae * Semaeostomeae * Scyphozoa * Philippines * Cebu * open water. The eight reddish, stinging tentacles, shown here contracted, can extend several feet and are used to stun and capture prey. The frilly lower portions of the animal are the food-gathering oral arms which draw the food into the mouth located on the underside of the bell. Most often encountered in warm offshore waters, these small (4 inches in length) jellyfish are sometimes carried inshore by currents.

286- *Phyllorhiza punctata* * Mastigiidae * Rhizostomae * Scyphozoa* Hawaii * coastal open water. This jellyfish occurs around the world in warm water. The polyps of this species seem able to survive in the ballast tanks of large ships or as fouling organisms. It has been reported from Pearl Harbor, San Diego Bay, San Juan, Puerto Rico, and Rio de Janeiro, all major ship harbors. This species can grow to almost three feet in diameter.

287- *Mastigias papua* * Mastigiidae * Rhizostomae * Scyphozoa * Papua New Guinea * Madang * mangrove * open water. This species appears to be quite variable in color and morphology. It occurs over the entire western Pacific and is easily confused with *P. punctata* above. It is possible there are several other species, closely related to *Phyllorhiza* and *Mastigias,* which occur in this region.

288- *Mastigias* sp. * Mastigiidae * Rhizostomae * Scyphozoa * Palau * Jellyfish Lake * midwater.
This species inhabits marine lakes in Palau. The most interesting difference between this species and *M. papua* above is its near inability to inflict a sting on bare skin. Both species have nematocysts and it appears those in the lake have lost almost all of their potency and ability to sting. This species is filled with zooxanthellae and swims to those parts of the lake with the best sun exposure.

289- *Cassiopea medusae* * Cassiopeidae * Rhizostomae * Scyphozoa * Philippines * Cebu * seagrass bed * 6 ft (2 m). *Cassiopea* is usually found in shallow bays and estuaries where the water is still. They are most often observed with the top of the bell laid against the substratum and the mouth and arms directed upwards. The margin of the bell pulses occasionally as if to help the jellyfish swim, however this effort is directed towards moving water and food across the oral region. The jellyfish is perfectly happy to be upside down.

288- ***Mastigias*** **sp. *** Palau

289-***Cassiopea medusae*** ***** Philippines

290- ***Cassiopea andromeda*** ***** Palau

291- ***Cephea cephea*** ***** Federated States of Micronesia

292- ***Crambione mastigophora*** ***** Federated States of Micronesia

293- ***Thystanostoma flagellatum*** ***** Palau

290- ***Cassiopea andromeda*** *** Cassiopeidae * Rhizostomae * Scyphozoa * Palau * Lighthouse Reef * sea grass bed * 6 ft (2m).** This jellyfish is similar in habit to *C. medusae* above, the principal difference being that this species prefers to inhabit seagrass areas.

291- ***Cephea cephea*** *** Cepheidae * Rhizostomae * Scyphozoa * Federated States of Micronesia * Chuuk * lagoon * open water.** This jellyfish varies a great deal in external appearance primarily as a function of size. There are several species described from the western Pacific, however we have only encountered this one in Chuuk and Palau. Large individuals are about one foot in diameter. Small individuals have numerous frilly tentacle-like extensions off their oral arms.

292- ***Crambione mastigophora*** *** Catostylidae * Rhizostomae * Scyphozoa * Federated States of Micronesia * Chuuk * lagoon * open water.** This jellyfish is common in the lagoon waters of Chuuk during late summer. The bell can be up to about one foot across. It is not a strong swimmer and large aggregations of up to several hundred individuals may be observed gently drifting with the tidal currents. *Crambione* can deliver a very painful sting if contact is made with exposed skin.

293- ***Thystanostoma flagellatum*** *** Thysanostomatidae * Rhizostomae * Scyphozoa * Palau * open water.** This is a small jellyfish for a rhizostome with a bell about four inches in diameter. The medusae is generally accompanied by a small fish (jack) which swims completely inside the bell. It seems to prefer open ocean, as we have only rarely seen it near shore.

Anthozoa is derived from a Greek word which means flower animal. Anthozoans are Cnidarians in which body form is based on the polyp. The medusa stage is absent. The Anthozoans are the most diverse class within the Cnidaria and that diversity is reflected in the wide variation in polyp morphology. They may be conveniently divided into octocorals and hexacorals, the former (as their name implies) have 8 branched tentacles while the latter usually have 6 unbranched tentacles or multiples thereof. Within the Anthozoa the octocorals (sea pens, soft corals, sea fans) are a fairly well-defined group, although taxonomy at the species level can be difficult. Within the hexacorals (hard corals, corallimorpharians, anemones, zoanthids, black corals and tube anemones) there is enough variation in nematocysts, skeletal structure and early life history to easily distinguish orders. In spite of the prominence of Anthozoans in shallow water tropical habitats, very little is positively known about the distribution (zoogeography) of species across the tropical Pacific due to inadequately and often times improperly identified specimens,.

In this book the Anthozoa are divided up into three picture and note sections; the octocorals, the stony hexacorals and the other hexacorals. A brief introduction of each group precedes the picture section for that group.

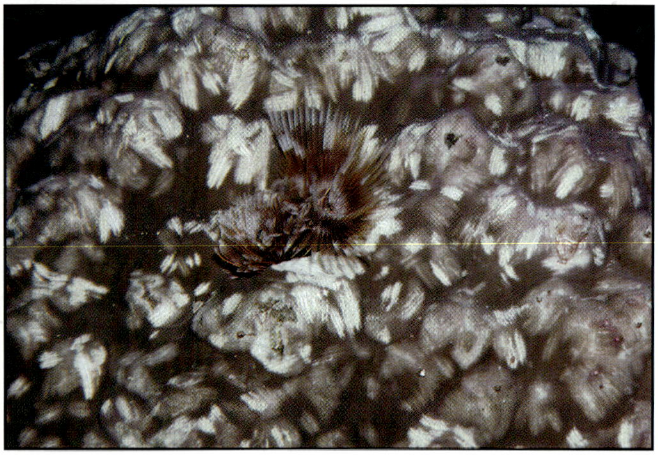

Above left- This close up photo of an octocoral shows typical polyps with eight pinately branched tentacles, also visble just beneath the red pigment are the spicules which make up the axial skeleton. Below left- This photo shows an ahermatypic coral, *Dendrophyllia* with its unbranched tentacles expanded at night. In general , only the octocorals have branched tentacles. The small orange spots on the transparent tentacles are clumps of nematocysts. Above right- This sea anemone *Heteractis magnifica* has commensal fish, crabs and shrimp. Many Anthozoans serve as partners in symbiotic relationships with other reef organisms. Below right- The coral, *Porites* has little defense against parrotfish which feed on the algae contained within the coral polyp. This photo is a good example of one way new substrate is created on the reef.

Four primary groups make up the octocorals: the stoloniferans, alcyonaceans (soft corals), gorgonians (seafans) and penatulaceans (sea pens). The stoloniferans are generally small polyps which, as their name implies, are often connected to each other by a thin runner or stolon. One species of stolonifera, *Tubipora,*(the organ pipe coral), however, secretes calcium carbonate and can form large colonies. The term "soft corals" usually refers to members of the Alcyonacea, which in shallow Pacific waters includes the brightly colored and often abundant *Dendronephthya* spp. Soft corals, particularly the species of *Sarcophyton, Lobophyton* and *Sinularia*, the leather corals, are opportunists when it comes to colonizing available substrate. They are fast to establish new individuals on the site and then grow rapidly to cover the available area. They are part of a complicated competition for space on the reef and in some cases, soft corals are able to overgrow stony corals. Gorgonian sea fans have an axial skeleton made of a horny scleroprotien substance called gorgonin. Gorgonian colonies can take the form of whips, fans, or bushy shrub-like colonies; they are often brightly colored and common in shallow water. The penatulacean seapens are highly modified octocorals. They usually consist of an elongate axial polyp which buds off secondary polyps along the sides of the stalk. Most sea pens have a thin horny central skeleton (the pen) and their tissues are reinforced with spicules which can be quite sharp. Some sea pens are bioluminescent and flash brilliant blue-green light if disturbed at night.

Octocorals have calcareous spicules within their body tissue which aid the support and maintenance of form in large colonies. The shape and size of these spicules, which differ from sponge spicules in shape, often determine classification of a species. The spicules are often times just under the outer surface of the octocoral and may cut or scratch a diver's hand if the animal is disturbed.

Most octocorals are filter feeders and inhabit areas where currents flow. Many shallow water species, especially the brown-colored ones, contain symbiotic zooxanthellae. Photosynthetic products from the zooxanthellae certainly augment the nutrient intake of the octocoral and enhance calcification in the stony corals.

Dendronephthya are some of the most spectacular organisms found on Pacific reefs, with brilliant colors, bizarre shapes and large sizes. They lack zooxanthellae so their sclerites are normally visible through the translucent body wall. Normally photographed when they are inflated with water, they often deflate to a small spiny lump on the bottom which is hard to identify as the same creature as the expanded individual.

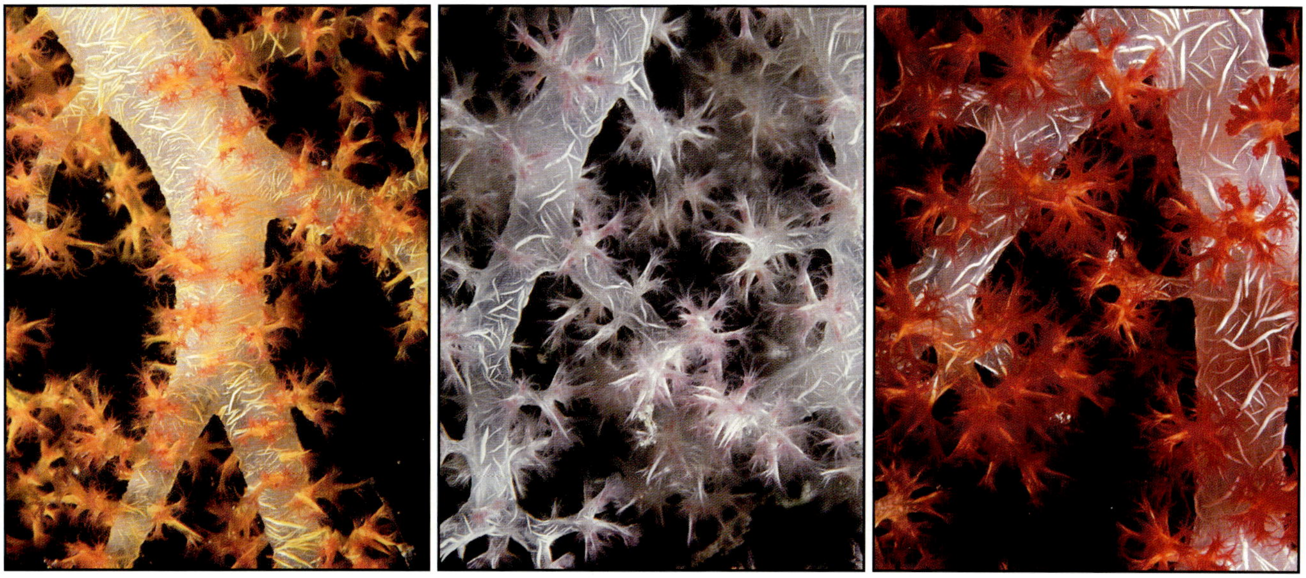

Above- Soft corals of *Dendronephthya* are admired for their delicate beauty. Because they have no zooxanthellae, the calcareous spicules are visible through the body wall. Their taxonomy is poorly known so they cannot be identified as a particular species.

294 - *Heliopora coerulea* * Philippines

295 - *Heliopora coerulea* * Marshall Islands

296 - *Clavularia* sp. * Papua New Guinea

294 - *Heliopora coerulea* * Helioporidae * Helioporacea * Octocorallia * Philippines * Pamalican Island * 13 ft (4 m). This abundant non-scleractinian coral, known as "blue coral", has the internal skeleton blue in color from iron salts. In the living colony this is visible only if a branch or plate has been broken off. It varies from delicate branches and vertical blades in shallow depths to horizontal plates in deeper water. It superficially resembles *Millepora* and some scleractinian corals, but once its characteristic appearance is learned, it is hard to confuse with anything else. In some areas, such as Ishigaki Island near Okinawa, it can be the dominant reef coral. Blue coral only occurs as far east as the Marshall and Gilbert Islands, and Samoa. It is not known from Fiji, French Polynesia or Hawaii.

295 - *Heliopora coerulea* * Helioporidae * Helioporacea * Octocorallia * Marshall Islands * Enewetak Atoll * Lojwa Island * 6 ft (2 m).

296 - *Clavularia* sp. * Clavulariidae * Stolonifera * Octocorallia * Papua New Guinea * Duke of York Islands * Ulu Pinnacle * 30 ft (9 m). These delicate soft corals with their eight branched tentacles look like little palm trees.

297 - *Pachyclavularia violacea* * Clavulariidae * Stolonifera * Octocorallia * Papua New Guinea * Duke of York Islands * Ulu Pinnacle * 50 ft (15 m).

298 - *Pachyclavularia* sp. * Clavulariidae * Stolonifera * Octocorallia * Philippines * Cebu * Mactan Island * 60 ft (18 m).

299 - *Carijoa* sp. * Clavulariidae * Stolonifera * Octocorallia * Federated States of Micronesia * Chuuk * Nematon Bay * 13 ft (4 m).

300 - *Carijoa* sp. * Clavulariidae * Stolonifera * Octocorallia * Marshall Islands * Enewetak Atoll * Cement Ship * 10 ft (3 m). This genus used to be called *Telesto*, but *Carijoa* was an earlier name. It is a very common fouling organism found on buoys, wharves and ship bottoms, plus turbid water reefs.

301 - *Paratelesto* sp. * Clavulariidae * Stolonifera * Octocorallia * Papua New Guinea * Madang * Planet Rock * 85 ft (25 m).

302 - *Tubipora musica* * Tubiporidae * Stolonifera * Octocorallia * Federated States of Micronesia * Nama Island * 20 ft (6 m). This picture shows a side view of a broken head of *Tubipora*, with tubes visible below polyps.

303 - *Tubipora musica* * Tubiporidae * Stolonifera * Octocorallia * Papua New Guinea * Duke of York Islands * Ulu Pinnacle * 10 ft (3 m). The dead skeleton of the organ pipe coral is unique and almost impossible to forget, the parallel red tubes bound together by horizontal platforms at regular intervals. The living colony, although often sizable, is much less distinctive and most divers never even notice it against a background of other corals and soft corals. *Tubipora musica* is an octocoral, the gray or greenish brown polyps having 8 tentacles. It is not known from east of the Marshall and Gilbert Islands, and is not found in Hawaii, Fiji or French Polynesia.

304 - *Cladiella* sp? * Alcyoniidae * Alcyoniina * Octocorallia * Federated States of Micronesia * Chuuk Atoll * Pizion Reef * 40 ft (12 m).

305 - *Eleutherobia* sp. * Alcyoniidae * Alcyoniina * Octocorallia * Indonesia * Manado * drop off * 66 ft (20 m).

306 - *Eleutherobia* sp. * Alcyoniidae * Alcyoniina * Octocorallia * Indonesia * Biak Island * dropoff * 60 ft (18 m).

297 - ***Pachyclavularia violacea*** * Papua New Guinea

298 - ***Pachyclavularia* sp.** * Philippines

299 - ***Carijoa* sp.** * Federated States of Micronesia

300 - ***Carijoa* sp.** * Marshall Islands

301 - ***Paratelesto* sp.** * Papua New Guinea

302 - ***Tubipora musica*** * Federated States of Micronesia

303 - ***Tubipora musica*** * Papua New Guinea

304 - ***Cladiella* sp.** * Federated States of Micronesia

305 - ***Eleutherobia* sp.** * Indonesia

306 - ***Eleutherobia* sp.** * Indonesia

307 - ***Minabea aldersladei*** * Papua New Guinea

307 - *Minabea aldersladei* * Alcyoniidae * Alcyoniina * Octocorallia * Papua New Guinea * Madang * reef * 66 ft (20 m).

308 - *Sinularia frondosa* * Alcyoniidae * Alcyoniina * Octocorallia * Papua New Guinea * Port Moresby * lagoon * 50 ft (15 m).

309 - *Sinularia* sp. * Alcyoniidae * Alcyoniina * Octocorallia * Papua New Guinea * Port Moresby * lagoon reef * 33 ft (10 m). The genus *Sinularia* has large sclerites densely packed into the tissues. This can form a hard material called "spicularite" and some species form columns of this hard rocky material which superficially resemble coral skeletons.

310 - *Sinularia* or *Sarcophyton* sp. * Alcyoniidae * Alcyoniina * Octocorallia * Federated States of Micronesia * Chuuk * lagoon * 30 ft (9 m).

311 - *Sarcophyton crassocaule* * Alcyoniidae * Alcyoniina * Octocorallia * Papua New Guinea * Madang * lagoon * 30 ft (9 m). This distinctive soft coral has anvil-like heads and when the polyps are expanded can form a continuous sheet of polyps. When the polyps retract, however, the individual heads are evident.

312 - *Sarcophyton* sp. * Alcyoniidae * Alcyoniina * Octocorallia * Federated States of Micronesia * Chuuk Atoll * Northeast Pass * 40 ft (12 m).

313 - *Sarcophyton* sp.* Alcyoniidae * Alcyoniina * Octocorallia * Papua New Guinea * Madang * lagoon * 20 ft (6 m).

314 - *Sarcophyton* sp. * Alcyoniidae * Alcyoniina * Octocorallia * Marshall Islands * Kwajalein Atoll * west barrier reef * night * 50 ft (15 m).

315 - *Sarcophyton* sp. * Alcyoniidae * Alcyoniina * Octocorallia * Palau * Kazia's Island * 10 ft (3 m).

316 - *Sarcophyton* sp. * Alcyoniidae * Alcyoniina * Octocorallia * Marianas Islands * Rota * fringing reef * 50 ft (15 m).

317 - *Sarcophyton* sp. * Alcyoniidae * Alcyoniina * Octocorallia * Indonesia * Biak Island * dropoff * 60 ft (18 m).

318 - *Sarcophyton* sp. * Alcyoniidae * Alcyoniina * Octocorallia * Philippines * Batangas * Pulang Buli * 20 ft (6 m).

319 - *Sarcophyton* sp. * Alcyoniidae * Alcyoniina * Octocorallia * Palau * barrier reef * 40 ft (12 m). Species of *Sarcophyton* can expand or retract their polyps which changes their appearance. They can also deflate the entire colony.

320 - *Sarcophyton* sp. * Alcyoniidae * Alcyoniina * Octocorallia * Federated States of Micronesia * Chuuk Atoll * lagoon reef * 40 ft (12 m).

321 - *Dendronephthya* sp. * Nephtheidae * Octocorallia * Palau * Ngerkuul Pass * 40 ft (12 m). Species of *Dendronephthya* have no zooxanthellae, which would mask the color of their sclerites. The sclerites, particularly in the branches of this genus, are close to the surface and can scratch human skin if touched, so contact should be avoided with them.

322 - *Dendronephytha* sp. * Nephtheidae * Octocorallia * Federated States of Micronesia * Chuuk Atoll * Fujikawa Maru * 30 ft (9 m). This is a contracted individual of *Dendronephthya*.

323 - *Dendronephthya* sp. * Nephtheidae * Octocorallia * Federated States of Micronesia * Chuuk Atoll * 30 ft (9 m). This is an expanded individual of the previous species.

308 - *Sinularia frondosa* * Papua New Guinea

309 - *Sinularia* sp. * Papua New Guinea

310 - *Sinularia* or *Sarcophyton* sp. * Papua New Guinea

311 - *Sarcophyton crassocaule* * Papua New Guinea

312 - *Sarcophyton* sp. * Federated States of Micronesia

313 - *Sarcophyton* * Papua New Guinea

314 - *Sarcophyton* sp. * Marshall Islands

315 - *Sarcophyton* sp. * Palau

316 - ***Sarcophyton* sp.** * Marianas Islands

317 - ***Sarcophyton* sp.** * Indonesia

318 - ***Sarcophyton* sp.** * Philippines

319 - ***Sarcophyton* sp.** * Palau

324 - *Dendronephthya* sp. * Nephtheidae * Octocorallia * Federated States of Micronesia * Chuuk Atoll * 60 ft (18m).

325 - *Dendronephthya* sp. * Nephtheidae * Octocorallia * Federated States of Micronesia * Chuuk Atoll * Fujikawa Maru * 30 ft (9 m). This soft coral was photographed on one of the wrecks in Chuuk (Truk) lagoon where they are extremely abundant and colorful. Many different color varieties exist together, and these consistent differences in the same environment implies that these color forms all represent separate species.

326 - *Dendronephthya* sp. * Nephtheidae * Octocorallia * Papua New Guinea * Madang * Rasch Passage * 66 ft (20 m).

327 - *Dendronephthya* sp. * Nephtheidae * Octocorallia * Papua New Guinea * Port Moresby * Horseshoe Reef * 50 ft (15 m).

328 - *Dendronephthya* sp. * Nephtheidae * Octocorallia * Papua New Guinea * Port Moresby * Horseshoe Reef * 40 ft (12 m).

329 - Genus species undetermined * Nephtheidae? * Octocorallia * Palau * Malakal Harbor * west entrance * sediment bottom * 36 m. This spiculose small soft coral is normally retracted during the day and looks like a little rounded stump sticking up out of the sediment. A few individuals, though, expand to their full extent even during the day, when this photo was taken, revealing the interesting form of the soft coral.

330 -*Dendronephthya* sp. * Nephtheidae * Octocorallia * Federated States of Micronesia * Chuuk Atoll * Northeast Pass * - 66 ft (20 m).

320 - ***Sarcophyton* sp.** * Federated States of Micronesia

321 - ***Dendronephthya* sp.** * Palau

322 - ***Dendronephytha* sp.** * Federated States of Micronesia

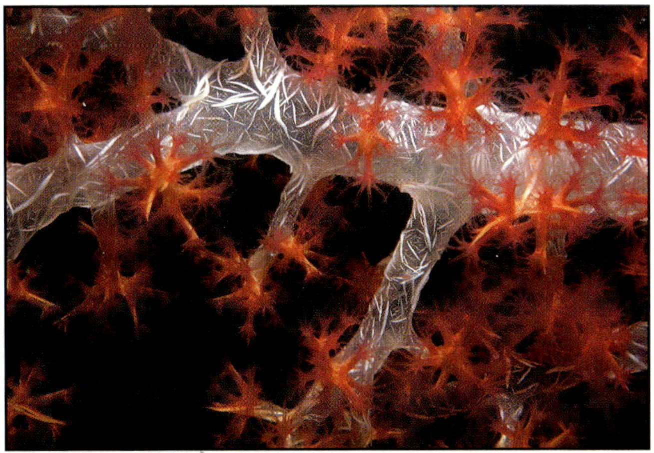

323 - ***Dendronephthya* sp.** * Federated States of Micronesia

324 - ***Dendronephthya* sp.** * Federated States of Micronesia

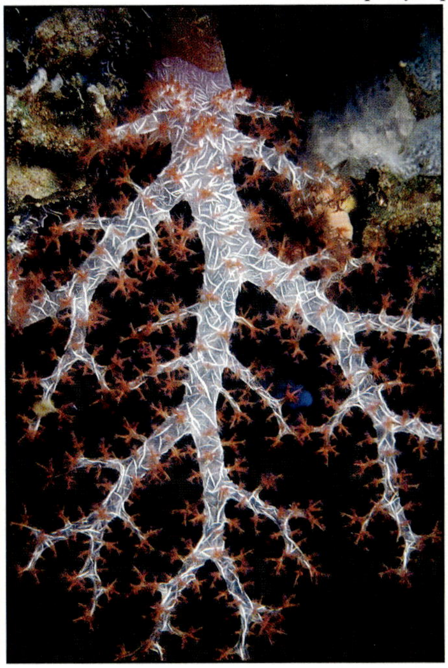

325 - ***Dendronephthya* sp.** * Federated States of Micronesia

326 - ***Dendronephthya* sp.** * Papua New Guinea

327 - ***Dendronephthya* sp.** * Papua New Guinea

328 - ***Dendronephthya* sp.** * Papua New Guinea

329 - **Unidentified Nephtheid** * Paalau.

330 - ***Dendronephthya* sp.** * Federated States of Micronesia

331 - ***Dendronephthya* sp.** * Federated States of Micronesia

332 - ***Nidalia occidentalis*** * Papua New Guinea

333 - ***Siphonogorgia* sp.** * Papua New Guinea

331- *Nidalia occidentalis* * Nidaliidae * Alcyoniina * Octocorallia * Papua New Guinea * Madang * harbor entrance * 100 ft (30 m). This photo was taken at night when ***Nidalia occidentalis*** is most active. At dusk the round head (capitulum) of the colony inflates with water and the polyps expand, exposing their tentacles. These are small octocorals, about four to five inches in length. This species is usually found on vertical faces near shaded ledges. It is often overlooked during the day.

332 -*Nidalia occidentalis* * Nidaliidae * Alcyoniina * Octocorallia * Papua New Guinea * Madang * harbor entrance * 100 ft (30 m). This photo shows *Nidalia* in its contracted state as it appears during the day.

333 - *Siphonogorgia* sp. * Nidaliidae * Alcyoniina * Octocorallia * Papua New Guinea * New Britain * Kimbe Bay * 100 ft (30 m). There are over twenty different, but very similar species of ***Siphonogorgia*** reported from the western Pacific. This octocoral has a semi-rigid skeleton due to spaces between the internal spicules. During periods of little current, species of *Siphonogorgia* appear to drupe. When the current increases, the softcoral inflates itself with water, expands all of its branches and faces the current to filter food.

334 - *Siphonogorgia* sp. * Nidaliidae * Alcyoniina * Octocorallia * Palau * Lighthouse Reef channel * 50 ft (15 m).

335 - *Siphonogorgia* sp. * Nidaliidae * Alcyoniina * Octocorallia * Papua New Guinea * New Ireland * Albatross Channel * 66 ft (20 m).

336 - *Cespitularia* sp * Xeniidae* Alcyoniina * Octocorallia * Papua New Guinea * Eastern Fields * reef wall * 50 ft (15 m).

337 - *Xenia* sp. * Xeniidae * Alcyoniina * Octocorallia * Papua New Guinea * Manam Island * 66 ft (20 m). Species of *Xenia* have small oval sclerites which diffract light and produce an opalescent sheen to the colonies. The polyps of *Xenia* are often observed opening and closing as if sweeping food toward the mouth. *Xenia* is common in shallow water reef habitsts.

338 - *Xenia* sp. * Xeniidae * Alcyoniina * Octocorallia * Federated States of Micronesia * Chuuk * Yamagiri Maru * 50 ft (15 m).

339 - *Xenia* sp. * Xeniidae * Alcyoniina * Octocorallia * Papua New Guinea * Madang * lagoon reef * 40 ft (12 m).

340 - *Briareum* sp * Briareidae * Scleraxonia * Octocorallia * Papua New Guinea * Madang * lagoon * 40 ft (12 m). *Briareum* is found in the Caribbean and Pacific. There are many species and many growth forms and it is difficult to classify a given specimen to species.

341 - *Briareum* sp. * Briareidae * Scleraxonia * Octocorallia * Papua New Guinea * Madang * Rasch Passage * 50 ft (15 m). This species of *Briareum* is similar to the previous one, but grows in a flat plate-like form oriented to capture the maximum amount of light. In this respect this growth form is similar to many stony corals which form plates in the deeper waters of the reef to capture light for their symbiotic zooxanthellae.

342 - *Ellisella* sp.? * Ellisellidae * Scleraxonia * Octocorallia * Papua New Guinea * West New Britain * 80 ft (22m).

343 - *Ellisella* sp.? * Ellisellidae * Scleraxonia * Octocorallia * Federated States of Micronesia * Chuuk * barrier reef * 100 ft (30 m).

344 - *Junceella* sp. * Ellisellidae * Holaxonia * Octocorallia * Papua New Guinea * Madang * Pig Island wall * 66 ft (20 m).

334 - ***Siphonogorgia* sp.** * Palau

335 - ***Siphonogorgia* sp.** * Papua New Guinea

336 - ***Cespitularia* sp** * Papua New Guinea

337 - ***Xenia* sp.** * Papua New Guinea

338 - ***Xenia* sp.** * Federated States of Micronesia

339 - ***Xenia* sp.** * Papua New Guinea

340 - ***Briareum* sp** * Papua New Guinea

341 - ***Briareum* sp.** * Papua New Guinea

342 - **Ellisella sp.?** * Papua New Guinea

343 - **Ellisella sp.?** * Federated States of Micronesia

344 - **Junceella sp.** * Papua New Guinea

The species of *Junceella* can reproduce by pinching off the top of the colony, which then falls to the bottom and starts a new individual.

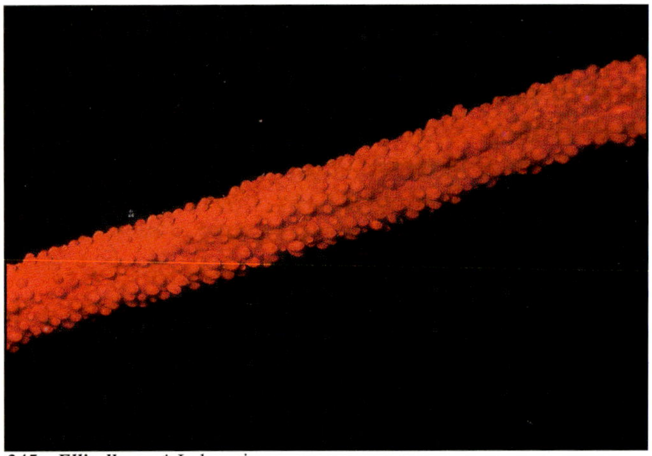
345 - **Ellisella sp.** * Indonesia

346 - **Nicella sp.** * Palau

345 - *Ellisella* sp. * Ellisellidae * Holaxonia * Octocorallia * Indonesia * Manado * fringing reef * 85 ft (25 m).

346 - *Nicella* sp. * Ellisellidae * Holaxonia * Octocorallia * Papua New Guinea * Madang * barrier reef * 66 ft (20 m).

347 - *Toeplitzella* sp. * Ellisellidae * Scleraxonia * Octocorallia * Palau * Ulong Channel * 23 m.

348 - *Alertigorgia* sp. * Anthothelidae * Scleraxonia * Octocorallia * Papua New Guinea * Madang * Pig Island * 66 ft (20 m).

349 - *Semperina* sp. * Anthothelidae * Scleraxonia * Octocorallia * Philippines * Batangas * 50 ft (15 m).

350 -*Semperina* sp. * Anthothelidae * Scleraxonia * Octocorallia * Papua New Guinea * Port Moresby * Basilisk Passage * 50 ft (15 m).

351 - *Solencaulon* sp. * Anthothelidae * Scleraxonia * Octocorallia * Indonesia * Manado * fringing reef * 60 ft (18 m). This gorgonian has broad flattened blades with polyps on one side only. The general form resembles some of the flattened sponges like *Phyllospongia*, but the polyps are an instant giveaway that this is a cnidarian, not a sponge.

352 - *Solenocaulon* sp. * Anthothelidae * Scleraxonia * Octocorallia * Papua New Guinea * Basilisk Passage * 50 ft (15 m).

353- *Subergorgia* sp.? * Subergorgiidae * Scleraxonia * Octocorallia * Papua New Guinea * New Britain * Kimbe Bay * 50 ft (15 m).

354 - *Subergorgia suberosa* * Subergorgiidae * Scleraxonia * Octocorallia * Federated States of Micronesia * East Fayu * reef wall * 85 ft (25 m).

347 - ***Toeplitzella* sp.** * Palau

348 - ***Alertigorgia* sp.** * Papua New Guinea

349 -***Semperina* sp.** * Philippines

350- ***Semperina* sp.** * Papua New Guinea

351 - ***Solencaulon* sp.** * Indonesia

352 - ***Solenocaulon* sp.** * Papua New Guinea

353 -***Subergorgia* sp.** * Papua New Guinea

354 - *Subergorgia suberosa* * Federated States of Micronesia

355 - *Subergorgia mollis* * Indonesia

356 - *Acabaria* sp. * Palau

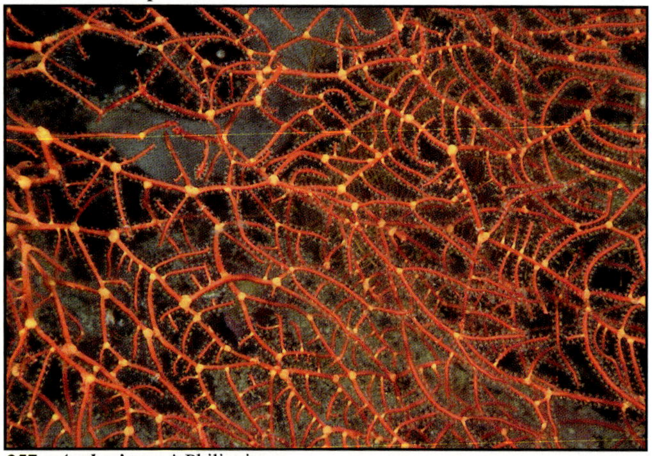

357 - *Acabaria* sp. * Philippines

358 - *Acabaria* sp. * Papua New Guinea

355 - *Subergorgia mollis* * Subergorgiidae * Scleraxonia * Octocorallia * Indonesia * Biak Island * reef wall * 85 ft (25 m). This sea fan can grow up to several feet in width.

356 - *Acabaria* sp. * Melithaeidae * Scleraxonia * Octocorallia * Palau * Babulukes * 66 ft (20 m).

357 - *Acabaria* sp. * Melithaeidae * Scleraxonia * Octocorallia * Philippines * Pamalican Island * 40 ft (12 m).

358 - *Acabaria* sp. * Melithaeidae * Scleraxonia * Octocorallia * Papua New Guinea * Eastern Fields * 40 ft (12 m).

359 - *Acabaria* sp. * Melithaeidae * Scleraxonia * Octocorallia * Marshall Islands * Kwajalein Atoll * patch reef * 40 ft (12 m).

360 - *Acabaria* sp. * Melithaeidae * Scleraxonia * Octocorallia * Papua New Guinea * Port Moresby * Basilisk Passage * 100 ft (30 m).

361 - *Melithaea* sp. * Melithaeidae * Scleraxonia * Octocorallia * Palau * Babulukes * 50 ft (15 m).

362 - *Melithaea* sp. * Melithaeidae * Scleraxonia * Octocorallia * Papua New Guinea * Eastern Fields * reef wall * 75 ft (23 m).

363 - *Melithaea* sp. *Melithaeidae * Scleraxonia * Octocorallia * Papua New Guinea * Bagabag Island * reef wall * 66 ft (20 m).

364 - *Acalycigorgia* sp. * Acanthogorgiidae * Holaxonia * Octocorallia * Papua New Guinea * Madang * offshore pinnacle * 90 ft (27 m). This genus and the two that follow are unique in that they have a tough horny axial skeleton, sharp glass spines and no calcium carbonate spicules embedded in the axis. *Acalycigorgia* lives on outer slopes in areas with current. This species forms tough flexible fans.

365 - *Acalycigorgia* sp. * Acanthogorgiidae * Holaxonia * Octocorallia * Federated States of Micronesia * Chuuk Atoll * barrier reef * 135 ft (40 m).

359 - *Acabaria* sp. * Marshall Islands

360 - *Acabaria* sp. * Papua New Guinea

361 - *Melithaea* sp. * Palau

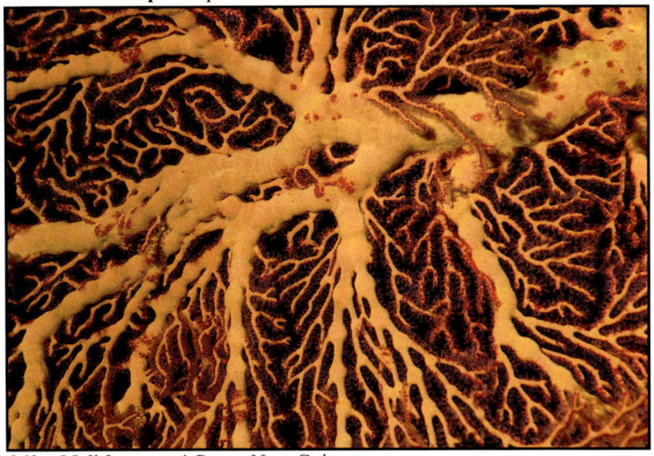
362 - *Melithaea* sp. * Papua New Guinea

363 - *Melithaea* sp. * Papua New Guinea

364 - *Acalycigorgia* sp. * Papua New Guinea

365 - *Acalycigorgia* sp. * Federated States of Micronesia

366 - *Acanthogorgia* sp. * Federated States of Micronesia

367 - ***Acanthogorgia* sp.** * Federated States of Micronesia

368 - ***Acanthogorgia* sp*** Papua New Guinea

369 - ***Acanthogorgia* sp.** * Marshall Islands

370 - ***Anthogorgia* sp.** * Papua New Guinea

366 - *Acanthogorgia* sp. * Acanthogorgiidae * Holaxonia * Octocorallia * Federated States of Micronesia * Chuuk * Anaw Reef wall * 165 ft (50 m).

367 - *Acanthogorgia* sp. * Acanthogorgiidae * Holaxonia * Octocorallia * Federated States of Micronesia * Chuuk * Pizion Reef * 165 ft (50 m).

368 - *Acanthogorgia* sp * Acanthogorgiidae * Holaxonia * Octocorallia * Marshall Island * Enewetak Atoll * 50 ft (15 m).

369 - *Acanthogorgia* sp. * Acanthogorgiidae * Holaxonia * Octocorallia * Marshall Islands * Kwajalein * west reef * 100 ft (30 m).

370 - *Anthogorgia* sp. * Acanthogorgiidae * Holaxonia * Octocorallia * Papua New Guinea * Yule Island * 66 ft (20 m).

371 - *Acanthogorgia* sp. * Acanthogorgiidae * Holaxonia * Octocorallia * Marshall Islands * Enewetak Atoll * lagoon reef * 50 ft (15 m).

372 - *Anthogorgia* sp. * Acanthogorgiidae * Holaxonia * Octocorallia * Papua New Guinea * Port Moresby * Bootless Bay * 75 ft (23 m).

373 - *Muricella* sp. * Acanthogorgiidae * Holaxonia * Octocorallia * Indonesia * Manado * reef drop off * 100 ft (30 m).

374 - *Astrogorgia* sp. * Plexauridae * Holaxonia * Octocorallia * Federated States of Micronesia * Chuuk * Northeast Pass * 50 ft (15 m),

375 - *Astrogorgia* sp. * Plexauridae * Holaxonia * Octocorallia * Papua New Guinea * Madang * barrier reef * 50 ft (15 m).

376 - *Astrogorgia* sp. * Plexauridae * Holaxonia * Octocorallia * Federated States of Micronesia * Chuuk * Falalu Island * reef * 50 ft (15 m).

377 - *Bebryce* sp. * Plexauridae * Holaxonia * Octocorallia * Federated States of Micronesia * Chuuk Atoll * Northeast Pass * 60 ft (18 m).

378 - *Bebryce* sp. * Plexauridae * Holaxonia * Octocorallia * Papua New Guinea * Duke of York Islands * Makada Reef * 50 ft (15 m).

379 -*Lophogorgia* sp. * Gorgoniidae * Holaxonia * Octocorallia * Federated States of Micronesia * Losap Atoll * East Channel * 50 ft (15 m).

380 - *Menella praelonga* * Plexauridae * Holaxonia * Octocorallia * Federated States of Micronesia * Chuuk Atoll * Northeast Pass * 85 ft (25 m).

381 - *Menella* sp. * Plexauridae * Holaxonia * Octocorallia * Indonesia * Manado * 50 ft (15 m).

382 - *Menella* sp. * Plexauridae * Holaxonia * Octocorallia * Palau * barrier reef * 40 ft (12 m).

383 -*Villogorgia* sp.? * Plexauridae * Holaxonia * Octocorallia * Indonesia * Biak Island * 50 ft (15 m).

384 - *Isis hippuris* * Isididae * Holaxonia * Octocorallia * Papua New Guinea * Madang * fringing reef * 10 ft (3 m). This species has an interesting internal skeleton, ribbed with alternating colors.

371 - ***Acanthogorgia* sp.** * Marshall Islands

372 - ***Anthogorgia* sp.** * Papua New Guinea

373 - ***Muricella* sp.** * Indonesia

374 - ***Astrogorgia* sp.** * Federated States of Micronesia

375 - ***Astrogorgia* sp.** * Papua New Guinea

376 - ***Astrogorgia* sp.** * Federated States of Micronesia

377 - ***Bebryce* sp.** * Federated States of Micronesia

378 - ***Bebryce* sp.** * Papua New Guinea

379 - ***Lophogorgia* sp.** * Federated States of Micronesia

380 - ***Menella praelonga*** * Federated States of Micronesia ***Pseudothesea* sp.**

381 - ***Menella* sp.** * Philippines

385 - *Rumphella* sp. * Gorgoniidae * Holaxonia * Octocorallia * Philippines * Cebu * Pescador Island * 30 ft (9 m).

386 - *Plumigorgia hydroides* * Ifalukellidae * Federated States of Micronesia * Satawan Atoll * reef wall * 100 ft (30 m).

387 - *Stephanogorgia* sp. * Chrysogorgiidae * Holaxonia * Octocorallia * Papua New Guinea * Eastern Fields * 100 ft (30 m).

388 - *Asterospicularia* sp? * Asterospiculariidae * Holaxonia * Octocorallia * Papua New Guinea * Karkar Island * 40 ft (12 m).

389 - *Rumphella* sp * Gorgoniidae * Holaxonia * Octocorallia * Federated States of Micronesia * Chuuk * Salat * 40 ft (12 m).

390 - *Cavernulina cylindrica* * Veretillidae * Pennatulacea * Octocorallia * Indonesia * Manado * Bangka Island * sandy slope * 33 ft (10 m).

391 - *Cavernularia* sp. * Veretillidae * Pennatulacea * Octocorallia * Papua New Guinea * Port Moresby * Motupore Island * 40 ft (12 m). The eight tentacles of the polyps can be clearly seen in this species of *Cavernulina*. This species would likely be identified as *C. obesa*, but that "species" is evidently a number of valid species, which are very similar in outward appearance. The identification of the photographed species is therefore in doubt.

392 - *Cavernularia* cf. *chuni*, * Veretillidae * Pennatulacea * Octocorallia * Papua New Guinea * Port Moresby * Bootless Bay * 60 ft (18 m).

382 - ***Menella* sp.** * Palau

383 - ***Villogorgia* sp?** * Indonesia

384 - ***Isis hippuris*** * Papua New Guinea

385 - *Rumphella* sp. * Philippines

386 - *Plumigorgia hydroides* * Federated States of Micronesia

387 - *Stephanogorgia* sp. * Papua New Guinea

388 - *Asterospicularia* sp? * Papua New Guinea

389 - *Rumphella* sp * Federated States of Micronesia

390 - *Cavernulina cylindrica* * Indonesia

391 - *Cavernulina* sp. * Papua New Guinea

392 - ***Cavernularia* cf. *chuni*** * Philippines

393 - ***Pteroeides* sp.** * Indonesia

394 - ***Pteroeides* sp.** * Philippines

395 - ***Pteroeides* sp.** * Papua New Guinea

393 - *Pteroeides* sp. * Pteroeididae * Pennatulacea * Octocorallia * Indonesia * Manado * fringing reef slope * 33 ft (10 m).

394 - *Pteroeides* sp. * Pteroeididae * Pennatulacea * Octocorallia * Philippines * Cebu * Olango Island * 50 ft (15 m).

395 - *Pteroeides* sp. * Pteroeididae * Pennatulacea * Octocorallia * Papua New Guinea * Port Moresby * Lion Island * 40 ft (12 m)

396 - *Pteroeides* sp. * Pteroeididae * Pennatulacea * Octocorallia * Papua New Guinea * Port Moresby * Horseshoe Reef * 66 ft (20 m).

397 - *Virgularia* sp. * Virgulariidae * Pennatulacea * Octocorallia * Philippines * Pamalican Island * 50 ft (15 m).

398 - *Virgularia* sp. * Virgulariidae * Pennatulacea * Octocorallia * Palau * Airai * mud bottom * 70 ft (21 m).

396 - ***Pteroeides* sp.** * Papua New Guinea

397 - ***Virgularia* sp.** * Philippines

398 - ***Virgularia* sp.** * Palau

As indicated previously, the stony corals fall into two general groups, the reef-building or hermatypic and the non-reef building or ahermatypic corals. Pacific Ocean stony corals are the most diverse coral fauna in the world, with some sites having as many as 300 to 400 species. The highest generic richness, about 90 genera, is believed to occur in the area formed by a triangle including the Philippines, northern Indonesia and New Guinea, and extends to the northern Great Barrier Reef. The taxonomy of Pacific Ocean reef corals is imperfectly known, although much progress has been made in the last few decades to sort out the variable and similar species. Identifications from dead skeletons may be difficult with many variations in growth form and skeletal characters depending on environmental conditions. Living colonies also pose problems since the colors of many corals can vary greatly. Additionally the living tissue of polyps masks the skeletal characteristics which are often necessary for accurate identification.

The stony corals fall into about 15 families. Assigning a particular coral to a specific family is the first step in identifying it. Some families contain genera which are easy to recognize. Others are much less distinctive and cannot practically be identified from photographs unless the colony photographed is also available as a specimen and primary literature is used.

The coral identifications presented in this book have been based on actual specimens in some cases, but more often have been made using available literature from the photographs alone. This is why many of the identifications are indefinite, often only to genus. It is hoped the species of stony corals included will assist in generally placing a particular coral amongst the many families and genera.

Tissues of most hermatypic stony corals contain symbiotic algae, called zooxanthellae, which give the coral polyps most of their color. In the presence of light these algal cells use nitrogen-containing waste products and carbon dioxide from the polyp to produce sugars and amino acids through the process of photosynthesis. Enzymes in the coral tissue cause some of these nutrients to leak out of the zooxanthellae; the nutrients are in turn used by the coral for its own growth. Obviously hermatypic corals grow best in shallow, clear, sunlit waters.

During the day most species of stony corals have the polyps retracted into cup-like calices, (fluted depressions that make up the upper part of the skeleton). However, at night the polyps expand, and the coral looks entirely different. To augment the nutrients produced by the zooxanthellae corals use their tentacles to capture food at night. Corals defend their space on the reef at night as well; some species have special sweeper tentacles which can reach distances of several inches to attack neighboring organisms to keep them from overgrowing the stony coral.

Some stony corals grow unattached to the bottom. Several genera of the family Fungiidae, the mushroom corals, are free living to the extent they can even move themselves along the bottom. Although these fungiid corals start out life as an attached polyp, early on the upper disc of the polyp breaks free. The disc settles on the bottom and grows. If the coral finds itself in an unsuitable spot, it is able to inflate itself with water and roll over along the bottom. Other families have fewer numbers of free-living stony corals; *Goniopora stokesi,* grows in small hemispherical colonies on sandy bottoms near reefs.

Corals reproduce both asexually and sexually. Colonies grow by two types of asexual division of the polyps, intratentacular (within the oral disc) and extratentacular (outside the oral disc) budding. Another form of asexual reproduction can occur by a method known as "polyp bailout" in which stressed coral polyps leave the skeleton and float away short distances and redevelop into new colonies. Some corals also form "satellite" colonies, such as *Goniopora stokesii*, in which small buds form off the original colony and eventually break off to form separate colonies.

Sexual reproduction also has several variations. Some corals have gonads of both sexes in each polyp (hermaphroditic). Other corals have colonies with separate sexes. Fertilization is either internal, in which case sperm released into the water swim to female polyps with eggs, or external, in which both eggs and sperm are released into the water. Self-fertilization, most likely, does not occur. Whether eggs are fertilized externally in the water or internally, they eventually produce larvae (planulae). The planulae are ciliated and able to swim, their oblong bodies are at most about 1/4 of an inch long. Planulae may remain in the water column for up to several months time and are the main means of medium to long distance dispersal for most corals. The species of *Pocillopora* have planulae which can be extremely long lived in the plankton and this is believed to be a major reason why this genus is found from the Red Sea to the western coast of North America.

In the past decade scientists have discovered that in some areas many species of stony corals spawn nearly simultaneously, resulting in a spectacular release of gametes over large areas of reef almost on cue. On the Great Barrier Reef the mass coral spawning occurs just after the full moon of November. In other areas timing is different; for example, off western Australia coral spawning occurs in April. In Micronesia coral spawning is less well known, but a number of species are believed to spawn during July.

When the planula larva settles to the bottom, it produces a single polyp which starts building the calcium carbonate skeleton. From that single polyp and its skeleton, the coral colony grows by budding of individual polyps, initially forming an area of firm attachment to the substrate. Once a sizeable base is established, the colony can begin to grow upward. *Acropora* species (one of the most common corals) have two types of polyps, axial (at the tips of the branches) and radial (along the sides of the branches). Growth is rapid at the tips of the branches, where the axial corallites occur, and members of this genus are among the fastest growing of all corals.

The genus *Acropora,* and other genera of corals occurs in many different growth forms and these forms are often times the best character for identifing species in the field. Color is not a good character to use, as it is often quite variable within a single species. Growth forms include: plate or table-like, tree-like (arboresecent), encrusting, corymbose (short branches arising from horizontal mass), pillow-like, digitate (short non-dividing, unconnected), bushy, and massive.

In just about any area of the tropical Pacific numerous species of *Acropora* are to be found. For example, in Australia, 73 species are reported from the eastern coast and 54 from western Australia. In Micronesia, 36 species are reported from Guam, and quite a few more occur in other Micronesian waters.

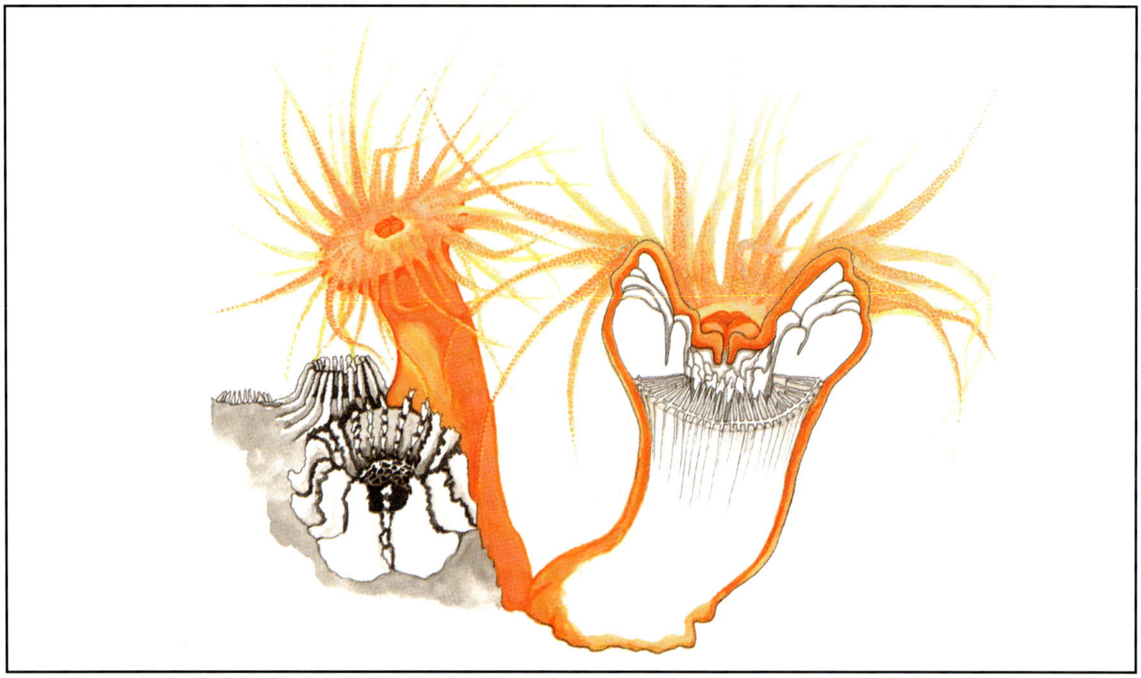

Above - This diagram of coral polyps is cut away to show the relationship of soft tissue to the coral skeletonn In the center of the cut away portion is a large gastrovascular cavity divided by mesentaries. The septa of the coral skeleton form against these folds.

399- *Stylocoeniella guentheri* * Federated States of Micronesia

400- *Pocillopora damicornis* * Palau

399- *Stylocoeniella guentheri* * Astrocoeniidae * Scleractinia * Federated States of Micronesia * Losap Atoll * east channel * 10 ft (3 m). This is the only genus in this family.

400- *Pocillopora damicornis* * Pocilloporidae * Scleractinia * Palau * Babulukes * 12 ft (4 m). This is one of the most common corals found in shallow waters of the western Pacific and occurs across the entire breadth of tropical and subtropical Pacific and Indian Oceans. The species is somewhat variable in growth form, depending on wave action. The genus *Pocillopora* has internal fertilization of eggs and releases planulae larvae which drift in the plankton. In *P. damicornis* these larvae are believed to be capable of surviving for several months in the open ocean, a major reason the species has such a broad distribution.

401- *Pocillopora damicornis* * Palau

401- *Pocillopora damicornis* * Pocilloporidae * Scleractinia * Palau * Kazia's Island * 20 ft (6 m). This coral is capable of withstanding a wide variety of conditions, however, as with any organism, the individuals at the geographic limits of the species are often found in marginal environments, where slight changes in conditions, such as water temperature, can have disastrous effects. In the eastern Pacific, *P. damicornis* was very abundant on reefs off the coasts of Costa Rica and Panama, but an "El Nino" phenomenon in the 1980's caused mass mortality of *P. damicornis* on many of the reefs where it formerly dominated.

402- *Pocillopora danae* * Federated States of Micronesia

402- *Pocillopora danae* * Pocilloporidae * Scleractinia * Chuuk * Weno * 6 ft (2 m). The species is common in shallow water throughout Micronesia, although it may not be present in Australia. Five species of the genus are known from Australia, although there is a total of 7-10 valid species in the genus.

403- *Pocillopora eydouxi* * Pocilloporidae * Scleractinia * Federated States of Micronesia* Fujikawa Maru * 50 ft (15 m). This species of *Pocillopora* is quite distinctive with upright flattened branches with pale ends. A number of crabs and fishes are associated with *P. eydouxi*, the crabs nestled deep in the crooks of the branches. This species occurs from the Red Sea and east Africa to Hawaii.

403- *Pocillopora eydouxi* * Federated States of Micronesia

404- *Pocillopora verrucosa* * Pocilloporidae * Scleractinia * Federated States of Micronesia * Chuuk * barrier reef * 15 ft (5 m). The genus *Pocillopora* is distinguished from other stony corals by the presence of small wart-like structures with polyps called verrucae which, obviously, is the source of the specific name of this species. This is a fine example of the ideal where a scientific name is descriptive of the species to which it is given. These verrucae give all the members of *Pocillopora* a distinctive look.

404- ***Pocillopora verrucosa*** * Federated States of Micronesia

405- ***Pocillopora*** **sp.** * Federated States of Micronesia

406- ***Seriatopora aculeata*** * Federated States of Micronesia

407- ***Seriatopora caliendrum*** * Federated States of Micronesia

405- *Pocillopora* sp. * Pocilloporidae * Scleractinia * Federated States of Micronesia * Chuuk Atoll * barrier reef * 30 ft (9 m).

406- *Seriatopora aculeata* * Pocilloporidae * Scleractinia * Federated States of Micronesia Chuuk Atoll * South Field Ramp * 6 ft (2 m). *Seriatopora* is quite a distinctive genus and the species with sharply pointed branch tips, *S. aculeata* and *S. hystrix*, are readily recognized. There are believed to be five species in the genus.

407- *Seriatopora caliendrum* * Pocilloporidae * Scleractinia * Federated States of Micronesia * Chuuk * Ozen Island * 15 ft (5 m). This species of *Seriatopora* does not have the sharp pointed tips to the branches. Rather, they are more rounded, but the overall colonies are delicately branched and fragile. This species is known from the Red Sea to Australia, New Caledonia and the Philippines.

408- *Seriatopora hystrix* * Pocilloporidae * Scleractinia * Papua New Guinea * Madang * barrier reef * 10 ft (3 m). This species is often called "needle coral" because of the fine sharp points on the branches. It is found from east Africa across the Indian Ocean and the Pacific to as far east as Micronesia and Samoa.

409- *Seriatopora* sp. * Pocilloporidae * Scleractinia * Philippines * Pamalican Island * 20 ft (6 m). This coral may be different than the previous two species. It resembles *S. hystrix* in color, but the branches are blunt, like those of *S. caliendrum*. There are over twenty described species of this genus, but reportedly only about 5 valid species. Much still remains to be done on the taxonomy of this genus.

410- *Stylophora mordax* * Pocilloporidae * Scleractinia * Federated States of Micronesia * Chuuk * Ozen Island * 20 ft (6 m). The genus *Stylophora* is believed to have about four species and occurs from the western Indian Ocean east through Micronesia and much of French Polynesia. It does not occur in Hawaii or the Marquesas Islands.

411- *Acropora granulosa* * Acroporidae * Scleractinia * Federated States of Micronesia * Chuuk * Ozen Island * 75 ft (23 m). The grand genus *Acropora* is considered by many to be the king of stony corals. The most speciose genus of all stony corals (with over 350 nominal species) and extremely distinctive as a genus, they are, with a few notable exceptions, among the most difficult to identify to species. The genus occurs throughout the tropical Indian and Pacific Oceans, reaching the west coast of the Americas off Colombia, and in the tropical western Atlantic.

412- *Acropora palifera* * Acroporidae * Scleractinia * Federated States of Micronesia * Nama Island * 20 ft (6 m). This is an abundant member of *Acropora* in much of Micronesia, and in that region seems most common on outer reefs exposed to wave action. There it usually has robust branches and can be the dominant stony coral species in some areas of the reef. In shallow water with strong wave action, it may grow in the form of short rounded ridges. It (and two other species *A. cuneata* and *A. bruggemanni*) differs from most other *Acropora* in that each branch has more than one axial corallite, resulting in blunt, thick, often non-circular cross section branches. To the inexperienced, this coral may not immediately be recognized as a member of *Acropora*, but may seem more like a member of *Pocillopora*. It is found throughout the region east to the Marshall Islands and Samoa. *A. cuneata* also occurs from the Indian Ocean through Micronesia, while *A. bruggemanni* only comes as far east as Australia, Indo-Malay region and the Philippines.

413- *Acropora pruinosa* * Acroporidae * Scleractinia * Hong Kong * Breaker Reef * 20 ft (6 m). This species is the most common *Acropora* of the four species that occur in Hong Kong waters, but it is well on its way to local extinction in Hong Kong due to turbidity and silt from dredging, plus general environmental degradation.

408- ***Seriatopora hystrix*** * Papua New Guinea

409- ***Seriatopora* sp.** * Philippines

410- ***Stylophora mordax*** * Federated States of Micronesia

411- ***Acropora granulosa*** * Federated States of Micronesia

412- ***Acropora palifera*** * Federated States of Micronesia

413- ***Acropora pruinosa*** * Hong Kong

414- ***Acropora tenella*** * Federated States of Micronesia

415- ***Acropora robusta*** * Philippines

416- ***Acropora*** **sp.** * Federated States of Micronesia

417- ***Acropora*** **sp.** * Papua New Guinea

418- ***Acropora verweberi*** * Philippines

419- ***Acropora*** **sp.** * Papua New Guinea

414- ***Acropora tenella*** *** Acroporidae * Scleractinia * Federated States of Micronesia * Chuuk Atoll * Ozen Island * 133 ft (40 m).** This remarkable species of *Acropora* is branched like many other species, but is greatly flattened with calices only on its upper surface to maximize exposure to light. Not surprisingly, it is found in deeper reef waters, usually below 80 feet, although small specimens are occasionally encountered shallower. It is quite a distinctive species that is easy to recognize.

415- ***Acropora robusta*** *** Acroporidae * Scleractinia * Philippines * Pescador Island * 20 ft (6 m).** This species is quite variable in growth form, even within the same colony. It is typically found in the shallow water areas, often where there is heavy wave action.

416- ***Acropora*** **sp. * Acroporidae * Scleractinia * Federated States of Micronesia * Chuuk Atoll * Northeast Passage * 30 ft (9 m).**

417- ***Acropora*** **sp. * Acroporidae * Scleractinia * Papua New Guinea * Mait Reef * 10 ft (3 m).**

418- ***Acropora verweberi*** *** Acroporidae * Scleractinia * Philippines * Pescador Island * 20 ft (6 m).**

419- ***Acropora*** **sp. * Acroporidae * Scleractinia * Papua New Guinea * Madang * lagoon * 20 ft (6 m).**

420- ***Acropora*** **sp. * Acroporidae * Scleractinia * Papua New Guinea * Madang * lagoon * 20 ft (6 m).**

421- ***Acropora*** **sp. * Acroporidae * Scleractinia * Papua New Guinea * Port Moresby * Lion Island * 20 ft (6 m).**

422- ***Acropora*** **sp. * Acroporidae * Scleractinia * Marshall Islands * Enewetak Atoll * Medren Pinnacle * 33 ft (10 m).**

423- ***Acropora*** **sp. * Acroporidae * Scleractinia * Indonesia * Cebu * Mactan Island * 20 ft (6 m).**

424- ***Acropora*** **sp. * Acroporidae * Scleractinia * Federated States of Micronesia * Chuuk Atoll * Ozen Island * 40 ft (12 m).**

425- ***Acropora*** **sp. * Acroporidae * Scleractinia * Federated States of Micronesia * Chuuk Atoll * Ozen Island * 120 ft (36 m).**

426- ***Acropora*** **sp. * Acroporidae * Scleractinia * Federated States of Micronesia * Chuuk Atoll * Ozen Island * 15 ft (5 m).**

427- ***Anacropora*** **sp. * Acroporidae * Scleractinia * Palau * Rock Islands * 15 ft (5 m).** Species of *Anacropora* superficially resembles *Acropora*, but lack the terminal axial corallites of the latter genus. Careful examination of the branches, even underwater, will reveal this difference. *Anacropora* is not as common as *Acropora* and there are only about six species of *Anacropora*. The genus ranges from the western Indian Ocean (but not the African coast) through the Indo-Malayan archipelago and the Philippines to Fiji and some areas of Microncsia. For example, the genus occurs in Palau and the Marshall Islands, but is absent from the Marianas, including Guam. Members of *Anacropora* are most often found in muddy environments, and are seldom seen by divers.

420- *Acropora* sp. * Papua New Guinea

421- *Acropora* sp. * Papua New Guinea

422- *Acropora* sp. * Marshall Islands

423- *Acropora* sp. * Indonesia

424- *Acropora* sp. * Federated States of Micronesia

425- *Acropora* sp. * Federated States of Micronesia

426- *Acropora* sp. * Federated States of Micronesia

427- *Anacropora* sp. * Palau

428- ***Astreopora gracilis*** * Federated States of Micronesia

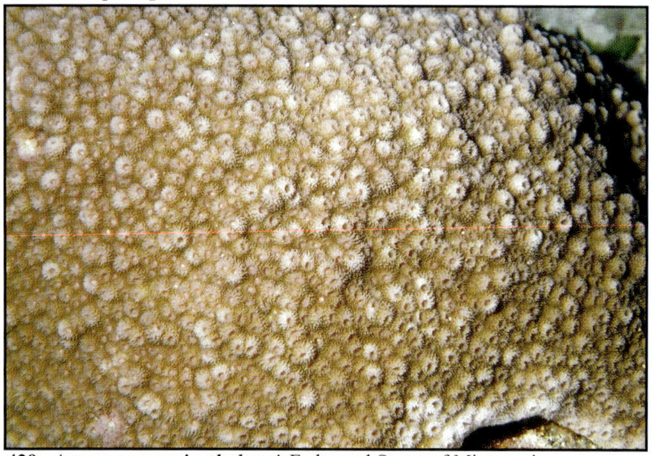
429- ***Astreopora myriopthalma*** * Federated States of Micronesia

430- ***Montipora aequituberulata*** * Philippines

431- ***Montipora* sp.** * Federated States of Micronesia

428- *Astreopora gracilis* * Acroporidae * Scleractinia * Federated States of Micronesia * Chuuk Atoll * lagoon * 30 ft (9 m).

429- *Astreopora myriopthalma* * Acroporidae * Scleractinia * Federated States of Micronesia * Nama Island * 30 ft (9 m).

430- *Montipora aequituberulata* * Acroporidae * Scleractinia * Philippines * Pescador Island * 20 ft (6 m). *Montipora* is second only to *Acropora* in the number of species occurring in a genus of stony coral, but as a whole are not distinctive nor easy to identify to species. There are over 200 nominal species, but many fewer valid species. In Australia, for example, there are 38 recognized species and Guam has 26 species. The corallites of *Montipora* are the smallest of all stony coral genera, not providing much in the way of structural clues to their identity. Species of *Montipora* are notoriously difficult to identify from underwater photographs, due to great variation in form and color. While alive, they are most easily confused with species of *Porites*, but can be readily distinguished when examining the skeleton under a microscope. The genus ranges from the east African coast and Red Sea through the Hawaiian Islands and Polynesia.

431- *Montipora* sp. * Acroporidae * Scleractinia * Federated States of Micronesia * Chuuk * Ozen Island * 50 ft (15 m).

432- *Montipora* sp. * Acroporidae * Scleractinia * Federated States of Micronesia * Chuuk Atoll * barrier reef * 50 ft (15 m).

433- *Montipora* sp. * Acroporidae * Scleractinia * Federated States of Micronesia * Chuuk Atoll * Ozen Island * 100 ft (30 m).

434- *Alveopora* sp. * Poritidae * Scleractinia * Federated States of Micronesia * Chuuk * Pou Bay * 20 ft (6 m).

435- *Goniopora fructicosa* * Poritidae * Scleractinia * Federated States of Micronesia * Chuuk * lagoon reef * 50 ft (15 m). The polyps of *Goniopora* are often expanded during the day and can reach surprising lengths. They have been difficult species for taxonomists to deal with and the final word on *Goniopora* taxonomy remains unsaid. The number of true species may number in the teens, with 14 recorded from Australia alone.

436- *Goniopora stokesii* * Poritidae * Scleractinia * Papua New Guinea * Karkar Island * 60 ft (18 m). This species is well known for often growing as unattached free-living colonies which have polyps growing on both the upper and lower surfaces. In deep areas of Chuuk lagoon, *G. stokesi* forms hemispherical colonies on sediment bottoms. The skeletons, for the size of the colonies, are surprisingly light and porous.

437- *Goniopora* sp. * Poritidae * Scleractinia * Papua New Guinea * Madang * lagoon reef * 66 ft (20 m).

438- *Goniopora* cf. *tenuidens* * Poritidae * Scleractinia * Federated States of Micronesia * Chuuk Atoll * 10 ft (3 m).

439- *Goniopora* sp. * Poritidae * Scleractinia * Federated States of Micronesia * Chuuk Atoll * Sankisan Maru * 60 ft (18 m).

440- *Goniopora* sp. * Poritidae * Scleractinia * Federated States of Micronesia * Chuuk Atoll * Eten Island * 40 ft (12 m).

441- *Porites cylindrica* * Poritidae * Scleractinia * Philippines * Pamalican Island * 10 ft (3 m). This coral can form stands of stubby branches several feet across in lagoon waters. During a tropical storm which produced large swells in the lagoon at Enewetak Atoll, large patches of *P. cylindrica* were broken. Pieces of the coral tumbled to new areas, where they survived and started growing as

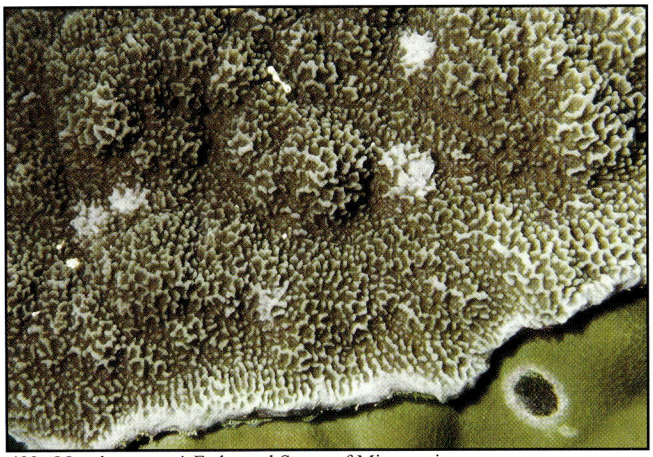
432- ***Montipora* sp.** * Federated States of Micronesia

433- ***Montipora* sp.** * Federated States of Micronesia

434- ***Alveopora* sp.** * Federated States of Micronesia

435- ***Goniopora fructicosa*** * Federated States of Micronesia

436- ***Goniopora stokesii*** * Papua New Guinea

437- ***Goniopora* sp.** * Papua New Guinea

438- ***Goniopora* cf. *tenuidens*** * Federated States of Micronesia

439- ***Goniopora* sp.** * Federated States of Micronesia

440- ***Goniopora* sp.** * Federated States of Micronesia

441- ***Porites cylindrica*** * Philippines

442- ***Porites lichen*** * Federated States of Micronesia

443- ***Porites lutea*** * Federated States of Micronesia

separate reef patches. Often, in Micronesia, *P. cylindrica* is yellow. It can be confused with *Palauastrea ramosa* which is superficially similar, but has distinct polyps and calices. *P. ramosa* is limited in its distribution, being found only in Palau among Micronesian islands.

442- *Porites lichen* * Poritidae * Scleractinia * Federated States of Micronesia * Chuuk * Pizion Reef * 30 ft (9 m).

443- *Porites lutea* * Poritidae * Scleractinia * Federated States of Micronesia * Chuuk * Eten Island * 12 ft (4 m). Heads of *Porites lutea* can reach great size, several feet across and ages of hundreds of years. Because of this, it is often used to sample past climatic and environmental conditions by examining the yearly growth layers of the colony, often by drilling a core from the outside to the central part of the head. It is found from the Red Sea to the Tuamotus, but not in Hawaii. Underwater it is difficult to distinguish from *Porites lobata*, another massive species which can form large heads.

444- *Porites rus* * Poritidae * Scleractinia * Papua New Guinea * Madang * barrier reef * 60 ft (18 m).

445- *Porites nigrescens* * Poritidae * Scleractinia * Federated States of Micronesia * Chuuk * Anaw Reef * 20 ft (6 m). This species is superficially similar to *Porites cylindrica*. *P. nigrescens* has deeper calices which give it a more pitted appearance and serve to distinguish it on close examination.

446- *Porites* sp. * Poritidae * Scleractinia * Federated States of Micronesia * Chuuk * Anaw Wall * 66 ft (20 m).

447- *Porites* sp. * Poritidae * Scleractinia * Federated States of Micronesia * Chuuk * Yamagiri Maru * 33 ft (10 m).

448- *Porites* sp. * Poritidae * Scleractinia * Federated States of Micronesia * Chuuk * Moen Island * 3 ft (1 m).

449- *Porites lutea* or ***lobata*** *** Poritidae * Scleractinia * Federated States of Micronesia * Chuuk Atoll * 20 ft (6 m).**

450- *Porites* sp. * Poritidae * Scleractinia * Philippines * Tubbataha Reef * 3 ft (1 m). *Porites* sp., the coral on the far right in the photograph, is living next to several other coral species on the reef crest.

451- *Psammocora contigua* * Siderastreidae * Scleractinia * Federated States of Micronesia * Chuuk Atoll * 6 ft (2 m). This species of *Psammocora* differs from *P. digitata* in that it is relatively delicate and occurs in much smaller colonies. It is similar to *P. obtusangula*.

452- *Psammocora digitata* * Siderastreidae * Scleractinia * Federated States of Micronesia * Chuuk * Eten Is. * 12 ft (4 m). This species is common in lagoon areas of Chuuk where it forms upright columns with rounded tops growing out of a basal mass of coral.

453- *Leptoseris gardineri* * Agariciidae * Scleractinia * Papua New Guinea * Port Moresby * Pt. Osbourne * 40 ft (12 m). The thin unifacial blades of *Leptoseris* are hard to confuse with any other stony coral with the exception of *Pavona cactus*. The latter species, however, is bifacial (polyps on both sides of the blades). There are about 14 species of *Leptoseris*, including one in the western Atlantic Ocean. Not all have frond-like blades, like the two species included here. Some are encrusting, others are foliaceous or plate-like.

444- ***Porites rus**** Papua New Guinea

445- ***Porites nigrescens*** * Federated States of Micronesia

446- ***Porites* sp.** * Federated States of Micronesia

447- ***Porites* sp.** * Federated States of Micronesia

448- ***Porites* sp.** * Federated States of Micronesia

449- ***Porites lutea or lobata*** * Federated States of Micronesia

450- ***Porites* sp.** * Philippines

451- ***Psammocora contigua*** * Federated States of Micronesia

452- ***Psammocora digitata*** * Federated States of Micronesia

453- ***Leptoseris gardineri*** * Papua New Guinea

454- ***Leptoseris papyracea*** * Federated States of Micronesia

455- ***Pachyseris rugosa*** * Federated States of Micronesia

454- *Leptoseris papyracea* * Agariciidae * Scleractinia * Federated States of Micronesia * Chuuk * lagoon bottom * 112 ft (34 m). This small coral can occur in dense masses in moderate lagoon depths. The photograph was taken in an area of the northwest Chuuk lagoon where this coral dominated the bottom, the dead skeleton building small hills on the bottom. It is found from the western Indian Ocean to Hawaii, and is much more delicate than the similar *Leptoseris gardineri*.

455- *Pachyseris rugosa* * Agariciidae * Scleractinia * Chuuk * Dublon Island * 40 ft (12 m).

456- *Pachyseris speciosa* * Agariciidae * Scleractinia * Papua New Guinea * Madang * barrier reef * 40 ft (12 m).

457- *Pavona cactus* * Agariciidae * Scleractinia * Papua New Guinea * Laing Island * 40 ft (12 m). At the opposite extreme from *Pavona minuta* is *P. cactus*, a delicate form which is most common in turbid water. Stands of *P. cactus* can be quite extensive, measuring many meters across. The species is bifacial, with polyps on both sides of the fronds. It occurs throughout the region as far east as the Marshall Islands.

458- *Pavona clavus* * Agariciidae * Scleractinia * Papua New Guinea * Manam Island * 40 ft (12 m).

459- *Pavona decussata* * Agariciidae * Scleractinia * Hong Kong * Breaker Reef * 33 ft (10 m). The form shown here is common in Hong Kong waters where it forms boxy plates.

460- *Pavona decussata* * Agariciidae * Scleractinia * Federated States of Micronesia * Chuuk * 10 ft (3 m). This is a different growth form than above. In this form the coral has fingers and is bifacial (polyps on both sides of blades or fingers).

461- *Pavona minuta* * Agariciidae * Scleractinia * Federated States of Micronesia * Chuuk * south barrier reef * 27 ft (8 m). Colonies of *Pavona minuta* can reach remarkable size. This photograph was taken on the southern barrier reef of Chuuk Atoll where a shallow pass empties out on the outer reef slope between Otta and Mesegon Islands. This particular colony of *P. minuta* was estimated to be over forty feet across and extended from about ten feet toforty feet in depth down the slope. The irregular lobate form is normal for Micronesia, but the species can also form a series of parallel ridges. The species occurs throughout the region, east to the Marshall Islands. This species is close to and perhaps the same as *Pavona duerdini*.

462- *Cycloseris* sp.* Fungiidae * Scleractinia * Federated States of Micronesia * Chuuk * lagoon bottom * 115 ft (35 m). *Cycloseris* are small fungiids, usually found on sediment bottoms away from reefs. They are nearly circular in shape. Some similar species belong to the genus *Diaseris*, although occurring in the same sort of habitats, differ in being irregular, not round, in shape.

463- *Fungia fungites* * Fungiidae * Scleractinia * Federated States of Micronesia * Chuuk * 15 ft (5 m). *Fungia* corals, or mushroom corals, are most easily recognized from the solitary species that form circular skeletons about 6-10 inches in diameter. These skeletons resemble the underside of large mushrooms, hence their common and scientific names. A number of species in this genus are colonial and form large domes or oblong plates several feet in length. These corals are common down to a depth of about 60 feet. They are not attached to the substrate as adults. Initially the larvae of this species attaches to the bottom and develops into a small coral polyp. Eventually the disc of the polyp breaks free and falls to the bottom. Most fungiid corals are able to right themselves by inflating the tissue of the polyp with water in order to push the coral right side

456- ***Pachyseris speciosa*** * Papua New Guinea

457- ***Pavona cactus*** * Papua New Guinea

458- ***Pavona clavus*** * Papua New Guinea

459- ***Pavona decussata*** * Hong Kong

460- ***Pavona decussata*** * Federated States of Micronesia

461- ***Pavona minuta*** * Federated States of Micronesia

462- ***Cycloseris*** * Federated States of Micronesia

463- ***Fungia fungites*** * Federated States of Micronesia

464- *Fungia* sp. * Papua New Guinea

465- **Undescribed fungiid** * Federated States of Micronesia

466- ***Fungia echinata*** * Federated States of Micronesia

467- ***Halomitra pileus*** * Federated States of Micronesia

up. In this manner they are also able to move from one area to another. The ridges on the surface of the coral are very sharp and handeling these corals is not advised.

464- *Fungia* sp. * Fungiidae * Scleractinia * Papua New Guinea * Mait reef * 30 ft (9 m).

465- Undescribed fungiid * Fungiidae * Scleractinia * Federated States of Micronesia * Chuuk * algal flat * 165 ft (50 m).

466- *Fungia echinata* * Fungiidae * Scleractinia * Federated States of Micronesia * Chuuk * Anaw Point * 50 ft (15 m).

467- *Halomitra pileus* * Fungiidae * Scleractinia * Federated States of Micronesia * Chuuk * Anaw Reef * 40 ft (12 m).

468- *Heliofungia actiniformis* * Fungiidae * Scleractinia * Papua New Guinea * Madang * lagoon * 30 ft (9 m). At first glance, due to the long white tipped tentacles, this coral looks more like a sea anemone. It is a species which is closely relate to the other fungiid corals.

469- *Herpolitha limax* * Fungiidae * Scleractinia * Marshall Islands * Enewetak Atoll * 20 ft (6 m).

470- *Lithophyllon edwardsi* * Fungiidae * Scleractinia * Hong Kong * 10 ft (3 m).

471- *Polyphyllia talpina* * Fungiidae * Scleractinia * Federated States of Micronesia * Chuuk * lagoon bottom * 100 ft (30 m).

472- *Podabacia crustacea* * Fungiidae * Scleractinia * Philippines * Pescador Island * 20 ft (6 m).

473- *Sandolitha robusta* * Fungiidae * Scleractinia * Palau * Ulong Channel * 20 ft (6 m).

474- *Acrhelia horrescens* * Oculinidae * Scleractinia * Papua New Guinea * Madang * barrier reef * 60 ft (18 m).

475- *Galaxea* sp. * Oculinidae * Scleractinia * Papua New Guinea * Wanganam * 30 ft (9 m).

476- *Galaxea paucisepta* * Oculinidae * Scleractinia * Palau * Ngchesar * 40 ft (12 m).

477- *Echinophyllia aspera* * Pectiniidae * Scleractinia * Federated States of Micronesia * Chuuk * Ozen Island * 100 ft (30 m) This coral is known from the Red Sea to Tahiti, and can be common in a wide range of habitats. It occurs in a wide variety of colors.

468- ***Heliofungia actiniformis*** * Papua New Guinea

469- *Herpolitha limax* * Marshall Islands

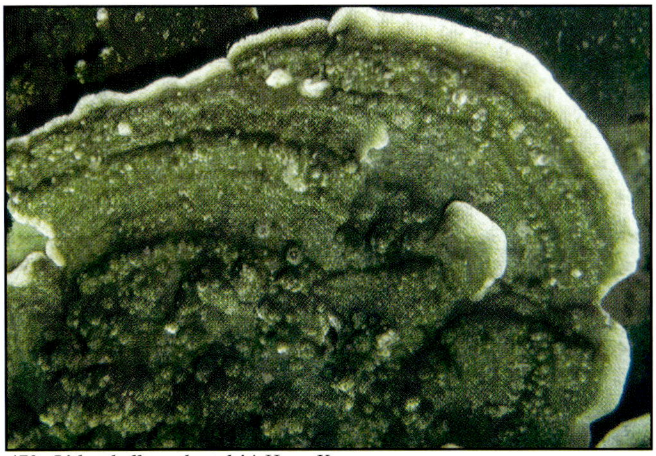

470- *Lithophyllon edwardsi* * Hong Kong

471- *Polyphyllia talpina* * Federated States of Micronesia

472- *Podabacia crustacea* * Philippines

473- *Sandolitha robusta* * Palau

474- *Acrhelia horrescens* * Papua New Guinea

475- *Galaxea* sp. * Papua New Guinea

476- *Galaxea paucisepta* * Palau

477- ***Echinophyllia aspera*** * Federated States of Micronesia

478- ***Mycedium elephantotus*** * Philippines

479- ***Mycedium elephantotus*** * Papua New Guinea

478- *Mycedium elephantotus* * Pectiniidae * Scleractinia * Philippines * Pescador Island * 20 ft (6 m).

479- *Mycedium elephantotus* * Pectiniidae * Scleractinia * Papua New Guinea * Eastern Fields * 100 ft (30 m).

480- *Oxypora glabra* * Pectiniidae * Scleractinia * Palau * Ngechesar * 90 ft (27 m).

481- *Pectinia lactuca* * Pectiniidae * Scleractinia * Papua New Guinea * Kavieng * Rai Island * 40 ft (12 m).

482- *Pectinia peonia* * Pectiniidae * Scleractinia * Marshall Islands * Kwajalein Atoll * 40 ft (12 m).

483- *Acanthastrea echinata* * Mussidae * Scleractinia * Papua New Guinea * Madang Channel * 40 ft (12 m).

484- *Cynarina lacrymalis* * Mussidae * Scleractinia * Papua New Guinea * Kavieng * Albatross Channel Wall * 100 ft (30 m).

485- *Lobophyllia* cf. *corymbosa* * Mussidae * Scleractinia * Papua New Guinea * Kranket Island slope * 66 ft (20 m).

486- *Lobophyllia corymbosa* * Mussidae * Scleractinia * Federated States of Micronesia * Chuuk Atoll * north lagoon * 100 ft (30 m).

487- *Lobophyllia hataii* * Mussidae * Scleractinia * Federated States of Micronesia * Chuuk * Weno * lagoon reef * 50 ft (15 m).

488- *Lobophyllia hemprichii* * Mussidae * Scleractinia * Philippines * Batangas * Pulangbuli reef * 33 ft (10 m).

489- *Lobophyllia hemprichii* * Mussidae * Scleractinia * Papua New Guinea * New Ireland * Kalihi harbor * 50 ft (15 m).

490- *Lobophyllia* sp. * Mussidae * Scleractinia * Federated States of Micronesia * Chuuk * lagoon reef * 50 ft (15 m).

491- *Lobophyllia* cf. *pachysepta* * Mussidae * Scleractinia * Federated States of Micronesia * Chuuk * lagoon * 50 ft (15 m).

492- *Scolymia vitiensis* * Mussidae * Scleractinia * Papua New Guinea * Dyaul Island * 60 ft (18 m).

493- *Symphyllia recta* * Mussidae * Scleractinia * Federated States of Micronesia * Chuuk * south field ramp * 12 ft (4 m).

480- ***Oxypora glabra*** * Palau

481- ***Pectinia lactuca*** * Papua New Guinea

482- *Pectinia peonia* * Marshall Islands

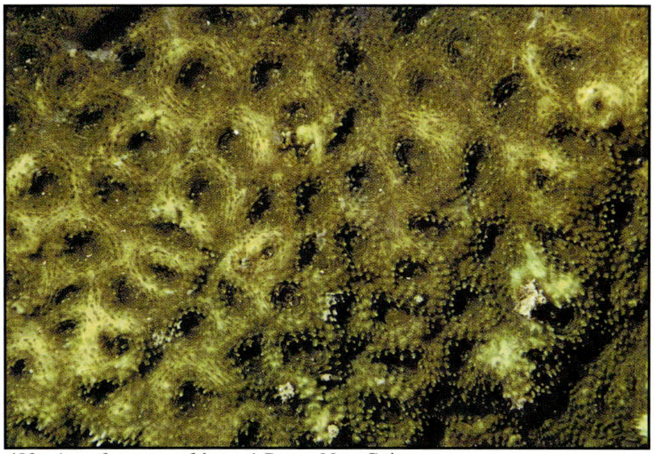
483- *Acanthastrea echinata* * Papua New Guinea

484- *Cynarina lacrymalis* * Papua New Guinea

485- *Lobophyllia* cf. *corymbosa* * Papua New Guinea

486- *Lobophyllia corymbosa* * Federated States of Micronesia

487- *Lobophyllia hataii* * Federated States of Micronesia

488- *Lobophyllia hemprichii* * Philippines

489- *Lobophyllia hemprichii* * Papua New Guinea

490- ***Lobophyllia* sp.** * Federated States of Micronesia

491- ***Lobophyllia* cf. *pachysepta*** * Federated States of Micronesia

492- ***Scolymia vitiensis*** * Papua New Guinea

493-***Symphyllia recta*** * Federated States of Micronesia

494- *Symphyllia* sp. * Mussidae * Scleractinia * Philippines * Pescador Island * 66 ft (20 m).

495- *Symphyllia* sp. * Mussidae * Scleractinia * Indonesia * Manado * 66 ft (20 m).

496- *Symphyllia* sp. * Mussidae * Scleractinia * Federated States of Micronesia * Chuuk * Ozen Island * 66 ft (20 m).

497- *Symphyllia* sp. * Mussidae * Scleractinia * Federated States of Micronesia * Chuuk * lagoon reef * 50 ft (15 m).

498- *Hydnophora exesa* * Merulinidae * Scleractinia * Papua New Guinea * Port Moresby * Lion Island * 33 ft (10 m).

499- *Hydnophora rigida* * Merulinidae * Scleractinia * Palau * Ngerkul Pass * 20 ft (6 m).

500- *Merulina amplicata* * Merulinidae * Scleractinia * Federated States of Micronesia * Chuuk * Ozen Island * 80 ft (25 m). This is a common species found from the Red Sea through Micronesia, Fiji and Samoa. It is quite variable in color.

501- *Scapophyliia cylindricus* * Merulinidae * Scleractinia * Federated States of Micronesia * Nama Island * 100 ft (30 m).

502- *Barabattoia amicorum* * Faviidae * Scleractinia * Federated States of Micronesia * Chuuk * Ozen Island * 40 ft (12 m).

503- *Caulastrea curvata* * Faviidae * Scleractinia * Papua New Guinea * Manam Island * volcanic sand slope * 80 ft (25 m).

494- ***Symphyllia* sp.** * Philippines

495- ***Symphyllia* sp.** * Indonesia

496- *Symphyllia* sp. * Federated States of Micronesia

497- *Symphyllia* sp. * Federated States of Micronesia

498- *Hydnophora exesa* * Papua New Guinea

499- *Hydnophora rigida* * Palau

500- *Merulina amplicata* * Federated States of Micronesia

501- *Scapophyliia cylindricus* * Federated States of Micronesia

502- *Barabattoia amicorum* * Federated States of Micronesia

503- *Caulastrea curvata* * Papua New Guinea

504- *Caulastrea furcata* * Palau

505- *Diploastrea heliopora* * Papua New Guinea

506- *Echinopora mammiformis?* * Papua New Guinea

507- *Favia stelligera* * Federated States of Micronesia

504- *Caulastrea furcata* * Faviidae * Scleractinia * Palau * Medusae Pass * 20 ft (6 m).

505- *Diploastrea heliopora* * Faviidae * Scleractinia * Papua New Guinea * Dyaul * 30 ft (9 m).

506- *Echinopora mammiformis?* * Faviidae * Scleractinia * Papua New Guinea * New Ireland * Kalili Harbor * 66 ft (20 m).

507- *Favia stelligera* * Faviidae * Scleractinia * Federated States of Micronesia * Chuuk * Nama Island * 20 ft (6 m).

508- *Favia stelligera* * Faviidae * Scleractinia * Federated States of Micronesia * Chuuk * Ozen Island * 40 ft (12 m).

509- *Favites flexuosa* * Faviidae * Scleractinia * Federated States of Micronesia * Chuuk * lagoon reef * 50 ft (15 m).

510- *Favites* cf. *halichora* * Faviidae * Scleractinia * Papua New Guinea * Madang Channel * 66 ft (20 m).

511- *Goniastrea actinata* * Faviidae * Scleractinia * Federated States of Micronesia * Chuuk * Anaw Reef * 20 ft (6 m).

512- *Goniastrea pectinata* * Faviidae * Scleractinia * Federated States of Micronesia * Chuuk * Weno * south field * 15 ft (5 m).

513- *Leptoria ptrygia* * Faviidae * Scleractinia * Federated States of Micronesia * Chuuk * Otta Island * 60 ft (18 m).

514- *Montastrea curta* * Faviidae * Scleractinia * Federated States of Micronesia * Nama Island * 40 ft (12 m).

508- *Favia stelligera* * Federated States of Micronesia

509- *Favites flexuosa* * Federated States of Micronesia

510- *Favites* cf. *halichora* * Papua New Guinea

511- *Goniastrea actinata* * Federated States of Micronesia

512- *Goniastrea pectinata* * Federated States of Micronesia

513- *Leptoria ptrygia* * Federated States of Micronesia

514- *Montastrea curta* * Federated States of Micronesia

515- *Moseleya latistellata* * Papua New Guinea

516- *Oulophyllia crispa* * Federated States of Micronesia

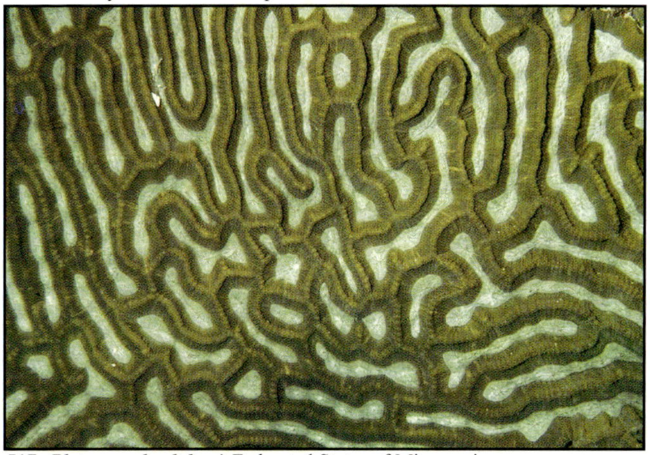

517- *Platygyra daedalea* * Federated States of Micronesia

515- *Moseleya latistellata* * Faviidae * Scleractinia * Papua New Guinea * Eastern Fields *40 ft (12 m).

516- *Oulophyllia crispa* * Faviidae * Scleractinia * Federated States of Micronesia * Chuuk * lagoon reef * 60 ft (18 m).

517- *Platygyra daedalea* * Faviidae * Scleractinia * Federated States of Micronesia * Chuuk * Eten Island * 40 ft (12 m).

518- *Platygyra lamellina* * Faviidae * Scleractinia * Papua New Guinea * Eastern Fields * 50 ft (15 m).

519- *Platygyra sinensis* * Faviidae * Scleractinia * Hong Kong * Hoi Ho Wan * 6 ft (2 m).

520- *Platygyra* sp. * Faviidae * Scleractinia * Federated States of Micronesia * Nama Island * 40 ft (12 m).

521- *Platygyra* sp. * Faviidae * Scleractinia * Federated States of Micronesia * Chuuk * Otta Island * 50 ft (15 m).

522- *Euphyllia ancora* * Caryophyllidae * Scleractinia * Papua New Guinea * New Ireland * Kalili Harbor * slope * 66 ft (20 m).

523- *Euphyllia divisa?* * Caryophyllidae * Scleractinia * Indonesia * Manado * fringing reef * 33 ft (10 m).

524- *Euphyllia glabrescens* * Caryophyllidae * Scleractinia * Papua New Guinea * Madang * barrier reef * 60 ft (18 m).

525- *Euphyllia parancora* * Caryophyllidae * Scleractinia * Marshall Islands * Kwajalein Atoll * patch reef * 30 ft (9 m).

526- *Euphyllia paranchora* * Caryophyllidae * Scleractinia * Papua New Guinea * Kalili Harbor * 33 ft (10 m).

527- *Euphyllia paradivisa* * Caryophyllidae * Scleractinia * Papua New Guinea * Port Moresby * Lion Island * 40 ft (12 m).

528- *Physogyra lichtensteini* * Caryophyllidae * Scleractinia * Palau * Mutremdiu Wall * 60 ft (18 m).

529- *Physogyra lichtensteini* * Caryophyllidae * Scleractinia * Federated States of Micronesia * Chuuk Atoll * lagoon * night * 50 ft (15 m).

530- *Pleurogyra synplex?* * Caryophyllidae * Scleractinia * Palau * rock islands * 40 ft (12 m). This coral is known from the Philippines, Cook Islands, eastern Papua New Guinea, Palau and New Caledonia.

518- ***Platygyra lamellina*** * Papua New Guinea

519- ***Platygyra sinensis*** * Hong Kong

520- ***Platygyra* sp.** * Federated States of Micronesia

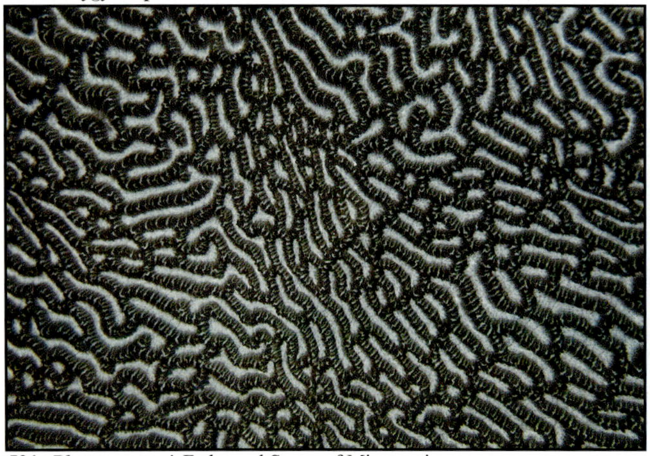

521- ***Platygyra* sp.** * Federated States of Micronesia

522- ***Euphyllia ancora*** * Papua New Guinea

523- *Euphyllia divisa* possibly * Papua New Guinea

524- *Euphyllia glabrescens* * Indonesia

525- *Euphyllia parancora* * Marshall Islands

526- *Euphyllia parancora* * Papua New Guinea

527- *Euphyllia paradivisa* * Papua New Guinea

528- *Physogyra lichtensteini* * Palau

529- *Physogyra lichtensteinii* * Palau

530- *Pleurogyra synplex*? * Palau

531- ***Pluerogyra sinuosa*** * Marshall Islands

532- ***Dendrophyllia* sp.** * Marshall Islands

533- ***Dendrophyllia* sp.** * Papua New Guinea

534- ***Tubastrea diaphana*** * Hong Kong

531- *Pluerogyra sinuosa* * Caryophyllidae * Scleractinia * Marshall Islands * Kwajalein Atoll * patch reef * 20 ft (6 m).

532- *Dendrophyllia* sp. * Dendrophyllidae * Scleractinia * Papua New Guinea * Madang * barrier reef cave * 60 ft (18 m).

533- *Dendrophyllia* sp. * Dendrophyllidae * Scleractinia * Papua New Guinea * Madang * barrier reef * 60 ft (18 m).

534- *Tubastrea diaphana* * Dendrophyllidae * Scleractinia * Hong Kong * Breaker Reef * 33 ft (10 m). This is the most common species of *Tubastrea* in Hong Kong.

535- *Tubastrea micrantha* * Dendrophyllidae * Scleractinia * Papua New Guinea * Madang * barrier reef * 60 ft (18 m).

536- *Tubastrea micrantha* * Dendrophyllidae * Scleractinia * Federated States of Micronesia * Chuuk * Fujikawa Maru * 80 ft (25 m).

537- *Tubastrea* sp * Dendrophyllidae * Scleractinia * Philippines * Pulang Buli * 3 ft (1 m).

538- *Tubastrea* sp. * Dendrophyllidae * Scleractinia * Philippines * Pescador Island * 60 ft (18 m). This coral is covered with unidentified acoela flatworms.

539- *Tubastrea* sp. * Dendrophyllidae * Scleractinia * Federated States of Micronesia * Chuuk * lagoon shipwreck * 50 ft (15 m).

540- *Turbinaria bifrons* * Dendrophyllidae * Scleractinia * Papua New Guinea * Port Moresby * Pt. Osbourne * 33 ft (10 m). The genus *Turbinaria* has species which exhibit extremely variable growth forms. Shallow and deep water forms of the same species often bear little resemblance to each other. Species of this genus are often found in murky water where they develop beautiful foliose growth forms. This species is unusual in that the polyps occur on both sides of the blade. In most species of the genus, polyps are found on one side. The genus also occurs in areas outside the tropics.

541- *Turbinaria peltata* * Dendrophyllidae * Scleractinia * Federated States of Micronesia * Chuuk Atoll * 40 ft (12 m).

542- *Turbinaria retiformis* * Dendrophyllidae * Scleractinia * Papua New Guinea * Madang * barrier reef * 50 ft (15 m).

543- *Turbinaria retiformis* * Dendrophyllidae * Scleractinia * Federated States of Micronesia * Nama Island * 50 ft (15 m).

535- ***Tubastrea micrantha*** * Papua New Guinea

536- ***Tubastrea micrantha*** * Federated States of Micronesia

537- ***Tubastrea*** **sp** * Philippines

538- ***Tubastrea*** **sp.** * Philippines

539- ***Tubastrea*** **sp.** * Federated States of Micronesia

540- ***Turbinaria bifrons*** * Papua New Guinea

541- ***Turbinaria peltata*** * Papua New Guinea

542- ***Turbinaria retiformis*** * Federated States of Micronesia

543- ***Turbinaria retiformis*** * Federated States of Micronesia

Subclass Corallimorpharia - Corallimorpharians

For all practical purposes corallimorpharians are corals without a skeleton. Their internal anatomy, nematocysts and tentacles are identical to the scleractinia. Many species are brightly colored. One species of *Pseudocorynactis*, which stays closed all day, spreads its disk open at night to reveal a beautiful red column, transparent tentacles and bright orange capitate spheres at the tentacle ends. Often times species of the genus *Rhodactis* cover large areas of bottom. We saw one entire patch reef north of New Britain in Papua New Guinea that was covered densely with nothing but *Rhodactis* over an area of thousands of square yards. Interestingly, other patch reefs nearby had very few. Corallimorpharians reproduce asexually through a process called transverse fission where the animal literally pulls itself apart and reforms into two individuals. Sexual reproduction is probably similar to that observed in corals.

Order Actiniaria - Sea Anemones

Sea Anemones are always solitary polyps. As in other Cnidarian polyps, the mouth is situated on an oral disc surrounded by tentacles. The body is columnar in shape with a flattened "foot" at the base for attachment. The main difference between anemones and the Cnidarians covered previously is the former have a well developed structure called the siphonoglyph located along one or both sides of the pharynx (octocorals have a reduced siphonglyph). The siphonoglyph is lined with cilia which beat in rhythmic fashion to bring water into the gastrovascular cavity or reverse direction to help expel waste material. By aiding circulation of oxygenated water, the siphonoglyph, enables anemones to attain larger sizes than most cnidarian polyps. Actinians resemble corallimorpharians, however they are generally firmer, larger and more variable in morphology.

There is great diversity in the body plan of different sea anemone species in the tropical Pacific. Sea anemones are common in shallow water where they attach to rocks and other hard substrate. Many species live on sandy and mud bottoms and several species live in association with other animals. The genus *Calliactis* has several species which are only found on the shells of hermit crabs. A number of species of large shallow water anemones harbor anemone fishes which only live in association with sea anemones. Much has been written about this association and it is generally accepted that the anemone fish develop a mucous coating that renders them unrecognizable as prey to the anemone. The association is well developed and there is certainly behavior on the part of the fish and probably on the part of the anemone which is not fully understood.

Order Ceriantharia - Tube-dwelling anemones

Cerianthids are solitary, tube-dwelling anemones. They differ morphologically from the sea anemones by having tentacles around the mouth (oral tentacles) as well as around the margin of the oral disc. In cerianthids, the bottom of the column is rounded and not modified into an attachment structure (there are a few anemones like this as well). They live mostly buried in the sand in tubes which they secrete. The tubes are made up of fired nematocysts and encrusted sand particles. The tubes of large cerianthids may be up to several feet in length. When disturbed the cerianthid quickly retracts into the safety of its tube. There are several species of shallow water cerianthids in the tropical Pacific. Cerianthids have planktonic larvae that may live for six months in the water column. It is quite possible that many species of cerianthids are widely distributed. Unfortunately most of the species are poorly known.

Order Zoanthidea - zoanthids

Zoanthids are generally colonial although a few are solitary. Colonial zoanthids may superficially resemble coral heads. Colonial polyps are connected by stolons or, in species forming large mats, the polyps are embedded in a tissue-like body (coenenchyme). Certain species of *Palythoa* and *Zoanthus* are capable of covering many square yards of reef flat or rocky bottom. When large areas of reef become overgrown with zoanthids it is usually a good indication that the water quality of the area has changed for the worse. Many species of zoanthids live in association with other Cnidarians. The genus *Parazoanthus* lives almost exclusively with other invertebrates such as sponges, hydroids and black corals. As previously mentioned, most zoanthids contain palytoxin and care should be taken to avoid contact with open cuts.

Order Antipatharia - Black Corals

In the black corals only the horny material of the axial skeleton is black. Polyps are colonial and-cover the surface of the black skeleton; they may be white, yellow, orange or brown. The skeleton often has small hooks or thorns on it which enable the polyps to grip the surface. The polyps are non-retractile and there are no "cups" into which they can retract as in stony corals or gorgonians. Members of the genus *Cirrhipathes* have a single whip-like skeleton. Other species, such as those of *Antipathes*, can be delicately branched on one or more planes or can be extremely bushy. Some black corals in deep water can reach several feet in height and more than four inches in diameter at the base of the skeleton.

The name antipatharian (Greek for "against disease") was given at a time when it was believed black corals possessed medicinal properties. To date, however, no pharmaceutical products have been obtained from black corals. Black coral, certain octocorals and even stony corals are used to make jewelry, but their true commercial potential is probably best realized by leaving them on the reef for tourist divers to observe.

Right- This photo shows a large black coral colony (antipatharian) on a reef outcrop near Madang in Papua New Guinea. Black corals grow slowly and a large individual such as this may be fifty years old or more. These corals are attached firmly to the reef by a modification of the central axis. Superficially, black coral seafans like this one resemble octocoral sea fans. Both types of seafans are filter feeders; they can be distinguished by the number and type of polyp tentacles. Black coral polyps have only six unbranched tentacles. Numerous other invertebrates, such as, shrimp, fish, oysters, worms, and other cnidarians are often found in association with black corals.

Below left- This photo of the reef flat at Enewetak Atoll in the Marshall Islands shows two very large aggregations of the zoanthid *Palythoa*. Chances are good that the initial asexual efforts of a single polyp gave rise to all the individual colonies in the photograph. Older colonies are at the center of each aggregation the youngest colonies are at the margin. Colonies of this species of *Palythoa* attain a maximum size of about four inches, only the colonies on the outer edge of the aggregation bud off new polyps.

Below Right- The seafan in this photo resembles both an octocoral and a black coral, actually it is a zoanthid.

544- ***Isaurus tuberculatus*** * Indonesia

545- ***Palythoa psammophila*** * Hawaii

546- ***Palythoa toxica*** * Papua New Guinea

547- ***Palythoa tuberculosa*** * Marshall Islands

544- *Isaurus tuberculatus* * Isauridae * Zoanthidea * Indonesia * Manado * fringing reef * 20 ft (6 m). Often inconspicuous, this zoanthid lives firmly attached to pieces of coral rubble which are sometimes partially covered with sand. During the day the tentacles are withdrawn and the animal appears as a bent, broken stem from an alga or soft coral. It is best observed at night when feeding with the tentacles extended. The body can vary in color from a dull brown to mottled yellow.

545- *Palythoa psammophila* * Zoanthidae * Zoanthidea * Hawaii * Kaneohe Bay * 3 ft (1 m). This zoanthid is usually buried in sand to the level of the oral disk. Its body wall may be heavily embedded with sand. The oral disk is green to light brown. It is known from Hawaii, particularly from sand flats of Kaneohe Bay, Oahu where the photograph was taken, however it probably occurs elsewhere.

546- *Palythoa toxica* * Zoanthidae * Zoanthidea * Papua New Guinea * New Britain * Kimbe Bay * 50 ft (15 m). This species may be identical to *Palythoa grandis*. In Hawaii *Palythoa toxica* lives in tide pools in a limited number of areas, but is widely distributed elsewhere in the Pacific. A strong toxin, called palytoxin, is found in the mucous and the gonads and is highly dangerous to humans. The ancient Hawaiians used to coat their spear points with this material to make them more deadly.

547- *Palythoa tuberculosa* * Zoanthidae * Zoanthidea * Marshall Islands * Enewetak Atoll * Lojwa Island * 20 ft (6 m). This species is known from Hawaii, and throughout the Indo-Pacific region.

548- *Palythoa tuberculosa* * Zoanthidae * Zoanthidea * Federated States of Micronesia * Chuuk Atoll * Fujikawa Maru * 60 ft (18 m). The genus *Palythoa* has the body wall heavily encrusted with sand.

549- *Palythoa vestitus* * Zoanthidae * Zoanthidea * Papua New Guinea * New Britain * 60 ft (18 m). This zoanthid grows over coral rubble in shallow water and may form large mats up to many square feet in area.

550- *Palythoa* sp. * Zoanthidae * Zoanthidea * Marshall Islands * Enewetak Atoll * reef flat * intertidal. This species can form large mats on atoll reef flats in Micronesia as seen in this photograph.

551- *Sphenopus* * Zoanthidae * Zoanthidea * Papua New Guinea * Port Moresby * 60 ft (18 m). This is a solitary zoanthid that lives on muddy bottoms.

552- *Parazoanthus* sp. * Zoanthidae * Zoanthidea * Papua New Guinea * Madang * barrier reef * 60 ft (18 m). This is a *Parazoanthus* which grows on abandoned worm tubes. This may possibly be a species of *Epizoanthus* or *Acrozoanthus australiae*. It can be fairly abundant, locally on outer reef slopes and elsewhere.

553- *Parazoanthus* sp. * Zoanthidae * Zoanthidea * Papua New Guinea * Madang * barrier reef * 60 ft (18 m). This zoanthid is believed to grow on dead worm tubes, but it may well form its own tube-like structure on which it grows.

554- *Parazoanthus* sp? * Zoanthidae * Zoanthidea * Papua New Guinea * New Britain * inshore reef * 60 ft (18 m). This species may not be *Parazoanthus*. It either grows on a dead hydroid or has an axial skeleton which resembles a sea fan. Further investigation is needed to see if the zoanthid secretes the skeleton or if the zoanthid as it overgrows a living hydroid stimulates growth of the later.

555- *Parazoanthus* sp. * Zoanthidae * Zoanthidea * Indonesia * Manado * fringing reef * 50 ft (15 m). This species grows in association with ascidians and sponges, in "turf balls" (see the sponge *Plakinastrella* sp. for more information).

556- *Parazoanthus* sp. * Zoanthidae * Zoanthidea * Papua New Guinea * New Britain * inshore reef * 30 ft (9 m). Intense competition for living space has provided the impetus for some organ-

548- ***Palythoa tuberculosa*** ***** Federated States of Micronesia

549- ***Palythoa vestitus*** ***** Papua New Guinea

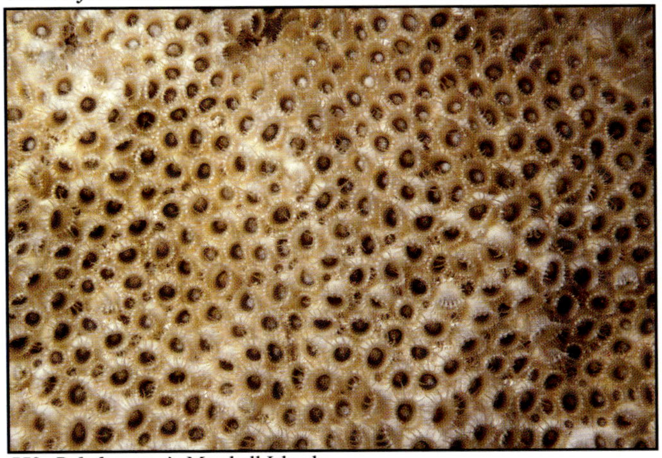
550- ***Palythoa* sp. *** Marshall Islands

551- ***Sphenopus*** ***** Papua New Guinea

552- ***Parazoanthus* sp. *** Papua New Guinea

553- ***Parazoanthus* sp. *** Papua New Guinea

554- ***Parazoanthus* sp. *** Papua New Guinea

555- ***Parazoanthus* sp. *** Indonesia

556- ***Parazoanthus* sp.** * Papua New Guinea

557- ***Parazoanthus* sp.** * Indonesia

558- ***Epizoanthus* sp.** * Philiipines

isms to overgrow other organisms. Several species of *Parazoanthus* live on sponges and hydroids. Studies have indicated that the zoanthid makes the host unpalatable, while the host provides substrate for the zoanthid. The color of the zoanthid usually contrasts with that of the host which makes it easier for predator species to recognize and avoid (aposymatic coloration). This species of *Parazoanthus* is commonly found growing on the surface of the orange sponge *Stylissa flabelliformis*. This is a typical growth form for *Parazoanthus*, similar to that of *P. axinella* from the Caribbean and Mediterranean.

557- *Parazoanthus* sp. * Zoanthidae * Zoanthidea * Indonesia * Manado * offshore island * 6 ft (2 m). This *Parazoanthus* is growing with ascidians and sea grass.

558- *Epizoanthus* sp. * Zoanthidae * Zoanthidea * Philippines * Cebu * Mactan Island * cave * 78 ft (23 m).

559- *Zoanthus pacificus* * Zoanthidae * Zoanthidea * Marshall Islands * Enewetak Atoll * patch reef * 20 ft (6 m). The genus *Zoanthus* does not have encrustations of the body wall. It is known from Hawaii, Samoa and Tahiti.

560- *Zoanthus mantoni* * Zoanthidae * Zoanthidea * Palau * marine lake * 3 ft (1 m). This zoanthid is very common in Chuuk Atoll in protected waters. On the northern barrier reef is a large pool about 600 feet across on the shallow reef, a remnant of lower sea level, in which much of the bottom is carpeted with this zoanthid.

561- *Zoanthus* sp. * Zoanthidae * Zoanthidea * Marshall Islands * Enewetak Atoll * lagoon patch reef * 16 ft (5 m).

562- *Zoanthina* larva of *Zoanthus* * Zoanthidae * Zoanthidea * open water.

563- *Zonathella* larva of *Palythoa* * Zoanthidae * Zoanthidea * open water.

564- *Anthopluera nigrescens?* * Actiniidae * Actinaria * Hong Kong * Cape d'Aguilar * 6 ft (2 m). In Hawaii the species is common in intertidal holes and crevices on rocky shores and was similarly found subtidally on rocky shores around Hong Kong.

565- *Condylactis* sp. * Actiniidae * Actinaria * Palau * lagoon reef * 20 ft (6 m).

566- *Dofleinia armata* * Actiniidae * Actinaria * Palau * sand slope * 60 ft (18 m).

567- *Dofleinia* sp. * Actiniidae * Actinaria * Indonesia * Manado * sand * 50 ft (15 m).

568- *Entacmea quadricolor* * Actiniidae * Actinaria * Federated States of Micronesia * Chuuk Atoll * fringing reef * 40 ft (12 m). *Entacmea quadricolor* lives among coral outcrops. It is easily recognized by the subterminal swelling of the tentacle tips and by its

559- ***Zoanthus pacificus*** * Marshall Islands

560- ***Zoanthus mantoni*** * Palau

561- ***Zoanthus* sp.** * Marshall Islands

562- ***Zoanthina* larvae** * open water

563- ***Zonathella* larve** * open water

564- ***Anthopluera nigrescens?*** * Hong Kong

565- ***Condylactis* sp.** * Palau

566- ***Dofleinia armata*** * Palau

567- ***Dofleinia* sp.** * Indonesia

568- ***Entacmea quadricolor*** * Federated States of Micronesia

habit of asexual division. The crevices on a coral head inhabited by this anemone will practically be filled with the tentacles of this anemone. *E. quadricolor* occurs in many areas of the Indo-West Pacific. This anemone harbors commensal fish (anemone fish).

569- *Boloceroides mcmurrichii* * Boloceroididae* Actinaria* Bahrain * seagrass bed * 6 ft (2 m). This anemone often attaches to sea grass blades and is easily dislodged. Amazingly, when knocked off the blades, they swim quite well with rhythmic pulsations of the tentacular crown. The shape and striped tentacles are distinguishing. It also lives on sandy and muddy bottoms. It is known from the Red Sea to Hawaii, Japan and Australia.

570- *Actinodendron arboreum* * Actinodendriidae * Actinaria * Marshall Islands * Enewetak Atoll * lagoon slope * 20 ft (6 m). Species of *Actinodendron* have been referred to as the "hells-fire" anemones due to the painful sting they can inflict. They live exposed, on sandy bottoms, and resemble a soft coral or clump of algae. Often they have commensal shrimp living on them.

571- *Actinodendron arboreum* * Actinodendriidae * Actinaria * Papua New Guinea *Madang * lagoon * night * 60 ft (18 m).

572- *Actinodendron plumosum* * Actinodendriidae * Actinaria * Palau * lagoon * sand bottom * 6 ft (2 m).

573- *Actinostephanus haechkeli* * Actinodendriidae * Actinaria * Marshall Islands * Enewetak Atoll * 30 ft (9 m).

574- *Megalactis hemprichii* * Actinodendriidae * Actinaria * Palau * lagoon * 30 ft (9 m).

575- *Aiptasia diaphana* * Aiptasiidae * Actinaria * Papua New Guinea * Madang * mangroves * 3 ft (1 m). This species of anemone is part of fouling communities, living on boat bottoms, pilings and other man-made objects that have other abundant growth.

576- *Aiptasia pulchella* * Aiptasiidae * Actinaria * Palau * marine lake * 3 ft (1 m). This is a fairly small, soft and flaccid anemone which usually has two distinct forms, a small greenish brown form marked with white flecks and a larger brown or pale form without markings. This species is known from Japan, French Polynesia, Hawaii, and the Central American coast of the eastern Pacific. It is probably identical to ***Aiptasia diaphana*** above.

577- *Alicia pretiosa* * Aliciidae * Actinaria * Federated States of Micronesia * Chuuk Atoll * Onang Island * 50 ft (15 m). This small anemone and other larger ones like it appear as shriveled clumps on rocks and seaweed by day, but at night they inflate to full height, which may be 2 feet, and extend their tentacles to capture food. Many species of anemones produce nematocysts that do not harm man, but others, like *Alicia*, are toxic to humans, causing severe pain and extreme illness. Especially painful are the nematocysts from the berry-like appendages on the column. These anemones should not be handled!

569- *Boloceroides mcmurrichii* * Bahrain

570- *Actinodendron arboreum* * Marshall Islands

571- *Actinodendron arboreum* * Papua New Guinea

572- *Actinodendron plumosum* * Palau

573- *Actinostephanus haechkeli* * Marshall Islands

574- ***Megalactis hemprichii*** * Palau

575- ***Aiptasia diaphana*** * Papua New Guinea

576- ***Aiptasia pulchella*** * Palau

577- ***Alicia pretiosa*** * Federated States of Micronesia

578- ***Phyllodiscus semoni*** * Philiipines

579- ***Phyllodiscus semoni*** * Philiipines

580- ***Triactis producta*** * Philippines

581- ***Edwardsia pudica*** * Marshall Islands

582- *Calliactis miriam* * Papua New Guinea

583- *Nemanthus annamensis* * Philippines

584- *Nemanthus annamensis* * Palau

585- *Phymanthus muscosus* * Papua New Guinea

578- ***Phyllodiscus semoni*** *** Aliciidae * Actinaria * Philippines * Cebu * Mactan Island * 50 ft (15 m).** This is a very cryptic anemone that is difficult to see on the reef. Its colors make it appear to be a rock coated with coralline and green algae. It is reported to be a bad stinger also.

579- ***Phyllodiscus semoni*** *** Aliciidae * Actinaria * Philippines * Cebu * Mactan Island * 50 ft (15 m).** This anemone is like something from a science fiction movie in that it has a smaller polyp segment which arises at night from the larger oral disk.

580- ***Triactis producta*** *** Aliciidae * Actinaria * Philippines * Cebu * Pescador Island * 40 ft (12 m).** This is the small anemone carried on the claws of the crab *Lybia.* In some areas the anemone occurs without the crab, but in others, such as Hawaii, it has not been found except in association with the crab. It is known from the Red Sea and India to Hawaii and eastern Australia.

581- ***Edwardsia pudica*** *** Edwardsiidae * Actinaria * Marshall Islands * Enewetak Atoll * 20 ft (6 m).** The species of *Edwardsia* are sand-dwelling anemones. Their taxonomy is poorly known, but at least two species are known to occur in Hawaii.

582- ***Calliactis miriam*** *** Hormathiidae * Actinaria * Papua New Guinea * Papua New Guinea * Port Moresby * Bootless Bay * 33 ft (10 m).** These anemones are found most often on the shells carried by hermit crabs. This relationship is believed to benefit both the crab and anemone by increased protection for the crab and mobility for the anemone. There is usually a second species of anemone, *Anthothoe* sp., attached to the inner surface (collumela) of the hermit crab's shell.

583- ***Nemanthus annamensis*** *** Nemanthidae * Actinaria * Philippines * Batangas * Pulang Buli * 66 ft (20 m).** *Nemanthus* is generally found living on whip corals. It can divide asexually, which allows it to overgrow large sections of the coral. Whether the anemone actually kills the coral or opportunistically settles on dead areas is not known.

584- ***Nemanthus annamensis*** *** Nemanthidae * Actinaria * Palau * barrier reef * 60 ft (18 m).** This anemone ranges in color from white to orange, and one apparent variety is mottled in pattern.

585- ***Phymanthus muscosus*** *** Phymanthidae * Actinaria * Papua New Guinea * Port Moresby * Bootless Bay * 33 ft (10 m).**

586- ***Stichodactyla haddoni*** *** Stichodactylidae * Actinaria * Papua New Guinea * Port Moresby * Lion Island * 50 ft (15 m).**

587- ***Stichodactyla tapetum?*** *** * Stichodactylidae * Actinaria * Indonesia * Manado * 60 ft (18 m).** This anemone has an unidentified *Periclimenes* shrimp on it.

588- ***Actineria villosa*** *** Thalassianthidae * Actinaria * Papua New Guinea * Kavieng * Dyaul Island * 6 ft (2 m).**

589- ***Cryptodendrum adhesivum*** *** Thalassianthidae * Actinaria * Papua New Guinea * Port Moresby * Bootless Bay * 60 ft (18 m).**

590- ***Heterodactyla hemprichii*** *** Thalassianthidae * Actinaria * Federated States of Micronesia * Chuuk Atoll * northern barrier reef * 78 ft (23 m).**

591- ***Heteractis aurora*** *** Stichodactylidae * Actinaria * Marshall Islands * Enewetak Atoll * sand flat * 40 ft (12 m).**

592- ***Heteractis malu*** *** Stichodactylidae * Actinaria * Papua New Guinea * Madang * patch reef * 30 ft (9 m).**

593- ***Heteractis magnifica*** *** Stichodactylidae * Actinaria * Papua New Guinea * Madang * 40 ft (12 m).**

594- ***Heteractis magnifica*** *** Stichodactylidae * Actinaria * Papua New Guinea * Bagabag Island * 50 ft (15 m).**

586- *Stichodactyla haddoni* * Papua New Guinea

587- *Stichodactyla tapetum?* * Indonesia

588- *Actineria villosa* * Papua New Guinea

589- *Cryptodendrum adhesivum* * Papua New Guinea

590- *Heterodactyla hemprichii* * Federated States of Micronesia

591- *Heteractis aurora* * Marshall Islands

592- *Heteractis malu* * Papua New Guinea

593- *Stichodactylus gigantea* * Papua New Guinea

594- ***Heteractis magnifica*** * Papua New Guinea

595- ***Heteractis magnifica*** * Papua New Guinea

596- ***Heteractis* sp.** * Papua New Guinea

597- ***Heteractis crispa*** * Papua New Guinea

598- ***Heteractis* sp.** * Marshall Islands

595- *Heteractis magnifica* * Stichodactylidae * Actinaria * Papua New Guinea * Madang * 40 ft (12 m).

596- *Heteractis* sp. * Stichodactylidae * Actinaria * Papua New Guinea * Madang * lagoon reef * 30 ft (9 m). This anemone only occasionally has anemonefish associated with it. Normally it is unoccupied. Its color varies considerably. The tentacles, like those of most hexacorals, have specialized nematocysts called spirocysts which make the tentacles extremely sticky. It is found throughout the tropical Pacific and Indian Oceans.

597- *Heteractis crispa* * Stichodactylidae * Actinaria * Papua New Guinea * Madang * 40 ft (12 m).

598- *Heteractis* sp. * Stichodactylidae * Actinaria * Marshall Islands * Enewetak Atoll * lagoon * 30 ft (9 m).

599- *Macrodactyla doreensis* * Stichodactylidae * Actinaria * Indonesia * Biak Island * 50 ft (15 m).

600- *Actinodiscus neglectus* * Actinodiscidae * Corallimorpharia * Marshall Islands * Enewetak Atoll * 40 ft (12 m).

601- *Amplexidiscus fenestrafer* * Actinodiscidae * Corallimorpharia * Papua New Guinea * West New Britain * Kimbe Bay * 60 ft (18 m). This is the largest corallimorpharian

599- ***Macrodactyla doreensis*** * Indonesia

600- ***Actinodiscus neglectus*** * Marshall Islands

601- ***Amplexidiscus fenestrafer*** * Papua New Guinea

602- ***Amplexidiscus fenestrafer*** * Philippines

603- ***Discosoma nummiformis*** * Indonesia

604- ***Discosoma nummiformis*** * Federated States of Micronesia

605- ***Discosoma* sp.** * Papua New Guinea

606- ***Discosoma* sp.** * Federated States of Micronesia

607- ***Discosoma* sp.** * Marshall Islands

608- *Discosoma* sp. * Federated States of Micronesia

609- "*Rhodactis*" * Papua New Guinea

610- *Rhodactis* sp. * Palau

611- *Corallimorphus* sp. * Philippines

612- *Pseudocorynactis* sp. * Marshall Islands

found on Indo-Pacific reefs, known from the Caroline Islands (Chuuk, Palau), Papua New Guinea and Indonesia. It has short conical tentacles on the inner part of the oral disk and only a few marginal tentacles. There is an area of the outer disk without tentacles. This remarkable corallimorph can capture and feed on fishes which rest on the disk at night. They are trapped by envelopment in the oral disk which bends upward and closes in a drawstring fashion.

602- *Amplexidiscus fenestrafer* * Actinodiscidae * Corallimorpharia * Philippines * Cebu * Mactan Island * 40 ft (12 m). This photo shows the disk folded in, the behavior used to capture fishes. This can be done quickly, in only a few seconds, and it is believed the captured fish is then killed quickly by toxic secretions and ingested. A few other large corallimorphs may be able to capture prey in this manner, but none as quickly as *A. fenestrafer*.

603- *Discosoma nummiformis* * Actinodiscidae * Corallimorpharia * Indonesia * Manado * 20 ft (6 m).

604- *Discosoma nummiformis* * Actinodiscidae * Corallimorpharia * Federated States of Micronesia * Chuuk Atoll * Fujikawa Maru * 50 ft (15 m).

605- *Discosoma* sp. * Actinodiscidae * Corallimorpharia * Papua New Guinea * Madang * lagoon reef * 50 ft (15 m).

606- *Discosoma* sp. * Actinodiscidae * Corallimorpharia * Federated States of Micronesia * Chuuk Atoll * barrier reef * 60 ft (18 m).

607- *Discosoma* sp. * Actinodiscidae * Corallimorpharia * Marshall Islands * Kwajalein Atoll * patch reef * 40 ft (12 m).

608- *Discosoma* sp. * Actinodiscidae * Corallimorpharia * Federated states of Micronesia * Chuuk * Shinkoku Maru * 40 ft (12 m).

609- *Rhodactis* sp. * Actinodiscidae * Corallimorpharia * Papua New Guinea * Madang * Wanganam Reef * 40 ft (12 m).

610- *Rhodactis* sp. * Actinodiscidae * Corallimorpharia * Palau * Mutremdiu Wall * 66 ft (20 m).

611- *Corallimorphus* sp. * Corallimorphidae * Corallimorpharia * Philippines * Batangas * Pulang Buli * 60 ft (18 m).

612- *Pseudocorynactis* sp. * Corallimorphidae * Corallimorpharia * Marshall Islands * Kwajalein Atoll * barrier reef * night * 50 ft (15 m).

613- *Pseudocorynactis* sp. * Corallimorphidae * Corallimorpharia * Philippines * Cebu * Mactan Island * 33 ft (10 m).

613- ***Pseudocorynactis* sp.** * Philippines

614- ***Ricordea* sp.** * Indonesia

615- ***Arachnanthus oligopodus*** * Marshall Islands

616- ***Arachnanthus* sp.** * Marshall Islands

617- ***Arachnanthus* sp.** * Papua New Guinea

618- ***Cerianthus* sp.** * Indonesia

619- ***Cerianthus* sp.** * Federated States of Micronesia

620- ***Cerianthus* sp.** * Philippines

621- *Cerianthus* sp. * Papua New Guinea

622- *Cerianthus* sp. * Bahrain

623- *Cerianthus* larva * open water

624- *Antipathes abies* * Papua New Guinea

614- *Ricordea* sp. * Corallimorphidae * Corallimorpharia * Indonesia * Biak Island * 60 ft (18 m).

615- *Arachnanthus oligopodus* * Arachnanthidae * Ceriantharia * Marshall Islands * Kwajalein Atoll * night * 60 ft (18 m).

616- *Arachnanthus* sp. * Arachnanthidae * Ceriantharia * Marshall Islands * Enewetak Atoll * quarry * night * 3 ft (1 m).

617- *Arachnanthus* sp. * Arachnanthidae * Ceriantharia * Papua New Guinea * Madang * 50 ft (15 m).

618- *Cerianthus* sp. * Cerianthidae * Ceriantharia * Indonesia * Biak Island * night * 60 ft (18 m).

619- *Cerianthus* sp. * Cerianthidae * Ceriantharia * Federated States of Micronesia * Chuuk Atoll * lagoon * 100 ft (30 m).

620- *Cerianthus* sp. * Cerianthidae * Ceriantharia * Philippines * Batangas * Pulangbuli * 60 ft (18 m).

621- *Cerianthus* sp. * Cerianthidae * Ceriantharia * Papua New Guinea * Kavieng * Albatross Channel * 66 ft (20 m).

622- *Cerianthus* sp. * Cerianthidae * Ceriantharia * Indonesia * Manado * 60 ft (18 m).

623- Cerianthid larvae * Cerianthidae * Ceriantharia * open water.

624- *Antipathes abies* * Antipathidae * Antipatharia * Papua New Guinea * Yule Island * 66 ft (20 m). This species is the epitome of a "bottle brush" antipatharian, a single filament with dense side branches. The small brown crab *Quadrella maculosa* is often found with this antipatharian.

625- *Antipathes bifaria* * Antipathidae * Antipatharia * Indonesia * Manado * Talisei * 100 ft (30 m).

626- *Antipathes elegans*? * Antipathidae * Antipatharia * Papua New Guinea * Port Moresby * Pt. Osbourne * 60 ft (18 m). This black coral forms a network of fine branches which forms an almost mist-like structure on lagoonal bottoms at 60 ft or more. At first glance it is hard to realize this is actually an antipatharian, as it seems more like a large hydroid.

627- *Antipathes* cf. *recticulata* * Antipathidae * Antipatharia * Palau * Mutremdiu wall * 90 ft (27 m).

628- *Antipathes* cf. *reticulata* * Antipathidae * Antipatharia * Federated States of Micronesia * Chuuk * Polle reef * 115 ft (35 m).

629- *Antipathes ulex* * Antipathidae * Antipatharia * Palau * Mutremdiu Wall * 100 ft (30 m).

630- *Antipathes* sp. * Antipathidae * Antipatharia * Marshall Islands * Enewetak Atoll * Cement Ship Reef * 66 ft (20 m). The genus *Antipathes* is poorly known, although there are many described species. The skeleton does not have calices where the polyps occur, such as the stony corals, so there are few characters in the skeleton on which to base descriptions and identifications. The general pattern of branching and the various spines and projections are used to differentiate species, but the variation within recognized species has made this work difficult. These cnidarians are an excellent case where careful observation of the living colony can provide taxonomic insights unobtainable from the dead skeleton alone.

631- *Antipathes* sp. * Antipathidae * Antipatharia * Federated States of Micronesia * Chuuk * Fujikawa Maru * 85 ft (25 m). This interesting black coral has densely packed, twisted branches and grows as a small tree. It is likely this is an undescribed species.

625- *Antipathes bifaria* * Indonesia

626- *Antipathes elegans*? * Papua New Guinea

627- *Antipathes* cf. *recticulata* * Palau

628- *Antipathes* cf. *reticulata* * Federated States of Micronesia

629- *Antipathes ulex* * Palau

630- *Antipathes* sp. * Marshall Islands

631- *Antipathes* sp. * Federated States of Micronesia

632- *Antipathes* sp. * Philippines

633- *Antipathes* sp. * Indonesia

634- *Antipathes* sp. * Papua New Guinea

635- *Antipathes* sp. * Papua New Guinea

636- *Antipathes* sp. * Federated States of Micronesia

632- *Antipathes* sp. * Antipathidae * Antipatharia * Philippines * Pescador Island * 40 ft (12 m). In some areas black corals can occur in clear water openly exposed to light, such as is seen here from a reef in the Philippines.

633- *Antipathes* sp. * Antipathidae * Antipatharia * Indonesia * Biak Island * 100 ft (30 m). The polyps of many black corals are actually white in color as seen in this species. This particular antipatharian forms large bushy trees on dropoffs.

634- *Antipathes* sp. * Antipathidae * Antipatharia * Papua New Guinea * Madang * Rasch Passage * 100 ft (30 m). There are four basic forms of antipatharians; single filament (whips), single whip with side branches (bottle brush), one to a few plane networks (fans) and bush or tree-like shapes. All have only a single holdfast.

635- *Antipathes* sp. * Antipathidae * Antipatharia * Papua New Guinea * Madang * Rasch Passage * 85 ft (25 m).

636- *Antipathes* sp. * Antipathidae * Antipatharia * Federated States of Micronesia * Chuuk * Fujikawa Maru * 85 ft (25 m).

637- *Antipathes* sp. * Antipathidae * Antipatharia * Papua New Guinea * Madang * Bagabag Island * 90 ft (27 m). This large black coral occurs on deep dropoffs in Papua New Guinea and elsewhere. The individual colonies become so large they are difficult to capture on film with a flash.

638- *Antipathes* sp. * Antipathidae * Antipatharia * Palau * Lighthouse Reef channel * 66 ft (20 m). This photo shows the reddish brown color found in some species of antipatharians. Along the thicker branches the black of the skeleton shows through the thin layer of tissue in places.

639- *Cirrhipathes* sp. * Antipathidae * Antipatharia * Federated States of Micronesia * Chuuk * Fujikawa Maru * 66 ft (20 m). The whip-like black corals of the genus *Cirrhipathes* are unlike anything else, yet are poorly known at the species level. They consist of a single filament, sheathed in polyps. Often some portion of the polyps will die and that section of the whip will become overgrown with various other invertebrates and algae. Small gobies of the genus *Bryanopsis* live on *Cirrhipathes.* The goby clears a section of polyps a few inches long on the whip to lay its eggs and, with luck, the goby tending its eggs can be found. Such damage to the sheath of polyps may help start the process of death and overgrowth of whips by other organisms.

640- *Cirrhipathes* sp. * Antipathidae * Antipatharia * Federated States of Micronesia * Chuuk * Fujikawa Maru * 85 ft (25 m). The polyps of this *Cirrhipathes* are clearly seen in this photograph, showing how they do not penetrate into the skeleton of the antipatharian. The living tissue is a thin film on the surface of the skeleton with the polyps protruding out of it.

641- *Cirrhipathes* sp. * Antipathidae * Antipatharia * Papua New Guinea * Madang * barrier reef * ledge * 85 ft (25 m). There are several different varieties of *Cirrhipathes*, with different colored polyps, which may represent different species. There are also differences in diameter of the whips consistent with color variations.

642- *Cirrhipathes* sp. * Antipathidae * Antipatharia * Palau * Mutremdiu Wall * 90 ft (27 m). Whips of *Cirrhipathes* can be extremely long, such as the individual photographed here, and reach many feet away from reef walls.

643- *Cirrhipathes* sp. * Antipathidae * Antipatharia * Palau * barrier reef * 66 ft (20 m). This photo shows the corkscrew characteristic that long specimens of *Cirrhipathes* often display.

637- ***Antipathes* sp.** * Papua New Guinea

638- ***Antipathes* sp.** * Palau

639- ***Cirrhipathes* sp.** * Federated States of Micronesia

640- ***Cirrhipathes* sp.** * Federated States of Micronesia

641- ***Cirrhipathes* sp.** * Papua New Guinea

642- ***Cirrhipathes* sp.** * Palau

643- ***Cirrhipathes* sp.** * Palau

The Ctenophores are a relatively small taxonomic group with 100 or so species. They are marine animals that superficially resemble jellyfish. Most ctenophores are transparent and pelagic, and they are found drifting in open water. Unlike most cnidarians, they have only a single body form during their life history and they are never colonial. The name Ctenophore comes from the eight rows of ciliary combs (ctenes) which traverse the body. The common name for the group, "comb jellies", refers to the appearance of these specialized structures. The fused cilia making up an individual comb diffract light and account for the irridescence seen in color photographs of ctenophores. Ctenophores have tentacles armed with adhesive cells called colloblasts which aid in capturing prey from the surrounding water. Other species of ctenophores, which lack large tentacles, capture their prey by enveloping them, much as we would capture small fish underwater with a plastic bag.

Some ctenophores (order Platyctenida) are creeping forms which live primarily in association with other invertebrates such as soft corals and echinoderms. They are usually pigmented rather than clear and have two long tentacles which they extend to capture prey in the water column.

Left- This bright red starfish, possibly *Echinaster luzonicus,* in the Solomon Islands has some uninvited guests, ctenophores of the genus *Astricola,* living on the surface of its arms. The relationship between the asteroid and these ctenophores is poorly known. Whether the ctenophores are parasites, or simply harmless commensals is open to investigation. These ctenophores are believed to divide by fission every so often to form two separate individuals, and that may be why certain starfish have several of these ctenophores, while nearby individuals have none. This starfish is support for that suggestion since it has several of the gayly patterned *Astricola.*

646- *Beroe forskali* * Beroida * open Pacific. *Beroe* is a member of the Class Nuda which lacks tentacles, even as larvae. This species is cosmopolitan in warm waters.

647- *Pleurobranchia* sp. * Cydippida * open Pacific. This species has short tentacles, one of which is easily visible in the photo. Members of this order have globular forms and can retract the tentacles quickly.

648- *Pleurobranchia* sp. * Cydippida * open Pacific. In this view the tentacles are fully extended fishing for prey.

649- *Bolinopsis* sp. * Lobata * open Pacific

650- *Leucathea* sp. * Lobata * open Pacific. This ctenophore can become seasonally abundant in Micronesian lagoons such as Chuuk lagoon during January through April. During this time, *Leucathea* is quite abundant in the upper several feet of water, but few individuals occur below about forty feet.

651- *Coeloplana bannwarthi* * Coeloplanidae * Platyctenea * Philippines * Pamalican Island * 9 m. This species lives on sea urchin spines, in this case *Diadema* sp. There is also a similar species which lives on the Crown-of-thorns starfish, *Acanthaster planci*. The members of this order are sessile, living attached to another creature.

652- *Astricola* sp. * Platyctenea * Philippines* Batangas * 40 ft (12 m). The mottled blotches on the skin of this starfish are actually a sessile ctenophore. These creatures are poorly known, but interestingly these ctenophores are believed to multiply by a process called fragmentation that produces new individuals. Generally any starfish which has this ctenophore has several. It is unknown for certain whether these ctenophores cause any damage to the starfish. This species, as well as the one that follows, is most active at night.

653- unidentified benthic ctenophore. * Platyctenea * Marshall Islands * Kwajalein Atoll * reef * 20 ft (6 m). This ctenophore is attached to a soft coral. While it may appear that it is damaging the soft coral, it is actually fishing using its tentacles which are clearly visible. When prey are captured by the tentacles, they are drawn in and the prey enveloped by the animal.

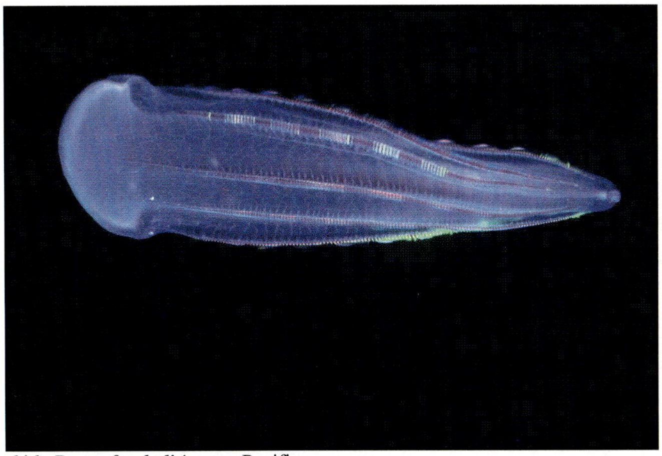

646- *Beroe forskali* * open Pacific.

647- *Pleurobranchia* sp. * open Pacific

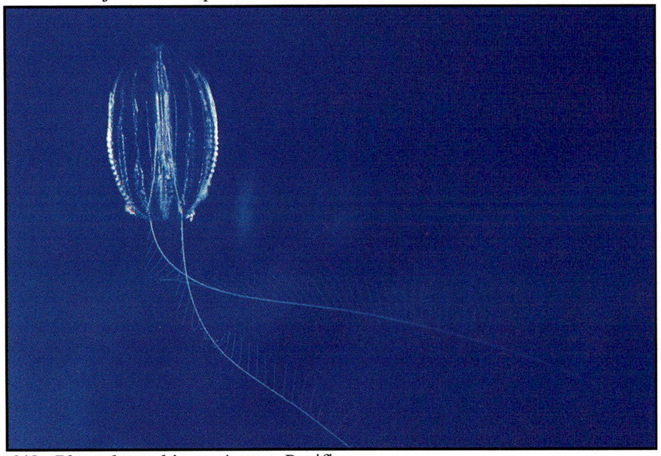

648- *Pleurobranchia* sp. * open Pacific.

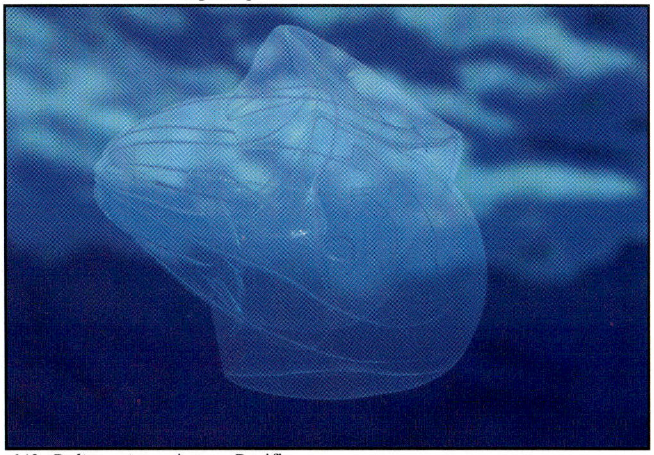

649- Bolinopsis sp. * open Pacific

650- *Leucathea* sp. * **o**pen Pacific.

651-*Coeloplana bannwarthi* * Philippines

652- *Astricola* sp. * Philippines

653- Unidentified benthic ctenophore * Marshall Islands

Phyla Platyhelminthes, Nemertea, Annelida, Sipuncula, Echiurida and Hemichordata

Marine Worms

The "worms", as grouped here, do not constitute a naturally related group, but because of their general form and the relatively few species that would be observed by non-specialists, they are considered together here for convenience.

Phylum Platyhelminthes

The flatworms are most commonly known for their parasitic members, which include the flukes (Class Trematoda) and tapeworms (Class Cestoda). The free-living flatworms (Class Turbellaria) found in shallow tropical waters however, are the most spectacular members of the group. They are among the most brightly colored of marine invertebrates, rivaled only by, and often confused with, the nudibranchs. They can swim with undulations of their thin, flattened bodies, but normally crawl along the bottom.

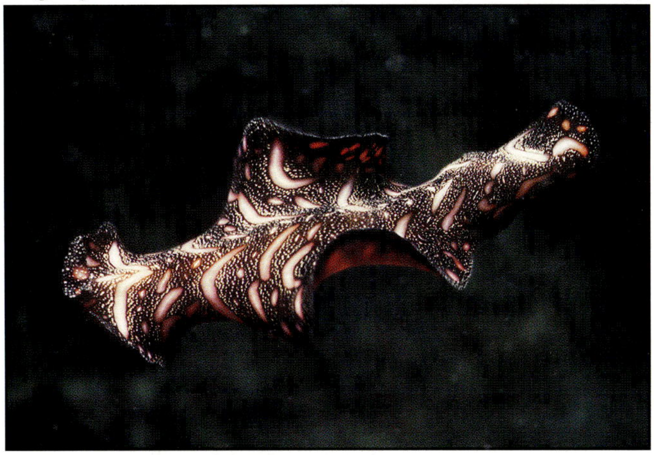

Above- The free-living flatworms found on reefs are excellent swimmers when removed from the bottom or disturbed. They swim by undulating the margins of their body in rhythmic waves.

The free-living flatworms are one of the most poorly known groups of tropical marine invertebrates. They are difficult to properly preserve for study by taxonomic specialists and even with well preserved specimens often require histological examination to identify. The polyclad, named because of their numerous gut branches, are the largest and most conspicuous. Their color patterns, once a species has been adequately described, are quite useful in identification. There are over 130 species of polyclad flatworms known from the Great Barrier Reef and a similar number from Papua New Guinea.

A number of cases of possible mimicry or evolutionary convergence exist between reef flatworms and nudibranchs, in which their color patterns closely resemble one another. The most visible distinction is the lack in the flatworms of the tuft of gill filaments found in the nudibranchs.

Opposite- The tube worm, *Protula* sp., is common near sheltered areas on outside reefs of Indonesia and other areas of the western Pacific. *Protula* uses the feathery cilliated structure, called the brachial plume, for filter feeding and as an aid in respiration. If disturbed they pull back into their tube in an instant.

Above- This photo shows a small nemertean crawling on coral rubble. Nemerteans are generally cryptic in habit. Some species forage on the reef and seagrass beds at night, others spend much of the time beneath rocks.

Above- The tube-dwelling polychaete *Salmacina dysteri* from Hawaii has numerous individuals entwined in a single mass. Orange sponge covers the tubes.

Above- The fragile white calcareous tubes of the serpulid worm *Filograna implexa* stand out on a *Melithaea* sea fan on which the worm is growing. It is common to see the tubes of this worm growing on other benthic organisms.

Phylum Nemertea - Ribbon worms

The ribbon worms are elongate, flattened worms that are unsegmented with a retractile proboscis in a fluid filled chamber above the gut. Many species attain a length of about three feet. Many coral reef species are brightly colored and some are known to contain chemical compounds of potential medicinal use. Some 22 species are known from the Great Barrier Reef, although about 1,200 species occur worldwide and an indeterminate number occur elsewhere in our region.

Phylum Annelida - Class Polychaeta - Segmented worms

The Phylum Annelida includes two predominantly terrestrial and freshwater groups (earthworms and leeches) and one large, primarily marine group, the polychaetes. Polychaetes are extremely abundant and widespread in the ocean. Some species are pelagic, but the vast majority live benthically as adults. These species burrow in the sediments, dwell under rocks, or live in tubes which they construct, and are not usually obvious to the untrained observer. Some, such as the Christmas tree worm, *Spirobranchus gigantea*, and at least the tentacles of terebellid worms (mentioned below) stand out, but otherwise polychaetes are generally not noticed.

The distinguishing characteristic of annelids is metamerism, a condition in which the body is divided into a linear series of similar segments (metameres) between the head and the tail (pygidium). The head typically bears sensory organs for processing environmental information. Growth involves the formation of new segments, which occurs just anterior to the pygidium. In polychaetes, each segment typically bears a lateral pair of paddle-like appendages called parapodia. The parapodia bear specialized bristles called setae, which aid in locomotion, environmental sensing, and defense.

In addition to burrowing forms, free-swimming, crawling and tube-dwelling life styles have also evolved. Modifications of the general polychaete body plan, such as fusion or differentiation of trunk segments and reduction of the parapodia, are associated with more sedentary habits. Various feeding modes have also developed in conjunction with the different lifestyles of polychaetes. Burrowing forms tend to be deposit feeders or raptorial feeders. Deposit feeders either ingest sediment directly and digest the nutritious particles it contains, or they send long, sticky palps or tentacles over the substratum to pick up food particles, which are then conveyed to the mouth on waves of cilia. Raptorial

Above- Some polychaete worms construct elaborate tubes, gluing together grains of sand and tiny pebbles for strength. This tube of an unidentified polychaete, family Pectinariidae, shows this construction. The tube is closed by the flat head of the worm.

Above- This is the epitoke of a "palolo" polychaete swimming free in the water. The epitoke holds the egg and sperm and breaks free from the rest of the body and swims towards the surface on a certain lunar phase during a limited season of the year.

feeders are equipped with jaws for catching small, mobile prey. Active crawlers tend to be scavengers. Many tube dwelling polychaetes have evolved specialized feeding structures, such as branchial or tentacular crowns, which remove suspended particles from the water and also double as respiratory organs.

Terebellid polychaetes live in tubes and spread their long tentacles over the surface around their tubes. The tentacles resemble white, red or green spaghetti and are drawn back to the tube when touched. The main body of the worm is lodged in fissures in the reef and is rarely seen. The worm feeds on algal and bacterial films with the food particles being brought back to the mouth via a ciliated groove on each tentacle.

Most polychaetes have separate sexes, although some are hermaphrodites and a few change sex. Fertilization of eggs takes place in the water column for those species which release gametes into the water. Other species mate and lay encapsulated eggs in the tube of the female (a few species retain fertilized eggs in the body of the female). Planktonic larvae develop from the fertilized eggs and eventually settle to the bottom to become juveniles. Among the most spectacular examples of polychaete spawning are the palolo worms. In Samoa, after sunset during the first lunar cycle of October or November, the reproductive portion of the body, the epitoke, breaks free and swims to the surface The epitokes are light sensitive and large numbers of them swarm on the surface, eventually breaking apart to release the eggs and sperm.

Phylum Sipuncula- peanut worms

Sipundulids do not belong phylogenetically with the previous three groups, and they are more closely related to the echiurids, below. They are unsegmented and live in sand, rocks and coral. Sipunculids are one of the major borers of coral skeletons, causing weakening of the skeleton and its eventual destruction. Their boring is believed to be a combination of mechanical and chemical action. They are eaten by some fishes and molluscs, especially *Mitra* spp. Some of the sand burrowers are almost a foot long, with unlined non-permanent burrows.

Diversity is moderate in the sipunculids, but a few species are usually present in most habitats. There are 23 species known from Great Barrier Reef while seven species have been collected from Enewetak Atoll.

Phylum Echiura- Echiurid worms

Echiurids are unsegmented worm-like animals, with a highly extendible proboscis. They live in burrows in mud, sand and rock. On reefs, echiurids are most often found in coral rock formed by overhangs of coral heads. The feeding proboscis, with a bifurcate tip, is occasionally seen extended out along the bottom for a distance of several feet or more from the body of the echiurid. If touched the proboscis is quickly drawn back to the worm.

Phylum Hemichordata - Acorn worms

Acorn worms live in sediment and the main evidence of their presence is a mound of coiled "castings" on the surface of the sand. The castings are sand, sheathed in a thin layer of mucous that has passed through the worm's gut. They live in sandy bottom habitats. Some polychaetes make similar castings.

654- Acanthozoon sp. * Papua New Guinea

655- ***Pseudoceros bimarginatus*** * Marshall Islands

656- ***Pseudoceros dimidiatus*** * Papua New Guinea

657- ***Pseudoceros dimidiatus*** * Philippines

654- *Acanthozoon* sp. * Pseudocerotidae * Polycladida * Platyhelminthes * Papua New Guinea * Madang * barrier reef * 30 ft (9 m). The most diverse and colorful group of flatworms in the western Pacific are members of the family Pseudocerotidae. The family includes several genera, four of which are included in this section.

655- *Pseudoceros bimarginatus* * Pseudocerotidae * Polycladida * Platyhelminthes * Marshall Islands * Enewetak Atoll * lagoon pinnacle * 20 ft (6 m). Little is known about the specific feeding habits of most polyclad flatworms, but they are reported, as a group, to feed on colonial ascidians.

656- *Pseudoceros dimidiatus* * Pseudocerotidae * Polycladida * Platyhelminthes * Papua New Guinea * New Britain * 20 ft 6 m).

657- *Pseudoceros dimidiatus* * Pseudocerotidae * Polycladida * Platyhelminthes * Philippines * 60 ft (20 m).

658- *Pseudoceros ferrugineus* * Pseudocerotidae * Polycladida * Platyhelminthes * Papua New Guinea * 10 ft (3m).

659- *Pseudoceros tritriastus* * Pseudocerotidae * Polycladida * Platyhelminthes * Philippines * Cebu * Mactan Is. * 10 ft (3 m).

660- *Pseudoceros zebra* * Pseudocerotidae * Polycladida * Platyhelminthes * Marshall Islands * Kwajalein Atoll * lagoon pinnacle * 30 ft (9 m).

661- *Pseudoceros* sp. * Pseudocerotidae * Polycladida * Platyhelminthes * Papua New Guinea * Madang * 10 ft (3m).

662- *Pseudoceros* sp. * Pseudocerotidae * Polycladida * Platyhelminthes * Papua New Guinea * Madang * 10 ft (3 m).

663- *Pseudoceros* sp. * Pseudocerotidae * Polycladida * Platyhelminthes * Marshall Islands * Enewetak Atoll * lagoon pinnacle * 32 ft (10 m).

664- *Pseudoceros* sp. * Pseudocerotidae * Polycladida * Platyhelminthes * Marshall Islands * Enewetak Atoll * lagoon pinnacle * 30 ft (9 m).

665- *Pseudoceros* sp. * Pseudocerotidae * Polycladida * Platyhelminthes * Marshall Islands * Enewetak Atoll * lagoon pinnacle * 30 ft(9 m).

666- *Pseudobiceros affinis* * Pseudocerotidae * Polycladida * Platyhelminthes * Papua New Guinea * New Britain * 60 ft (20 m).

667- *Pseudobiceros bedfordi* * Pseudocerotidae * Polycladida * Platyhelminthes * Papua New Guinea * Madang * 60 ft (20 m).

658- ***Pseudoceros ferrugineus*** * Papua New Guinea

659- ***Pseudoceros tritriastus*** * Philippines

660- ***Pseudoceros zebra*** * Marshall Islands

661- ***Pseudoceros* sp.** * Papua New Guinea

662- ***Pseudoceros* sp.** * Papua New Guinea

663- ***Pseudoceros* sp.** * Marshall Islands

664- ***Pseudoceros* sp.** * Marshall Islands

665- ***Pseudoceros* sp** * Marshall Islands

666- ***Pseudobiceros affinis*** * Papua New Guinea

667- ***Pseudobiceros bedfordi*** * Papua New Guinea

668- ***Pseudobiceros damawan*** * Pseudocerotidae * Polycladida * Platyhelminthes * Marshall Islands * Enewetak Atoll * channel * 32 ft (10 m).

669- ***Pseudobiceros* cf. *fulgor*** * Pseudocerotidae * Polycladida * Platyhelminthes * Federated States of Micronesia * Chuuk * lagoon patch reef * 30 ft (9 m).

670- ***Pseudobiceros paralaticlavus*** * Pseudocerotidae * Polycladida * Platyhelminthes * Philippines * 3 ft (1 m).

671- ***Pseudobiceros gloriosus*** * Pseudocerotidae * Polycladida * Platyhelminthes * Federated States of Micronesia * Mortlock Islands * 60 ft (20 m).

672- ***Pseudobiceros gratus*** * Pseudocerotidae * Polycladida * Platyhelminthes * Papua New Guinea * Madang * reef * 30 ft (9 m).

673- ***Pseudobiceros* sp.** * Pseudocerotidae * Polycladida * Platyhelminthes * Papua New Guinea * West New Britain * sand slope * 40 ft (12m).

674- ***Pseudoceros* sp.** * Pseudocerotidae * Polycladida * Platyhelminthes * Hawaii * Pupukea * 25 ft (8 m).

675- ***Pseudobiceros* sp.** * Pseudocerotidae * Polycladida * Platyhelminthes * Palau * seagrass * 5 ft (2 m).

676- ***Pseudobiceros* sp.** * Pseudocerotidae * Polycladida * Platyhelminthes * Philippines * Cebu * Moalboal * 32 ft (10 m).

677- ***Thysanozoon* sp.** * Pseudocerotidae * Polycladida * Platyhelminthes * Philippines * Cebu * Mactan Is. * 60 ft (20 m).

668- ***Pseudobiceros damawan*** * Marshall Islands

669- ***Pseudobiceros* cf. *fulgor*** * Federated States of Micronesia

670- ***Pseudobiceros paralaticlavus*** * Philippines

671- ***Pseudobiceros gloriosus*** * Federated States of Micronesia

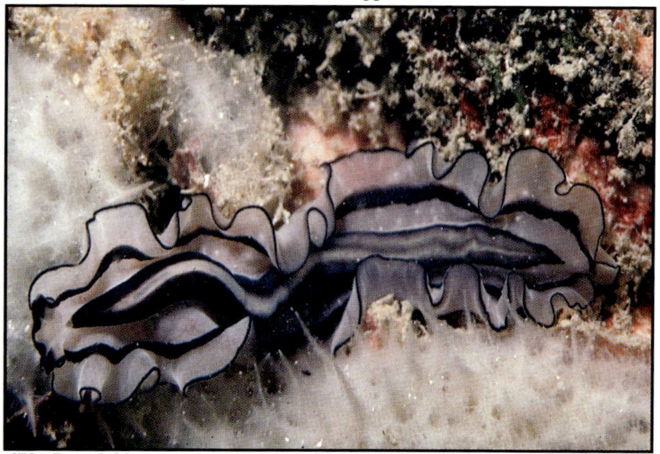
672- ***Pseudobiceros gratus*** * Papua New Guinea

673- *Pseudobiceros* sp. * Papua New Guinea

674- *Pseudoceros* sp. * Hawaii

675- *Pseudobiceros* sp. * Palau

676- *Pseudobiceros* sp. * Philippines

677- *Thysanozoon* sp. * Philippines

678- Acoela * Palau

679- Callioplanidae * Marshall Islands

680- Euryheptidae * Papua New Guinea

681- ***Eurlepta* cf. *punctata*** * Palau

682- ***Paraplanocera oligoglena*** * Marshall Islands

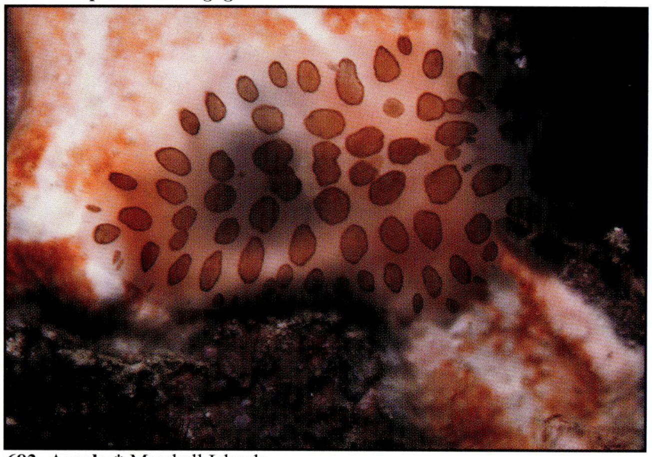

683- **Acoela** * Marshall Islands

678- Acoela * Platyhelminthes * Palau * Malakal Harbor * 20 ft (6 m). These small, brown flatworms are often abundant on various hard and soft corals in shallow water.

679- Callioplanidae * Polycladida * Platyhelminthes * Marshall Islands * Lagoon reef * 30 ft (9 m).

680- Euryleptidae * Polycladida * Platyhelminthes * Papua New Guinea * Reef flat * 10 ft (3 m). The euryleptids are less diverse than the Pseudocerotids. but are still often brightly colored.

681- *Eurylepta* cf. *punctata* * Euryleptidae * Polycladida * Platyhelminthes * Palau * Rock Islands * 15 ft (5 m).

682- *Paraplanocera oligoglena* * Paraplanoceridae * Polycladida * Marshall Islands * Kwajalein Atoll * pass * 30 ft (10 m).

683- Acoela * Marshall Islands * Kwajalein Atoll * west reef * 35 ft (12 m). This species is similar to number 678 above.

684- *Baseodiscus delineatus* * Heteronemertea * Nemertea * Philippines * Cebu * Mactan Island * 6 ft (2 m).

685- *Baseodiscus hemprichii* * Heteronemertea * Nemertea * Philippines * Cebu * Mactan Island * 6 ft (2 m).

686- *Baseodiscus mexicanus* * Heteronemertea * Nemertea * Lighthouse Reef * seagrass bed * night * 3 ft (1 m).

687- *Baseodiscus quinquelineata* * Heteronemertea * Nemertea * Indonesia * Manado * night * 15 ft (5 m).

688- *Cerebratulus* sp. * Heteronemertea * Nemertea * Palau * Lighthouse Reef * sea grass * night * 3 ft (1 m).

689- *Paralepidonotus ampulliferus* * Polynoidae * Polychaeta * Marshall Islands * Kwajalein Atoll * oceanside reef * 20 ft (6 m). This type of polychaete is known as a scale worm, due to the resemblence of its upper surface to fish scales.

690- *Gastrolepideela clavigera* * Polynoidae * Polychaeta * Palau * Lighthouse Reef * 6 ft (2 m). This worm lives on the holothurian *Bohschadsia argus*, it also occurs with *Stichopus horrens*.

691- *Pherecardia striata* * Amphinomidae * Polychaeta * Hawaii * Pupukea * 40 ft (9 m). An Indo-west Pacific species, found among coral and coral rubble from the intertidal and subtidal.

692- *Notopygos albiseta* * Amphonomidae * Polychaeta *Marshall Islands * Enewetak Atoll * 40 ft (9 m).

693- *Leocrates* sp. * Hesionidae * Polychaeta * Enewetak Atoll * Medren Pinnacle * 60 ft (20 m).

684- ***Baseodiscus delineatus*** * Philippines

685- ***Baseodiscus hemprichii*** * Philippines

686- *Baseodiscus mexicanus* * Palau

687- *Baseodiscus quinquelineata* * Indonesia

688- *Cerebratulus* sp. * Palau

689-*Paralepidonotus ampulliferus* * Marshall Islands

690- *Gastrolepideela clavigera* * Palau

691- *Pherecardia striata* * Hawaii

692- *Notopygos albiseta* * Marshall Islands

693- *Leocrates* sp. * Marshall Islands

694- Unidentified Sabellarid * Palau

695- Terrebellidae * Marshall Island

696- Sabellidae * Federated States of Micronesia

694- Unidentified Sabellarid * Sabellaridae * Polychaeta * Palau * Malakal Harbor entrance * 100 ft (35 m). We are uncertain exactly what this tube dwelling polychaete is, but it is unusual in appearance. Twin "antennae" protrude from its tube and the worm pulls back deep into the tube at the slightest disturbance.

695- Unidentified Terrebellid * Terrebellidae * Polychaeta * Marshall Islands * Enewetak Atoll * Medren Pinnacle * 30 ft (10 m). Terebellids are tube dwelling, deposit-feeding worms commonly known as spaghetti worms. They extend tentacles over the substratum and food particles are carried to the mouth by ciliary action of the grooved tentacles or carried to the mouth after capture by the tentacles. Their tubes are constructed of sand and shell fragments and are buried in sediments and crevices or attached to the undersurfaces of rocks.

696- Unidentified Sabellid * Sabellidae * Polychaeta * Federated States of Micronesia * Chuuk * patch reef * 40 ft (9 m).

697- *Bispira* sp. * Sabellidae * Polychaeta * Philippines * Cebu * seagrass bed * 3 ft (1 m).

698- Unidentified Sabellid * Sabellidae * Polychaeta * Papua New Guinea * Madang * Barracuda Point * 30 ft (10 m). The marks on the coral below the worm are from parrotfish.

699- Unidentified Sabellid * Sabellidae * Polychaeta * Papua New Guinea * Manam Island * volcanic sand slope * 70 ft (23 m).

700- Unidentified Sabellid * Sabellidae * Polychaeta * Papua New Guinea * Manam Island * volcanic sand * 60 ft (20 m).

701- Unidentified Sabellid * Sabellidae * Polychaeta * Papua New Guinea * Manam Island * sand slope * 30 ft (9 m).

702- Unidentified Sabellid * Sabellidae * Polychaeta * Papua New Guinea * Manam Island * sand slope * 100 ft (30 m).

703- Unidentified Sabellid * Sabellidae * Polychaeta * Philippines * Batangas * Pulang Buli Island * 40 ft (12 m).

704- *Spirobranchus giganteus* * Serpulidae * Polychaeta * Philippines * Batangas * Pulang Buli Island * 40 ft (12 m).

705- *Spirobranchus gigantea* * Serpulidae * Polychaeta * Papua New Guinea * Madang Channel * 50 ft (15 m).

706- *Protula magnifica* * Serpulidae * Polychaeta * Federated States of Micronesia * Chuuk * lagoon reef * 6 ft (2 m). *Protula*, like other Serpulids, builds a tube of calcium carbonate. This species has a relatively large (2-3 inches across) and colorful brachial crown (the feathery part).

697- *Bispira* sp.* Philippines

698- *Sabellastarte* sp.* Papua New Guinea

699- Sabellidae * Papua New Guinea

700- Sabellidae * Papua New Guinea

701- Sabellidae * Papua New Guinea

702- Sabellidae * Papua New Guinea

703- Sabellidae * Philippines

704- *Spirobranchus giganteus** Philippines

705- *Spirobranchus giganteus** Papua New Guinea

706- *Protula magnifica** Philippines

707- ***Protula* sp.** * Papua New Guinea

707- ***Protula* sp. * Serpulidae * Polychaeta * Papua New Guinea * West New Britain * reef wall * 50 ft (15 m).**

708- ***Filogranella elatensis* * Serpulidae * Polychaeta * Philippines * Cebu * Mactan Island * cave * 55 ft (18 m).**

709- ***Filograna implexa* * Serpulidae * Polychaeta * Papua New Guinea * Madang * barrier reef * 50 ft (15 m).**

710- ***Filograna implexa* * Serpulidae * Polychaeta * Philippines * Pamalican Island * 40 ft (12 m).**

711- ***Eunice* sp.* Eunicidae * Polychaeta * Palau * Lighthouse Reef * seagrass bed * 3 ft (1 m).** This worm is long, over three feet, and it is probably capable of biting. Spectacular worms such as this are rarely seen on reefs.

712- ***Eunice* sp.* Eunicidae * Polychaeta * Marshall Islands * Enewatak * patch reef * 3 ft (1 m).** There are many species of eunicids, most are difficult to identify.

713- ***Eunice* sp.* Eunicidae * Polychaeta * Marshall Islands * Enewatak Atoll * patch reef * 10 ft (3 m).**

714- **Unidentified sipunculid * Sipuncula * Marshall Islands * Enewatak Atoll * patch reef *20ft (7 m).** Sipunculids, like many of the other worm-like animals are usually cryptic in habit. They have an extensible proboscis that bears the mouth and they feed on organic detritus.

715- **Unidentified sipunculid * Sipuncula * Philippines * Cebu * Mactan Island * mud bottom * 3 ft (1 m).**

716- ***Bonellia fuliginosa* * Bonellidae * Echiura * Marshall Islands * Kwajalien Atoll * lagoon reef * 30 ft (10 m).**

717- ***Achaetobonellia maculata* * Bonellidae * Echiura * Marshall Islands * Enewetak Atoll * lagoon reef * 30 ft (10 m).** This is all that is normally seen of echiurid worms on the reef, the feeding proboscis with its bifurcate end. If touched, the proboscis is quickly withdrawn back to the burrow or crevice where the main body of the worm is found.

718- **Unidentified hemichordate * Hemichordata * Marshall Islands * Enewetak Atoll* patch reef * 10 ft (3 m).** This hemichordate is usually living beneath the sand. It was found and photographed here after a large typhoon. It makes castings similar to the one in the next photograph.

719- **Hemichordata mound * Marshall Islands * Enewetak Atoll * lagoon bottom * 9 m.**

708- ***Filogranella elatensis*** * Philippines

709- ***Filograna implexa*** * Papua New Guinea

710- ***Filograna implexa*** * Philippines

711- ***Eunice* sp.** * Palau

712- *Eunice* sp. Marshall Islands

713- *Eunice* sp. * Marshall Islands

714- Unidentified Sipunculid * Marshall Islands

715- Unidentified Sipunculid * Philippines

716- *Bonellia fuliginosa* * Marshall Islands

717- *Achaetobonellia maculata* * Marshall Islands

718- Unidentified Hemichordata * Marshall islands

719- Hemichordata mound * Marshall Islands

Molluscs

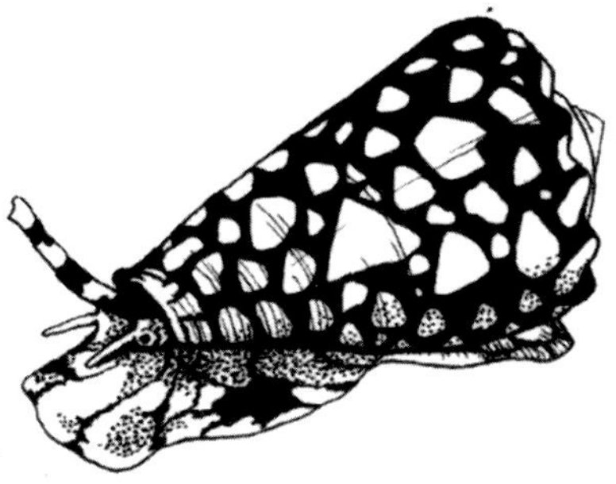

The historical popularity of shell collecting and the durable nature of the hard shells of specimens, with a wealth of characters available, are the reasons for molluscs having perhaps the best known taxonomy of all marine invertebrate groups. After the arthropods, the phylum Mollusca contains the greatest number of described living species (100,000), plus an additional 60,000 known fossil species. About half of the species are marine, the rest being freshwater or terrestrial.

The molluscs are extremely diverse in form and include polyplacophorans (chitons), gastropods (snails, nudibranchs, sea hares and relatives), bivalves (clams, oysters, mussels, and scallops), cephalopods (squids, octopods, and chambered *Nautilus*), and a few other small groups. The diversity of molluscan species reflects their great success in adapting to many different habitats and lifestyles.

Although the molluscs appear to be a heterogeneous assemblage, they are derived from the same fundamental body plan. The body is typically divided into a head, with well-developed sensory organs, a large, soft visceral mass, from which the phylum gets its name, and a muscular foot. As possession of a muscular foot is believed to be the primitive condition in molluscs, locomotion by crawling is common in the group. Octopods and squids can take in seawater, expel it from their mantle cavity, and spurt forward by means of jet propulsion, in addition to their ability to crawl over the substrate with their well-developed tentacles. These animals are collectively called cephalopods ("head-footed") because the tentacles are derived from the muscular foot and emanate from the head. In sessile forms, such as oysters, the muscular foot is greatly reduced.

Most molluscs possess an external calcareous shell which they secrete. Evolutionary change in some groups, like the nudibranchs, squids, and octopus, has resulted in reduction, internalization, or complete loss of the shell. These species have developed other means of protection and evasion, for example, the octopus confuses predators by emitting a cloud of ink.

Molluscs form the basis of many economically valuable fisheries. Squid and octopods are fished throughout the world. Various oysters produce both natural and cultured pearls. Many bivalves are highly prized for food, from the clams and oysters familiar in temperate climates to the giant clams of the tropical Pacific. Various gastropods are also valuable for food, from the queen conch of the Caribbean to many small neritids and limpets which are utilized in subsistence fisheries in the Pacific.

Opposite- The giant clam, *Tridacna gigas*, on a coral reef at Bagabag Island, offshore of Madang, Papua New Guinea. The giant clams were a regular component of most western Pacific coral reefs, but have been exploited in many areas, causing a great decrease in their abundance throughout the region.

Class Polyplacophora- Chitons

The chitons are strictly marine, but occur in all seas. They are flattened, with eight overlapping plates comprising the shell, surrounded by a girdle. Most tropical Pacific species areintertidal or shallow subtidal, and have a large foot by which they can clamp down to rock so tightly that they cannot be moved. Most feed on algae, grazing the surface of rocks, while a few also feed on assorted encrusting invertebrates. They are generally slow-moving, inconspicuous animals. There are approximately 500 living species of chitons world-wide.

Class Gastropoda- Snails and slugs

There are about 35,000 described species of gastropods. There are three subclasses, only two of which concern us here; the Prosobranchia (snails) and the Opisthobranchia (bubble shells, sea hares, nudibranchs and others). The third subclass, the Pulmonata, are land snails and slugs.

Not all gastropods have a shell. Many of the opisthobranchs have a reduced shell or no shell at all. Shells are made of calcium carbonate secreted by the mantle of the animal. Many species have an operculum, or trap door, which helps to seal the entrance of the shell when the animal draws inside. Gastropod shells are usually coiled in a counter-clockwise or right hand direction (looking with the spire pointed away from the observer). A limited number of species coil in a left hand direction, such as the common Lightning Whelk in the southeast United States, and some species have both right and left handed shells.

Most gastropods have a large, fleshy foot which is used for locomotion over a variety of substrates, propelled by either ciliary action or waves of fine muscular contraction along the surface of the foot. Mucous secreted at the foot helps the animals glide over the substrate. Some, such as members of the genus *Strombus*, use their claw-like operculum to "pole" themselves along the bottom with what appears to be a loping motion. Some species have an escape reaction when a potential predator is detected, which consists of a rapid rolling or leaping action, usually with the aid of the operculum.

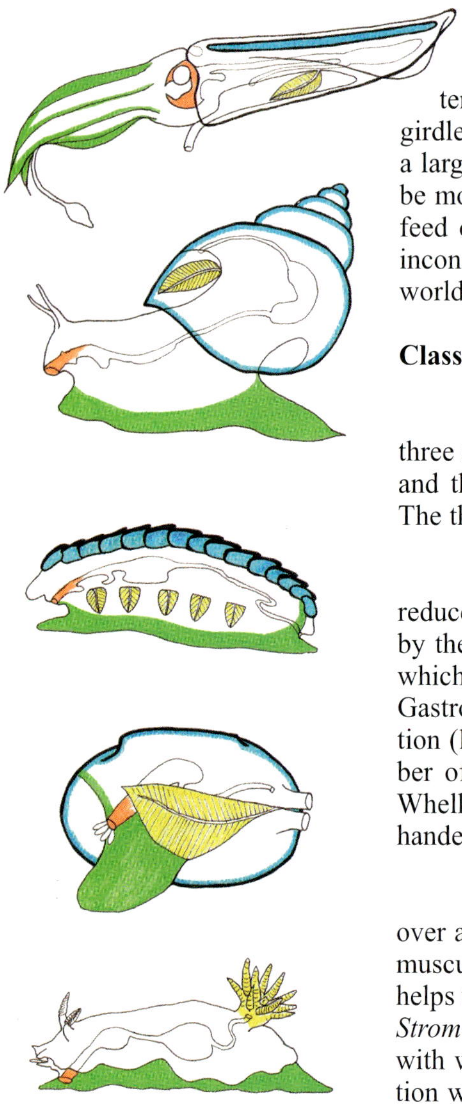

Above- The molluscs have their basic morphology modified in (from top to bottom) cephalopods, snails, chitons, clams and nudibranchs. In these drawings the foot is green, the shell is blue, the gills are yellow and the mouth is orange.

Shell shape can indicate a lot about the habitat of a gastropod. The spiny species of *Murex* are inhabitants of soft, muddy bottoms, their spines helping to protect them from predators or possibly to aid in prey capture (usually other molluscs). Limpets and abalones have low, broad shells which offer less resistance in high wave habitats.

Most gastropods have a unique file-like mouth part, called a radula. Herbivorous species use this to rasp or cut algae from rocks. In the carnivorous cone shells, the radula is modified into a barbed harpoon-like structure, which they use to inject a powerful toxin into their unsuspecting prey, usually other molluscs. Some families of gastropods are adapted to use the radula for drilling holes in the shells of their prey. The spined murexes are noted for this, where the animal alternates drilling and secreting an acid, to form a virtually round hole in the shell of the prey. A hole such as this is often seen in empty bivalve shells.

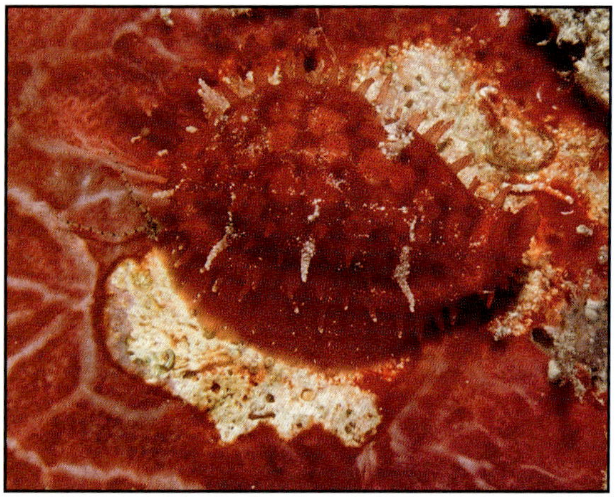

Below- The cowry *Cypreae cribraria* closely matches the color of the sponge it is feeding on in this photo taken at Enewetak Atoll.

Among molluscs there are grazers (herbivores) and a surprising number of carnivorous predators. The cone snails are perhaps the most extraordinary predators, and some species aggressively hunt fishes. Those that kill vertebrate prey, such as

Conus geographus, have neurotoxin venoms which can prove fatal to humans. Most cone shells, however, feed on invertebrate prey, such as worms and other mollusc-sand pose little danger to humans.

Among other gastropods, some smaller species can also be ruthless predators. The snails of the genus *Drupa* can attack and kill stony coral colonies. The snails line up around the living coral and literally suck the soft tissue away with their long proboscis. The trumpet triton, *Charonia tritonis*, is well-known for its ability to kill and eat the crown-of-thorns starfish. This gastropod is one of the few predators of this venomous echinoderm which has caused major disruption on reefs throughout the tropical Pacific.

Above- The cone shell *Conus textilus* is attempting to sting *Nasaarius papillosus* with its venomous harpoon. The prey snail is jumping out of the way in an escape response, trying to save its life.

At first glance, the Opisthobranchs, with some 2,000 species worldwide, hardly seem to be gastropods. Their often large bodies do not have sizable external shells, but they do exhibit the basic gastropod plan. The Cephalaspidea are the most primitive members of the opisthobranchs. Some have shells they can retract into (bubble shells), while others have only internal or vestigial shells. All members have a head shield over the head end of the body. The Anaspidea, or sea hares, have either a reduced shell buried in the mantle or it has been completely lost. Many release purple ink if disturbed. The Notaspidea, or pleurobranchs, have a single gill on the right side and reach medium to large sizes for opisthobranchs. The last group, Nudibranchia or sea slugs, are perhaps the most conspicuous molluscs. They are generally brightly colored, and can range in size from less than one eigth of an inch to over one foot in length. They have no shell and the body is bilaterally symmetrical. The head always has a pair of antennae-like rhinophores. Most have gills on the posterior part of the body and some can retract their gills into a branchial pocket. The upper surface of nudibranchs often have cerata, digitate or club-like projections of tissue, that can be brightly colored. These are used in respiration, defense and digestion.

Below- The trumpet triton, *Charonia tritonis*, is a predator of echinoderms. In this photo a large triton is feeding on a pincushion star, *Culcita novaeguineae*. The large horny operculum of the triton can be seen clearly.

Class Bivalvia- Clams and oysters

Bivalves are molluscs with two shells, or valves, hinged along one edge, with the animal in between. The shell's valves are held closed by two strong muscles. The foot of bivalves is compressed. In sand-dwelling bivalves the foot is adapted for burrowing. In sessile bivalves which have one valve firmly cemented to rock or other hard substratum (oysters), the foot is reduced in size.

Some bivalves, such as members of the genus *Lithophaga,* are among the most destructive organisms, excavating tunnels through coral skeletons. Sometimes corals are so riddled with these tunnels that they are almost hollow. The only evidence of their presence are the dark holes where the siphons of the clams are exposed on the surface of the coral.

The gastropod snail in this photograph, *Phallium* sp. is laying eggs. The column shaped mass of eggs is typical for this group of snails.

Above left- These oysters, *Pteria* sp., are growing upon a gorgonian in Palau. Winged oysters are common on gorgonians and black corals on outer reef drop offs. Above right- The strange tube-like mollusc is a "watering pot clam", *Brechites* sp. *Brechites* starts life with a normal pair of valves, but during development the valves fuse to form a long tube, perforated at the top and fluted at the bottom.

The black lip pearl oyster (*Pinctada margaritifera*) above, and the gold lip pearl oyster (*Pinctada maxima*) produce commercially valuable pearls. Other bivalves, including the giant clams, produce pearls of lesser quality .

Many bivalves are filter feeders, straining food items from the surrounding water brought across their enlarged gills by ciliated cells. As seawater is passed over the gills for gas exchange, food particles are also captured and passed to the mouth by ciliary action. Some, however, possess zooxanthellae, algal cells similar to those found in stony corals and other cnidarians, and apparently these animals also derive at least a portion of their nutrition from the alga. The best example of these are the giant clams of the Indo-Pacific, members of the genera *Tridacna* and *Hippopus*. This symbiotic relationship is the major reason these clams are able to reach such large sizes. Most zooxanthellae are found in the fleshy mantle, which is expanded when the clam opens to expose it to light. The photosynthetic requirements of the algae restrict giant clams to relatively shallow water, and they are seldom found below about forty feet in depth.

Bivalves also include the pearl oysters. Two bivalves in our region are sources of commercial pearls. The black-lip pearl oyster (*Pinctada margaritifera*) is found across the Pacific from Baja California to the Mediterranean Sea and produces black pearls. The gold-lip pearl oyster (*P. maxima*) occurs from Burma to the Solomon Islands and produces white pearls. There are other species of *Pinctada* in our region, but they are not significant pearl producers. Natural pearls are rare in these two species and most are now cultured. The shell of the black-lip is also of significant value for use largely in button production (mother of pearl).

The pearl is formed around a foreign object lodged between the mantle and the shell of the oyster. Often the oyster simply builds the area of the shell where the object occurs and produces a nacreous lump called a half pearl. For pearl culture, planktonic young oysters (spat) are usually obtained by having them settle on lines or other objects in the ocean (spat collectors). The juvenile oysters grow on the lines and are seeded with pearl nuclei when about 2 years of age. The pearls are harvested about 2 years later, they are about 4 mm diameter or more.

Pearls can be a significant money earner for island nations, for example producing several million dollars a year for the Cook Islands. Culture efforts are limited by overfishing of adult populations in some areas (resulting in low spat collection), pollution (causing production of low quality pearls) and diseases. Hatchery production of spat has helped some of these problems, but a clean non-polluted environment has proven to be the most important component in successful pearl culture.

Pearls are found in some other bivalves. The giant clams occasionally produce pearl-like growths of great size, but these lack the lustre of oyster pearls. The gastropod, *Strombus,* occasionally has pearls.

Class Cephalopoda- Nautilus, squids, cuttlefish and octopus

The cephalopods are generally adapted for swimming, with the exception of the octopods which have taken up a benthic existence. There are about 600 living species of cephalopods, but over ten times that many are known from the fossil record. The swimming abilities of squid are exceptional and some species are even capable of short gliding flights in the manner of flying fishes, using the lateral fins as temporary wings. The cold-water giant squid (more than fifty feet long) is the largest known invertebrate.

The cephalopods are divided into two subclasses, the Nautilida and the Coleoidea. *Nautilus*, with five species, is the last surviving genus of the subclass Natuloidea. While generally found at depths of 300-600 feet in the Pacific tropics, they come to within diving depths at night in areas with relatively cool surface waters, such as New Caledonia. Some dive boats catch *Nautilus* using deep water traps, then release the animals in shallow water so divers can observe and photograph them and allow the *Nautilus* to return to the depths. The shells are often found floating or washed ashore. *Nautilus* are slow swimmers and are largely scavengers with 38 tentacles. It is believed that early Nautiloids could not compete with the advanced bony fishes in ancient shallow tropical seas and are restricted to deep reef environments today because of reduced competition in such areas.

Cephalopods eyes are very similar in some ways to those of vertebrate animals, although the image is formed by pinhole optics rather than by a lens. They are adapted for finding and capturing prey using their arms and prehensile tentacles, which often have adhesive suckers. Squid and cuttlefish have 8 arms and 2 tentacles, while octopods have 8 arms, but lack tentacles. There is a pair of beak-like jaws in the oral (buccal) cavity, which often has salivary glands associated with it. These glands secrete enzymes and toxins, which include the venom of the blue-ringed octopus, *Hapalochlaena maculosa,* which can be fatal to humans. The radula, also found in gastropods, serves as a tongue. Prey of cephalopods includes fishes, crustacea, and other cephalopods.

Squids, cuttlefish and many other cephalopods propel themselves by expelling a jet of water from the mantle cavity through a funnel which can be directed to control the direction of movement of the animal. The streamlined squids can swim faster than any other invertebrate, up to 25 miles per hour.

Most cephalopods have ink glands, capable of releasing a quantity of black ink-like substance from the anus when disturbed. This dark fluid, high in the pigment melanin, is believed to confuse predators by creating a "dummy" cephalopod and it may also have anesthetic properties on the chemoreceptors of predators. Nautiloids lack an ink gland.

Cuttlefish bodies are shorter, broader and more flattened than those of squids. There are several species of cuttlefishes found in the tropical Pacific, ranging from small species only a few cm long, to the half meter long *Sepia latimanus*. Most cuttlefish swim in open water, but some bury themselves in sand lying in wait for prey. They are particularly known for their ability to change color and pattern rapidly, almost as though waves of color were passing up and down their bodies. The internal shells of cuttlefishes (cuttlebone) provide buoyancy in life. After death it is commonly found washed up on beaches. The cuttlebone differs between species and is an important character for species identification.

The most common squid on tropical Pacific reefs is *Sepioteuthis lessoniana*, which can occur in schools. There are numerous other species which might be encountered. Diving at night around a light hung over the side of a boat is often a good way to observe and closely approach squids, which normally are quite shy during the day. Some squids are also luminescent.

Octopods are fairly common in reef environments, usually hiding in crevices and holes during the day. One species found in the Pacific, *Octopus macropus*, also occurs around the world in tropical and subtropical waters, including the Mediterranean. All octopods have a beak capable of inflicting minor wounds.

Cephalopods are either male or female and reproduction occurs through copulation. The male has a modified arm, termed the hectocotylus, which is used to deposit a spermatophore into the mantle cavity or elsewhere on the body of the female. As the eggs are released, they are fertilized by sperm from the spermatophore and then are deposited on the bottom by most shallow water species. Some eggs are attached to objects such as rock or seagrass. Some octopods engage in parental care of the eggs until hatching. Many species die after reproducing and at most live only a few years. Others, such as *Nautilus*, are believed to live 20 years or more.

Above- This large cuttlefish, *Sepia* sp. is common on coral reefs of northern New Guinea. Cuttlefish can change color and texture dramatically and underwater it is difficult to identify many of them to species.

720 - Chiton * Papua New Guinea

721 - *Cryptoplax* sp. * Philippines

722 - *Acanthopleura* sp. * Indonesia

723 - *Stenoplax alata* * Philippines

720 - Chiton * Polyplacophora * Papua New Guinea * Dyaul Island * 3 ft (1 m).

721 - *Cryptoplax* sp. * Polyplacophora * Philippines * Cebu * Mactan Island * 10 ft (3 m). This chiton has minute shell plates embedded in the girdle. It is usually found under rocks on reefs.

722 - *Acanthopleura* sp. * Polyplacophora * Indonesia * Manado * intertidal under rocks. Chitons have eight transverse plates which fit into the sides of the girdle, a fold of the mantle.

723 - *Stenoplax alata* * Polyplacophora * Philippines * Cebu * Mactan Island * intertidal.

724 - *Haliotis asinina* * Haliotidae * Archaeogastropoda * Palau * Lighthouse Reef * 10 ft (3 m). While abalone are usually thought of as cold water creatures, there are about 10 species which occur in the tropical western Pacific. The tropical species are generally smaller (averaging about 2 inches or 50 mm) than their temperate relatives. The species illustrated is widespread in the region, but some others are restricted to a single island or group.

725 - *Trochus niloticus* * Trochidae * Archaeogastropoda * Marshall Islands * Enewetak Atoll * reef flat * night * 10 ft (3 m). *Trochus niloticus* is one of the most important commercial molluscs of the Pacific. While long used for food by local populations, its shells are used for button manufacturing. The natural distribution of *T. niloticus* was widespread, from Sri Lanka to Wallis Island, but spotty over the region. For example in the Caroline Islands, it naturally occurred in Palau and Yap only. Starting in the 1920's the Japanese successfully transplanted *T. niloticus* to many islands and populations rapidly grew to the point they could be commercially harvested within 20 years or so. It occurs in the intertidal or very shallow subtidal and is easily harvested by experienced divers or waders. The demand for "pearl" shell buttons is so great that most island states have had to institute protective measures to insure their *Trochus* stock are not decimated.

726 - *Turbo petholatus* * Turbinidae * Archaeogastropoda * Papua New Guinea * New Ireland * night * 3 ft (1 m). Commonly known as the tapestry turban, this species has a glossy attractively colored shell. It has a calcareous operculum which closes off the opening when the animal withdraws which is a thick convex multi-colored disk, known as a cat's eye.

727 - *Serpulorbis variabilis* * Vermetidae * Mesogastropoda * Papua New Guinea * West New Britain * fringing reef * 20 ft (6 m). Vermetids have the whorls of the shell completely separated, more or less, like a corkscrew. Their foot is reduced, but they have a circular operculum which effectively closes the opening. Vermetids lay down mucous nets which are used in capture of food.

728 - Vermetid snails * Vermetidae * Mesogastropoda * * Papua New Guinea * West New Britain * patch reef * 20 ft (6 m).

729 - *Cerithium aulco* * Cerithiidae * Mesogastropoda * Indonesia * Manado * fringing reef * 15 ft (5 m). This genus is common on sandy bottoms around reefs and seagrass beds. They live in the upper layer of sand and often leave trails on the surface as they plow through the sand.

730 - *Thyca crystallina* * Eulimidae * Mesogastropoda * Philippines * Cebu * Mactan Island 10 ft (3 m). These snail are parasites, living on starfishes where they either sit on the surface or actually burrow into the arms.

731 - *Lambis scorpius* * Strombidae * Mesogastropoda * Federated States of Micronesia * Chuuk Atoll * barrier reef * 40 ft (12 m). The species of *Lambis* are generally known as spider shells due to their long finger-like projections. This species has perhaps the most spectacular spines and colorful shell among *Lambis*, but its outer surface is often heavily encrusted and until the shell is turned over its true beauty cannot be seen.

732 - *Lambis truncata* * Strombidae * Mesogastropoda * Marshall Islands * Enewetak * 10 ft (3 m). This is the largest species of *Lambis*. Immature shells have undeveloped spines making them hard to recognize as members of *Lambis*.

724 - *Haliotis asinina* * Palau

725 - *Trochus niloticus* * Federated States of Micronesia

726 - *Turbo petholatus* * Papua New Guinea

727 - *Serpulorbis variabilis* * Papua New Guinea

728 - Vermetid snails * Papua New Guinea

729 - *Cerithium aulco* * Indonesia

730 - *Thyca crystallina* * Philippines

731 - *Lambis scorpius* * Federated States of Micronesia

732 - ***Lambis truncata*** * Marshall Islands

733 - ***Strombus dentatus*** * Marshall Islands

734 - ***Strombus gibberulus*** * Indonesia

735 - ***Strombus luhuanus*** * Marshall Islands

733 - *Strombus dentatus* * Strombidae * Mesogastropoda * Marshall Islands * Kwajalein Atoll * lagoon pinnacle * 33 ft (10 m). This is a small species of *Strombus* which has a glossy shell and reaches about 2.6 inches (60 mm) in length. Species of *Strombus* have their eyes on stalks.

734 - *Strombus gibberulus* * Strombidae * Mesogastropoda * Indonesia * Biak Island * lagoon * night * 30 ft (9 m). This is another small *Strombus* which can be extremely abundant on shallow sand and reef flats. Like other strombids, the eyestalks can be clearly seen in this individual.

735 - *Strombus luhuanus* * Strombidae * Mesogastropoda * Marshall Islands * Enewetak Atoll * 20 ft (6 m). Commonly known as the blood mouth conch due to the red color inside the shell opening, it is a very common gastropod of sandy bottoms around reefs. Often the snails are found in groups on the surface of the sand. They are a highly-prized food in Papua New Guinea.

736 - *Strombus sinuatus* * Strombidae * Mesogastropoda * Federated States of Micronesia * Chuuk Atoll * 60 ft (18 m). This strombid has a beautiful flared lip with up to 5 projections on its upper margin. This species can be well camouflaged, as the outer surface of its shell can be overgrown.

737 - *Coriocella nigra* * Lamellariidae * Mesogastropoda * Federated States of Micronesia * Chuuk * lagoon patch reef * 20 ft (6 m). These unusual gastropods look more like nudibranchs than mesogastropods, and have an internal shell.

738 - *Coriocella* sp. * Lamellariidae * Mesogastropoda * Palau * Lighthouse Reef * night * 10 ft (3 m). This mollusc is possibly just a variant of *Coriocella nigra*, but at least in Palau is always a brown, rather than black color.

739 - *Cypraea annulus* * Cypraeidae * Mesogastropoda * Federated States of Micronesia * Chuuk * Weno * 6 ft (2 m). The family Cypraeidae are commonly called cowries and they are especially beloved by shell collectors because of their smooth shiny shells with colorful bold patterns. This is known as the gold-ring cowry, and is used in shell handicrafts, especially those made in the Philippines.

740 -*Cypraea annulus* * Cypraeidae * Mesogastropoda * Federated States of Micronesia * Chuuk * Weno * 6 ft (2 m).

741 - *Cypraea arabica* * Cypraeidae * Mesogastropoda * Hong Kong * Breaker Reef * 15 ft (5 m). Many cowries are herbivores, grazing algae from rock surfaces with their radula. A smaller number of cowries are believed to eat sponges, but the feeding habits of these species are not accurately known. *C. arabica* is a variable, nocturnally active species known from the Red Sea to Hawaii, although it is believed that its presence in Hawaii was due to an accidental introduction.

742 - *Cypraea aurantium* * Cypraeidae * Mesogastropoda * Marshall Islands * Kwajalein Atoll * reef drop off * night * 50 ft (15 m). The golden cowry is one of the most highly prized of Pacific gastropods by shell collectors. Until people started night diving on Micronesian reefs their habitat was a mystery. They live in caves and crevices, particularly on steep outer reefs, hiding during the day.

743 - *Cypraea chinensis* * Cypraeidae * Mesogastropoda * Hawaii * Pupukea * 30 ft (9 m). This cowry has its vermilion mantle covered with papillae. It is known from Hawaii and the Tuamotus to Indonesia and Okinawa. There may be a disjunct isolated population off east Africa.

744 - *Cypraea cribraria* * Cypraeidae * Mesogastropoda * Marshall Islands * Kwajalein Atoll * lagoon pinnacle * 30 ft (9 m). The sieve cowry has a very colorful shell with a thin translucent mantle. It is known from Micronesia to east Africa. Cowries have a spiral shell, but this may only be obvious in its early life. As the cowry begins to mature, the outer edge of the spiral shell curls then thickens, forming a narrow aperture. The shell will not grow in length after the lip has formed, but the entire shell will continue to thicken.

736 - ***Strombus sinuatus*** * Federated States of Micronesia

737 - ***Coriocella nigra*** * Federated States of Micronesia

738 - ***Coriocella* sp.** * Palau

739 - ***Cypraea annulus*** * Federated States of Micronesia

740 - ***Cypraea annulus*** * Federated States of Micronesia

741 - ***Cypraea arabica*** * Hong Kong

742 - ***Cypraea aurantium*** * Marshall Islands

743 - ***Cypraea chinensis*** * Hawaii

744 - *Cypraea cribraria* * Marshall Islands

745 - *Cypraea humphreysi* * Marshall Islands

746 - *Cypraea mappa* * Indonesia

747 - *Cypraea onyx* * Hong Kong -

745 - *Cypraea humphreysi* * Cypraeidae * Mesogastropoda * Marshall Islands * Enewetak Atoll * lagoon pinnacle * 40 ft (12 m). The smooth outer surface of cowry shells is unusual for gastropods and exists because the shell-secreting mantle is normally extended like a cloak upward to completely enclose the shell. The line on the top of the shell where the sides of the mantle join is known as the mantle line. If a cowry is disturbed it will slowly pull in its mantle, uncovering the shell. Because of its large surface area, the mantle is thought to perhaps also function as an auxiliary respiratory organ.

746 - *Cypraea mappa* * Cypraeidae * Mesogastropoda * Indonesia * Biak Island * dock area * 40 ft (12 m). The map cowry is one of the most attractive and distinctive cowries, with the common name coming from the unusual pattern on the shell produced where the two sides of the mantle meet. This species is known from Micronesia and French Polynesia to east Africa.

747 - *Cypraea onyx* * Cypraeidae * Mesogastropoda * Hong Kong * Ngau Shek Chau * 33 ft (10 m). This cowry has the shell completely covered by the mantle, the two sides meeting along the back of the shell where the dark area occurs.

748 - *Cypraea scurra* * Cypraeidae * Mesogastropoda * Hawaii * Makua * 80 ft (23 m). The species is known from east Africa to the islands of the eastern Pacific, but is not believed to reach the continental coast of the Americas. In Hawaii it is reported to be common, living in the deeply-luted heads of the coral *Porites lobata*.

749 - *Cypraea talpa* * Cypraeidae * Mesogastropoda * Marshall Islands * Kwajalein Atoll * lagoon pinnacle * 40 ft (12 m). This is one of the cowries which reaches a large size in Hawaii, but is smaller in areas where water temperatures are higher and more consistent. The species is known on reefs from east Africa to Hawaii and Polynesia.

750 - *Cypraea tigris* * Cypraeidae * Mesogastropoda * Marshall Islands * Enewetak * 20 ft (6 m). The tiger cowry is a large species and fairly common in many areas. It is quite variable and it has been said that "no two are alike". The shell has dark spots and is a popular curio item. In Hawaii *C. tigris* is often found with the coral *Pocillopora meandrina;*, there it reaches its greatest length, over 5 inches (125 mm).

751 - *Cypraea vitellus* * Cypraeidae * Mesogastropoda * Indonesia * Biak Island * dock area * night * 40 ft (12 m). This and the following photo show the mantle extended and retracted in the same species. Commonly known as the Pacific deer cowry, it occurs from east Africa to Hawaii and is one of the species which has its greatest sizes away from the equator.

752 - *Cypraea vitellus* * Cypraeidae * Mesogastropoda * Federated States of Micronesia * Chuuk * lagoon bottom * night * 100 ft (30 m).

753 - *Cypraea* sp. * Cypraeidae * Mesogastropoda * Palau * Lighthouse Reef * night * 3 ft (1 m).

754 - *Ovula ovum* * Ovulidae * Mesogastropoda * Philippines * Pamalican Island * 20 ft (6 m). The egg cowry preys upon the octocoral *Sarcophyton* sp. and even lays its eggs upon the soft coral. There is a remarkable contrast between the black mantle and brilliant white shell of this animal.

755 - *Crenovolva renovata* * Ovulidae * Mesogastropoda * Hong Kong * Cape d'Aguilar * 30 ft (9 m). These small ovulids live on gorgonians, sea whips and soft corals. The color of their mantle and shell usually matches that of the gorgonian making them inconspicuous.

756 - *Phenacovolva rosea* * Ovulidae * Mesogastropoda * Hong Kong * Cape d'Aguilar * 30 ft (9 m). This purple species lives on gorgonians. When the polyps are expanded and the mantle of the gastropod is extended, it is difficult to detect the snails held tight against the branches of the gorgonian.

757 - *Natica stellata* * Naticidae * Mesogastropoda * Papua New Guinea * Port Moresby * Lion Island * night * 40 ft (12 m). The

748 - ***Cypraea scurra*** * Hawaii

749 - ***Cypraea talpa*** * Marshall Islands

750 - ***Cypraea tigris*** * Marshall Islands

751 - ***Cypraea vitellus*** * Indonesia

752 - ***Cypraea vitellus*** * Federated States of Micronesia

753 - ***Cypraea* sp.** * Palau

754 - ***Ovula ovum*** * Philippines

755 - ***Crenovolva renovata*** * Hong Kong

756 - ***Phenacovolva rosea*** * Hong Kong

757 - ***Natica stellata*** * Papua New Guinea

758 - ***Natica* sp.** * Papua New Guinea

759 - ***Polinices tumidus*** * Philippines

members of *Natica* are predatory molluscs which use their toothed radula to drill small holes in the shells of other molluscs to kill and consume them.

758 - *Natica* sp. * Naticidae * Mesogastropoda * Papua New Guinea * Madang * lagoon bottom * 80 ft (24 m). This species, probably a member of *Natica*, has a spectacularly colored mantle and foot. Although the shells of this and the previous species appear similar, the coloration of the live animal reveals that they are worlds apart.

759 - *Polinices tumidus* * Naticidae * Mesogastropoda * Philippines * Cebu * Mactan Island * reef flat * 6 ft (2 m). The moon snails have a huge mantle and foot which amazingly can be withdrawn into the shell when danger threatens. They cruise slowly over the bottom searching for potential food items. Many such species use the large foot to envelope and subdue prey.

760 - *Natica* egg case * Naticidae * Mesogastropoda * Papua New Guinea * Yule Island * 10 ft (3 m). This type of egg case, a conical whorl found on sand, is produced by members of the Naticidae.

761 - *Cassis cornuta* * Cassidae * Mesogastropoda * Indonesia * Manado * fringing reef * 33 ft (10 m). The helmet shells are found on sandy bottoms and are among the largest of gastropods. This species can reach 16 inches (40 cm) in length. They are generally predators on echinoderms, particularly sea urchins and sand dollars. The unfortunate echinoderms are paralyzed by a secretion from the *Cassis*. The helmet shells can bury almost completely in sand, so deep that only the upper spire of the shell remains exposed.

762 - *Casmaria erinaceus* * Cassidae * Mesogastropoda * Philippines * Cebu * Mactan Island * 10 ft (3 m). This small cassid is in a group called the bonnet shells. They are also found on sandy areas near reefs. In the Philippines, where the photograph was taken, such molluscs are relatively abundant, probably a result of the elimination of their fish predators by over-fishing.

763 - *Malea pomum* * Tonnidae * Mesogastropoda * Papua New Guinea * Duke of York Islands * Makada Reef * night * 20 ft (6 m). This species has a solid glossy shell and a large foot which spreads out over sand substrates.

764 - *Tonna cepa* * Tonnidae * Mesogastropoda * Palau * German Channel * night * 20 ft (6 m). Members of *Tonna* have thin, fragile shells and are nocturnally active.

765 - *Tonna perdix* * Tonnidae * Mesogastropoda * Marshall Islands * Kwajalein Atoll * reef drop off * night * 50 ft (15 m). The Pacific partridge tun is large, to about 5 inches (125 mm) in length, and has a huge animal compared to the size of the shell. They crawl around the reef at night and are predators of other molluscs.

766 - *Charonia tritonis* * Cymatiidae * Mesogastropoda * Marshall Islands * Kwajalein Atoll * offshore reef * 30 ft (9 m). The trumpet triton shell is one of the most spectacular and prized of the Pacific gastropods reaching 16 inches (40 cm) or more in length. They are predators on echinoderms, particularly asteroids, and are one of the few animals which prey upon the crown-of-thorns starfish, *Acanthaster planci*. The individual in the photograph is eating a young pin cushion star, *Culcita novaeguinea*, a known predator of stony corals.

767 - *Cymatium aquatile* * Cymatiidae * Mesogastropoda * Federated States of Micronesia * Chuuk Atoll * lagoon wreck * 20 ft (6 m). The various species of *Cymatium* are predators of molluscs and echinoderms. Some species, such as the hairy triton, *C. pileare*, have thick periostracum, a thick horny layer, with hair-like projections over the shell. A thin periostracum can be seen on the photographed species.

768 - *Chicoreus microphyllus* * Muricidae * Neogastropoda * Federated States of Micronesia * Chuuk * Anaw Channel *50 ft (15 m). The muricids are one of the best known and most distinctive families of gastropods with many remarkable feeding and reproductive adaptations. Most are voracious predators, feeding on other gastropods, bivalves and barnacles. Many are capable of drilling holes, using their toothed radula, in the shells of other molluscs. This

760 - ***Natica*** **egg case** * Papua New Guinea

761 - ***Cassis cornuta*** * Indonesia

762 - ***Casmaria erinaceus*** * Philippines

763 - ***Malea pomum*** * Papua New Guinea

764 - ***Tonna cepa*** * Palau

765 - ***Tonna perdix*** * Marshall Islands

766 - ***Charonia tritonis*** * Marshall Islands

767 - ***Cymatium aquatile*** * Federated States of Micronesia

768 - ***Chicoreus microphyllus*** * Federated States of Micronesia

769 - ***Chicoreus ramosus*** * Federated States of Micronesia

770 - ***Murex pecten*** * Papua New Guinea

771 - ***Nassarius* cf. *coronatus*** * Philippines

species is often seen with a thin orange periostracum covering the outside of the shell, which is not the shell's true color.

769 - *Chicoreus ramosus* * Muricidae * Neogastropoda * Federated States of Micronesia * Chuuk Atoll * lagoon bottom * 75 ft (23 m). This is the largest muricid, reaching about 12 inches in length, and is found throughout the entire Indo-west Pacific region.

770 - *Murex pecten* * Muricidae * Neogastropoda * Papua New Guinea * Yule Island * 66 ft (20 m). The delicate venus comb *Murex,* is found on muddy bottoms. It buries during the day and comes to the sand surface at night. Members of Muricidae are renowned for the intricate and elaborate sculpturing of their shells. There are about 400 species in the family and many produce clusters fibrous egg capsules which are occasionally seen on the reef and rocks.

771 - *Nassarius* cf. *coronatus* * Nassariidae * Neogastropoda * Philippines * Cebu * Mactan Island * 6 ft (2 m). The species of *Nassarius* are carnivorous scavengers which can move surprisingly fast over the bottom.

772 - *Nassarius glans* * Nassariidae * Neogastropoda * Solomon Islands * Guadalcanal Island * Ruaniu * 40 ft (12 m). The dog whelk is a very active gastropod with a large foot and a long slender siphon. The small brown operculum can be seen on the top of the rear of the foot and two tiny projections from the rear of the foot are believed to be false antennae to confuse predators as to which is the head end of the snail.

773 - *Nassarius papillosus* * Nassariidae * Neogastropoda * Marshall Islands * Kwajalein Atoll * 30 ft (9 m).

774 - *Pleuroploca trapezium* * Fasciolariidae * Neogastropoda * Indonesia * Lembeh Island * 33 ft (10 m).

775 - *Oliva reticulata* * Olividae * Neogastropoda * Indonesia * Manado * Lembeh Island * night * 66 ft (20 m). The olive shells are high gloss tubular gastropods which live in sand. They are nocturnally active and leave tracks in the sand. This species reaches about 2 inches (50 mm) in length and is a common shallow water species found throughout the Indo-Pacific.

776 - *Oliva tessellata* * Olividae * Neogastropoda * Indonesia * Manado * 50 ft (15 m). Olive shells have a large foot which they use to envelope prey, usually other molluscs. They are also general scavengers with a highly developed sense of smell, which they use to find prey. This species is found from the Bay of Bengal to the Solomon Islands. It is not found further east.

777 - *Harpa amouretta* * Harpidae * Neogastropoda * Marshall Islands * Enewetak Atoll * lagoon * night * 40 ft (12 m). The harp shells have an immense foot, with the shell seeming to be a tiny accessory on top. There are about 12 species of harp shells in the tropical/subtropical Pacific. This species reaches about 2.5 inches (60 mm) and is found from east Africa to Hawaii and French Polynesia.

778 - *Harpa harpa* * Harpidae * Neogastropoda * Indonesia * Manado * fringing reef * night * 50 ft (15 m). The harp shells are carnivores, feeding largely on crabs and other crustaceans. They suffocate their prey with their large foot. If threatened, they can shed the rear part of the foot, a behavior called autotomy, the writhing shed piece serves as an excellent decoy to occupy the attention of the predator.

779 - *Cymbiola vespertilio* * Volutidae * Neogastropoda * Philippines * Cebu * Mactan Island * seagrass bed * night * 10 ft (3 m). The Philippine bat volute has several subspecies from the Indian Ocean to the Philippines and northern Australia, where it is common in areas of mud. Most volutes do not have free swimming larvae, so local populations often evolve which differ from others geographically resulting in the subspecies.

772 - ***Nassarius glans*** * Solomon Islands

773 - ***Nassarius papillosus*** * Marshall Islands

774 - ***Pleuroploca trapezium*** * Indonesia

775 - ***Oliva reticulata*** * Indonesia

776 - ***Oliva tessellata*** * Indonesia

777 - ***Harpa amouretta*** * Marshall Islands

778 - ***Harpa harpa*** * Indonesia

779 - ***Cymbiola vespertilio*** * Philippines

780- *Melo broderipii* * Philippines

781 - *Melo melo* * Hong Kong

782 - *Mitra mitra* * Indonesia

783 - *Vexillum plicarium* * Philippines

780 - *Melo broderipii* * Volutidae * Neogastropoda * Philippines * 20 ft (6 m). The baler shells are often large and usually feed on sand dwelling molluscs. The word Baler is actually a misspelling of the word 'Bailer', dating back to when Europeans saw islanders use the shells to bail canoes.

781 - *Melo melo* * Volutidae * Neogastropoda * Hong Kong * 20 ft (6 m).

782-*Mitra mitra* * Mitridae * Neogastropoda * Indonesia * Biak Island * 50 ft (15 m). The mitres are sand-dwelling molluscs, again leaving trails in the sand as they push their way through the surface of the sediment. They are nocturnally active.

783 - *Vexillum plicarium* * Costellariidae * Neogastropoda * Philippines * Cebu * Mactan Island * seagrass bed * 3 ft(1 m). Two families of gastropods the Mitridae and the Costellariidae are considered to be mitre shells. This genus falls into the later family.

784 - Turridae * Neogastropoda * Indonesia * Manado * 60 ft (18 m). Turrids can be of various shapes, but all have a notch in the outer lip of the shell.

785 - *Conus aulicus* * Conidae * Neogastropoda * Papua New Guinea * Madang * Pig Island * 20 ft (6 m). This attractive cone shell is a large species, reaching over 6 inches (15 cm), which is a predator on other molluscs. It is found through the Indo-west Pacific. All members of *Conus* are predatory carnivores, using their venom glands and radular teeth to subdue their prey. Overall the genus has three types of feeding, on worms (vermivores), molluscs (molluscivores) and fish (piscivores).

786 - *Conus circumcisus* * Conidae * Neogastropoda * Papua New Guinea * West New Britain * 10 ft (3 m). This cone shell reaches 3.5 inches (9 cm) and is found from the Indian Ocean to the Philippines. It is possibly a piscivore, but its food habits are not really well known.

787 - *Conus crocatus* * Conidae * Neogastropoda * Marshall Islands * Enewetak Atoll * lagoon pinnacle * 33 ft (10 m). This is a fairly small species reaching only about 2.5 inches.

788 - *Conus eburneus* * Conidae * Neogastropoda * Federated States of Micronesia * Chuuk * northern lagoon * 40 ft (12 m). This is common cone shell, lives in sand. It is widespread, but does not occur in Hawaii.

789 - *Conus floccatus* * Conidae * Neogastropoda * Marshall Islands * Kwajalein Atoll * offshore reef * 50 ft (15 m). This species is believed to be a piscivore and is found from the Philippines and Solomon Islands to Fiji and the Marshall Islands.

790 - *Conus geographus* * Conidae * Neogastropoda * Papua New Guinea * Port Moresby * Lion Island * 40 ft (12 m). This is the species of cone snail most dangerous to humans, having been responsible for a number of deaths. Normally feeding on fish (a piscivore), it actively hunts at night, crawling over the bottoms in search of prey with its deadly proboscis. In such cone snails, the teeth of the radula, the rasping tongue found in gastropods, are modified into barbed "harpoons", which are shot at prey. The toxin of this cone can is so stable it can be dangerous to humans even after the animal is dead.

791 - *Conus imperialis Conidae * Neogastropoda * Federated States of Micronesia * Chuuk * 6 ft (2 m).**

792 - *Conus marmoreus* * Conidae * Neogastropoda * Marshall Islands * Enewetak Atoll * 40 ft (12 m). This is a common species, known as the marble cone, found throughout the Indo-west Pacific. It is a molluscivore.

793 - *Conus nussatella* * Conidae * Neogastropoda * Marshall Islands * Enewetak Atoll * lagoon pinnacle * 33 ft (10 m).

794 - *Conus striatus* * Conidae * Neogastropoda * Papua New Guinea * West New Britain * Witu Islands * 20 ft (6 m). This species can be locally abundant, found in sandy habitats.

784 - **Turridae** * Indonesia

785 - ***Conus aulicus*** * Papua New Guinea

786 - ***Conus circumcisus*** * Papua New Guinea

787 - ***Conus crocatus*** * Marshall Islands

788 - ***Conus eburneus*** * Federated States of Micronesia

789 - ***Conus floccatus*** * Marshall Islands

790 - ***Conus geographus*** * Papua New Guinea

791 - ***Conus imperialis*** * Federated States of Micronesia

792 - ***Conus marmoreus*** * Marshall Islands

793 - ***Conus nussatella*** * Marshall Islands

794 - ***Conus striatus*** * Papua New Guinea

795 - ***Conus textile*** * Philippines

795 - *Conus textile* * Conidae * Neogastropoda * Philippines * Cebu * Mactan Island * 20 ft (6 m). Most cones with a tented pattern on the shell, such as *C. textile,* feed on other molluscs. This species hunts at night. It buries in the sand during the day.

796 - *Terebra guttata* * Terebridae * Neogastropoda * Marshall Islands * Enewetak * lagoon bottom * 70 ft (21 m). The auger shells are sand-dwellers found around most reefs, their presence revealed by a furrow in the sand created as they plow along. Some species have venom glands and harpoon-shaped radular teeth. This species reaches about 7 inches (175 mm) in length and occurs from the intertidal to 330 ft (100 m) depths.

797 - *Terebra maculata* * Terebridae * Neogastropoda * Fiji * Kaimbu Island * reef sand * night * 30 ft (9 m). This is the largest of the auger shells, reaching 11 inches (275 mm) in length; commonly called the marlinspike shell. Because of its size it has often been used as a tool in Pacific cultures. It lacks a poison gland and feeds on polychaete worms.

798 - *Terebra* sp. * Terebridae * Neogastropoda * Papua New Guinea * Port Moresby * Lion Island * 50 ft (15 m). Auger shells feed on polychaetes, hemichordates and other "worms". It is believed that each species is prey specific. Species within this genus are often quite variable in their markings and other features.

799 - *Janthina janthina* * Janthinidae * Heterogastropoda * Central Pacific * surface. This is actually a pelagic mollusc found worldwide in the tropics which floats on a bubble raft produced by the snail. The photographed individual has several goose-neck barnacles living on it.

800 - Pyramidellidae * Entomotaeniata * Opisthobranchia * Papua New Guinea * Madang * lagoon * 40 ft (12 m). These tiny white shelled opisthobranchs, seen here on a sea cucumber, are often parasitic on a wide variety of invertebrates, such as echinoderms and bivalve molluscs, including giant clams. Their taxonomy is poorly known.

801 - *Chelidonura electra* * Aglajidae * Cephalaspidea * Opisthobranchia * Papua New Guinea * Port Moresby * Pt. Osbourne * 50 ft (15 m). The cephalaspideans have a cephalic (head) shield, a flattened flap of skin which extends up and back from the head region. This is believed to aid in burrowing. They also have parapodia, like the sacoglossans, and these can be used for swimming by flapping them. Many members of the order have a thin bulbous shell. Cephalaspideans can be both herbivores and carnivores.

802 - *Chelidonura hirundinina* * Aglajidae * Cephalaspidea * Opisthobranchia * Philippines * Cebu * Mactan Island * 10 ft (3 m). Members of this genus are predators of opisthobranchs. This species is common in the intertidal in Hawaii.

803 - *Chelidonura inornata* * Aglajidae * Cephalaspidea * Opisthobranchia * Federated States of Micronesia * Mortlock Islands * reef * 10 ft (3 m).

796 - ***Terebra gutatta*** * Marshall Islands

797 - *Terebra maculata* * Fiji

798 - *Terebra* sp. * Papua New Guinea

799 - *Janthina janthina* * Central Pacific

800 - Pyramidellidae * Papua New Guinea

801 - *Chelidonura electra* * Papua New Guinea

802 - *Chelidonura hirundinina* * Philippines

803 - *Chelidonura inornata* * Federated States of Micronesia

804 - *Chelidonura varians* * Indonesia

805 - ***Philinopsis cyanea*** * Philippines

804 - *Chelidonura varians* * Aglajidae * Cephalaspidea * Opisthobranchia * Indonesia * Manado * 30 ft (9 m).

805 - *Philinopsis cyanea* * Aglajidae * Cephalaspidea * Opisthobranchia * Philippines * Cebu * Mactan Island * algal flat * 3 ft (1 m).

806 - *Philinopsis pilsbryi* * Aglajidae * Cephalaspidea * Opisthobranchia * Philippines * Cebu * algal flat * 10 ft (3 m).

807 - *Phanaropthalmus smaragdinus* * Haminoeidae * Cephalaspidea * Opisthobranchia * Philippines * Cebu * Mactan Island * 10 ft (3 m).

808 - *Bulla ampulla* * Bullidae * Cephalaspidea * Opisthobranchia * Philippines * Cebu * Mactan Island * sea grass bed * 3 ft (1 m). The thin shell of this cephalaspidean is clearly visible and it is a common species is shallow areas throughout the region.

806 - ***Philinopsis pilsbryi*** * Philippines

809 - *Elysia ornata* * Elysiidae * Sacoglossa * Opisthobranchia * Federated States of Micronesia * Chuuk * lagoon patch reef * 40 ft (12 m). The sacoglossids are mostly herbivorous and lack both a shell and cerata. They possess parapodial lobes, flaps extending upward along the back which are evident in the photograph. Most species are similar in appearance. They usually live on algae.

810 - *Elysia* sp. * Elysiidae * Sacoglossa * Opisthobranchia * Federated States of Micronesia * Chuuk * Northeast Passage * 20 ft (6 m). Species of *Elysia* have functional chloroplasts in their tissues which give them their green color. The source of these chloroplasts (the cellular organelles responsible for photosynthesis in plant cells) is the algae preyed upon by sacoglossans. Their radula is adapted for piercing the cells of algae and sucking out the contents.

811 - *Plakobranchus ocellata* * Elysiidae * Sacoglossa * Opisthobranchia * Hawaii * Oahu * Coconut Island * 10 ft (3 m). Member of this genus are typically found in muddy and silty areas.

807 - ***Phanaropthalmus smaragdinus*** * Philippines

812 - *Plakobranchus* sp. * Elysiidae * Sacoglossa * Opisthobranchia * Philippines * Cebu * Mactan Island * 10 ft (3 m). The 'flaps' of this species are rolled inward; when extended the green color of the chloroplasts which cover their entire back (dorsal) surface can be seen.

813 - *Cyerce* sp. * Calyphyllidae * Sacoglossa * Opisthobranchia * Palau * Lighthouse Reef * night * 10 ft (3 m).

814 - *Aplysia dactylomela* * Aplysiidae * Anaspidea * Opisthobranchia * Palau * Lighthouse Reef * 10 ft (3 m). The anaspids lack a head shield and are commonly known as sea hares. They have long rhinophores, which are chemosensory tentacles. All are herbivores, feeding on algae. They have a small remnant internal shell. This species occurs circumtropically.

815 - *Aplysia* sp. * Aplysiidae * Anaspidea * Opisthobranchia * Philippines * Cebu * Mactan Island * algal flat * 3 ft (1 m). The

808 - ***Bulla ampulla*** * Philippines

809 - ***Elysia ornata*** * Federated States of Micronesia

810 - *Elysia* sp. * Federated States of Micronesia

811 - *Plakobranchus ocellata* * Hawaii

812 - *Plakobranchus* sp. * Philippines

813 - *Cyerce* sp. * Palau

814 - *Aplysia dactylomela* * Palau

815 - *Aplysia* sp. * Philippines

816 - *Dolabella auricularia* * Papua New Guinea

817 - *Dolabella auricularia* * Papua New Guinea

818 - ***Umbraculum umbraculum*** * Palau

819 - ***Berthella martensi*** * Papua New Guinea

820 - ***Berthellina citrina*** * Indonesia

821 - ***Pleurobranchus brockii*** * Hong Kong

species of *Aplysia* can be relatively large, up to 4-8 inches (10-20 cm). They release a purple secretion when disturbed which is an irritant and may deter predators.

816 - *Dolabella auricularia* * Aplysiidae * Anaspidea * Opisthobranchia * Papua New Guinea * Madang * 30 ft (9 m). There is believed to be only a single member of *Dolabella* and it reaches over 16 inches (40 cm) in length. The animal is flattened on the posterior end and is cryptically colored so it is often overlooked. It is nocturnally active, hiding under rocks or buried in sediment during the day. It is found from east Africa to the western Pacific.

817 - *Dolabella auricularia* * Aplysiidae * Anaspidea * Opisthobranchia * Papua New Guinea * Port Moresby * Bootless Bay * 50 ft (15 m). The highly cryptic coloration of *Dolabella auricularia* is evident in this photograph. The animal could easily be mistaken for a rock covered with algae. The truncated appearance of the posterior of the animal is also evident.

818 - *Umbraculum umbraculum* * Umbraculidae * Notaspidea * Opisthobranchia * Palau * Lighthouse Reef * night * 10 ft (3 m). This unusual opisthobranch has the shell reduced to an umbrella-like disk.

819 - *Berthella martensi* * Pleurobranchidae * Notaspidea * Opisthobranchia * Papua New Guinea * Madang * Rempi * 50 ft (15 m).

820 - *Berthellina citrina* * Pleurobranchidae * Notaspidea * Opisthobranchia * Indonesia * Manado * Ruang Island * 50 ft (15 m). This is the most common Hawaiian notaspid and appears to be circumtropical in distribution. It is active at night and eats sponges and hard corals, including the orange *Tubastrea*. The photograph shows a typical individual amid didemnid ascidians, hydroids and sponges.

821 - *Pleurobranchus brockii* * Pleurobranchidae * Notaspidea * Opisthobranchia * Hong Kong * Cape d'Aguilar * 85 ft (25 m).

822 - *Pleurobranchus forskali* * Pleurobranchidae * Notaspidea * Opisthobranchia * Indonesia * Manado * 40 ft (12 m). This large opisthobranch can be very common in some areas. The species is variable in color with the extremes being almost positive and negative images of the same pattern. This can be seen in this and the following photograph.

823 - *Pleurobranchus forskali* * Pleurobranchidae * Notaspidea * Opisthobranchia * Indonesia * Manado * night * 40 ft (12 m). This is an alternate color pattern of *P. forskali*. Compared with the previous photograph, this is essentially a negative image of the pattern.

824 - *Pleurobranchus grandis* * Pleurobranchidae * Notaspidea * Opisthobranchia * Palau * German Channel * night * 20 ft (6 m). This opisthobranch has a spectacular hard to forget color pattern. The small commensal shrimp *Periclimenes imperator* can be found riding on the surface of this species.

825 - *Pleurobranchus peroni* * Pleurobranchidae * Notaspidea * Opisthobranchia * Indonesia * Manado * Ruang Island * 66 ft (20 m).

826 - *Pleurobranchus* sp. * Pleurobranchidae * Notaspidea * Opisthobranchia * Philippines * Cebu * Santa Rosa * seagrass bed * 6 ft (2 m). This photograph shows very clearly the side gills of the notaspids.

827 - *Andeadoris egrettae* * Dorididae * Nudibranchia * Opisthobranchia * Opisthobranchia * Papua New Guinea * Biak Island * reef * 50 ft (15 m). The nudibranchs are the final group of opisthobranchs, so named because of their naked gills which are often carried on the back of the animal. They are similar in appearance to other opisthobranchs and are easily confused with them. Nudibranchs have anterior rhinophores and a large flat foot. They are all carnivorous, feeding on a broad spectrum of benthic animals, including some of the most noxious species found in nature.

828 - *Sebadoris nubilosa* * Dorididae * Nudibranchia * Opisthobranchia * Philippines * Cebu * Mactan Island * 9 ft (2

822 - ***Pleurobranchus forskali*** * Indonesia

823 - ***Pleurobranchus forskali*** * Indonesia

824 - ***Pleurobranchus grandis*** * Palau

825 - ***Pleurobranchus peroni*** * Indonesia

826 - ***Pleurobranchus* sp.** * Philippines

827 - ***Andeadoris egrettae*** * Papua New Guinea

828 - ***Sebadoris nubilosa*** * Philippines

829 - ***Trippa intecta*** * Indonesia

830 - ***Asteronotus* sp.** * Papua New Guinea

831 - ***Asteronotus caespitosus*** * Indonesia

832 - ***Asteronotus* sp.** * Federated States of Micronesia

m). This nudibranch was relatively common in seagrass beds where it was photographed.

829 - *Trippa intecta* * Dorididae * Nudibranchia * Opisthobranchia * Indonesia * Manado * 33 ft (10 m).

830 - *Asteronotus* sp. * Asteronotidae * Nudibranchia * Opisthobranchia * Papua New Guinea * Duke of York Islands * Mioko Island * lagoon bottom * 33 ft (10 m).

831 - *Asteronotus caespitosus* * Asteronotidae * Nudibranchia * Opisthobranchia * Indonesia * Manado * fringing reef * 15 ft (5 m). This large species is cryptically colored, and warty on the upper surface. It is known to be widespread in the Indo-west Pacific.

832 - *Asteronotus* sp. * Asteronotidae * Nudibranchia * Opisthobranchia * Federated States of Micronesia * Nama Island * night * 30 ft (9 m). This dark *Asteronotus* contrasts sharply with the previous species, not having the warty surface.

833 - *Halgerda* sp. * Asteronotidae * Nudibranchia * Opisthobranchia * Papua New Guinea * West New Britain * 13 ft (4 m). The species of *Halgerda* all have a characteristic appearance with large conules or ridges on their backs.

834 - *Jorunna funebris* * Kentrodorididae * Nudibranchia * Opisthobranchia * Palau * Short drop off * 60 ft (18 m). This nudibranch has been observed feeding on the sponge, *Haliclona* cf. *coerulescens*.

835 - *Platydoris cruenta* * Platydoridae * Nudibranchia * Opisthobranchia * Palau * Lighthouse Reef * 6 ft (2 m). This species is often found on intertidal reef flats, under rocks or algae during the day.

836 - *Platydoris scabra* * Platydoridae * Nudibranchia * Opisthobranchia * Papua New Guinea * Madang * Kranket Island * lagoon * 10 ft (3 m).

837 - *Platydoris* sp. * Platydoridae * Nudibranchia * Opisthobranchia * Papua New Guinea * New Ireland * Kavieng * 80 ft (24 m). This is an undescribed species of *Platydoris*, recently found.

838 - *Ceratosoma moloch* * Chromodorididae * Nudibranchia * Opisthobranchia * Solomon Islands * Guadalcanal * Pt. Cruz * 66 ft (20 m).

839 - *Ceratosoma trilobata* * Chromodorididae * Nudibranchia * Opisthobranchia * Papua New Guinea * New Britain * 100 ft (30 m).

840 - *Chromodoris albopunctata* * Chromodorididae * Nudibranchia * Marshall Islands * Enewetak Atoll * barrier reef * 60 ft (18 m). This is a large genus of nudibranchs, with unlimited color patterns. Many new species of *Chromodoris* are still being found.

833 - ***Halgerda* sp.** * Papua New Guinea

834 - ***Jorunna funebris*** * Palau

835 - *Platydoris cruenta* * Palau

836 - *Platydoris scabra* * Papua New Guinea

837 - *Platydoris* sp. * Papua New Guinea

838 - *Ceratosoma moloch* * Solomon Islands

839 - *Ceratosoma trilobata* * Papua New Guinea

840 - *Chromodoris albopunctata* * Marshall Islands

841 - *Chromodoris annae* * Papua New Guinea

842 - *Chromodoris annulata* * Papua New Guinea

843 - ***Chromodoris coi*** * Papua New Guinea

844 - ***Chromodoris elizabethina*** * Marshall Islands

841 - *Chromodoris annae* * Chromodorididae * Nudibranchia * Papua New Guinea * Madang * lagoon patch reef * 40 ft (12 m).

842 - *Chromodoris annulata* * Chromodorididae * Nudibranchia * Papua New Guinea * Madang * lagoon * 30 ft (9 m).

843 - *Chromodoris coi* * Chromodorididae * Nudibranchia * Papua New Guinea * Madang * lagoon reef * 50 ft (15 m).

844 - *Chromodoris elizabethina* * Chromodorididae * Nudibranchia * Marshall Islands * Enewetak Atoll * 20 ft (6 m).

845 - *Chromodoris kunei* * Chromodorididae * Nudibranchia * Papua New Guinea * New Britain * patch reef * 40 ft (12 m). This photo shows the nudibranch on *Padina* algae.

846 - *Chromodoris lineolata* * Chromodorididae * Nudibranchia * Papua New Guinea * Manam Island * 60 ft (18 m).

847 - *Chromodoris lochi* * Chromodorididae * Nudibranchia * Indonesia * Manado * 60 ft (18 m).

848 -*Chromodoris magnifica* * Chromodorididae * Nudibranchia * Philippines * Batangas * patch reef * 40 ft (12 m).

849 - *Chromodoris reticulata* * Chromodorididae * Nudibranchia * Papua New Guinea * Madang * 33 ft (10 m).

850 - *Chromodoris willani* * Chromodorididae * Nudibranchia * Philippines * Batangas * patch reef * 50 ft (15 m).

851 - *Chromodoris* cf. *tinctoria* * Chromodorididae * Nudibranchia * Marshall Islands * Kwajalein Atoll * harbor * 15 ft (5 m).

852 - *Chromodoris* sp. * Chromodorididae * Nudibranchia * Indonesia * Manado * 60 ft (18 m).

845 - ***Chromodoris kunei*** * Papua New Guinea

847 - ***Chromodoris lochi*** * Indonesia

846 - ***Chromodoris lineolata*** * Papua New Guinea

848 - ***Chromodoris magnifica*** * Philippines

849 - ***Chromodoris reticulata*** * Papua New Guinea

850 - ***Chromodoris willani*** * Philippines

851 - ***Chromodoris* cf. *tinctoria*** * Marshall Islands

852 - ***Chromodoris* sp.** * Indonesia

853 - ***Chromodoris* sp.** * Indonesia

854 - ***Chromodoris* sp.** * Papua New Guinea

855 - ***Chromodoris* sp.** * Philippines

856 - ***Chromodoris* sp.** * Philippines

857 - ***Dendrodoris tuberculosa*** * Papua New Guinea

858 - ***Glossodoris atromarginata*** * Philippines

859 - ***Hypselodoris festiva*** * Hong Kong

853 - *Chromodoris* sp. * Chromodorididae * Nudibranchia * Indonesia * Manado * 40 ft (12 m).

854 - *Chromodoris* sp. * Chromodorididae * Nudibranchia * Opisthobranchia * Papua New Guinea * 110 ft (33 m).

855 - *Chromodoris* sp. * Chromodorididae * Nudibranchia * Opisthobranchia * Philippines * Cebu * Mactan Island * 30 ft (9 m).

856 - *Chromodoris* sp. * Chromodorididae * Nudibranchia * Opisthobranchia * Philippines * Batangas * 40 ft (12 m).

857 - *Dendrodoris tuberculosa* * Chromodorididae * Nudibranchia * Opisthobranchia * Papua New Guinea * Duke of York Islands * Mioko Island * 30 ft (9 m). This species is large, firm and rubbery. It can produce a toxic secretion which is irritating to human eyes.

858 - *Glossodoris atromarginata* * Chromodorididae * Nudibranchia * Opisthobranchia * Philippines * Batangas * 60 ft (18 m).

859 - *Hypselodoris festiva* * Chromodorididae * Nudibranchia * Opisthobranchia * Hong Kong * Cape d'Aguilar * 50 ft (15 m).

860 - *Hypselodoris kanga* * Chromodorididae * Nudibranchia * Opisthobranchia * Federated States of Micronesia * Chuuk * lagoon bottom * 100 ft (30 m).

861 - *Hypselodoris mandadilus* * Chromodorididae * Nudibranchia * Opisthobranchia * Marshall Islands * Enewetak Atoll * lagoon * 10 ft (3 m).

862 - *Miamira sinuata* * Chromodorididae * Nudibranchia * Opisthobranchia * Palau * Ngerkuul Pass * 100 ft (30 m)

863 - *Risbecia imperialis* * Chromodorididae * Nudibranchia * Opisthobranchia * Papua New Guinea * West New Britain * Kimbe Bay * 100 ft (30 m).

864 - *Reticulidia fungia* * Chromodorididae * Nudibranchia * Opisthobranchia * Federated States of Micronesia * Chuuk * Rio de Janeiro Maru * 60 ft (18 m). On sponge.

865 - *Hexabranchus sanguineus* * Hexabranchidae * Nudibranchia * Opisthobranchia * Philippines * Cebu * Mactan Island * 33 ft (10 m). The spanish dancer is found from the Red Sea to the western Pacific. It is one of the largest nudibranchs, reaching over 12 inches (30 cm) and four pounds in weight. The species swims, if disturbed, by undulating its body in a front to rear motion.

866 - *Hexabranchus sanguineus* * Hexabranchidae * Nudibranchia * Opisthobranchia * Papua New Guinea * Manam Island * 85 ft (25 m). There is some question as to whether there might be more than one species of the spanish dancer because a significant amount of color variation exists. This photo, along with the

860 - ***Hypselodoris kanga*** * Federated States of Micronesia

861 - ***Hypselodoris mandadilus*** * Marshall Islands

862 - *Miamira sinuata* * Palau

863 - *Risbecia imperialis* * Papua New Guinea

864 - *Reticulidia fungia* * Federated States of Micronesia

865 - *Hexabranchus sanguineus* * Philippines

866 - *Hexabranchus sanguineus* * Papua New Guinea

867 - *Hexabranchus sanguineus* eggs * Papua New Guinea

868 - *Fryaria ruppeli* * Indonesia

869 - *Phyllidia babai* * Papua New Guinea

previous, demonstrates this. It is possible that much of the color variation can be attributed to changes during growth, rather than specific differences.

867 - *Hexabranchus sanguineus* eggs * Hexabranchidae * Nudibranchia * Opisthobranchia * Papua New Guinea * Manam Island * 66 ft (20 m). Nudibranch eggs are often brightly colored and laid in exposed areas in a spiral ring.

868 - *Fryeria ruppeli* * Phyllidiidae * Nudibranchia * Opisthobranchia * Indonesia * Manado * reef * 30 ft (9 m).

869 - *Phyllidia babai* * Phyllidiidae * Nudibranchia * Opisthobranchia * Papua New Guinea * Dyaul Island * 70 ft (21 m). Unlike most nudibranchs, *Phyllidia* have the gills located under the lateral mantle margins. This is evident here and in the following *Phyllidia* spp. photos. The rhinophores are still present and there is various sculpturing on the dorsal surface.

870 - *Phyllidia carlsonhoffi* * Phyllidiidae * Nudibranchia * Opisthobranchia * Fiji * Kaimbu Island * lagoon * 5 ft (1.5 m).

871 - *Phyllidia elegans* * Phyllidiidae * Nudibranchia * Opisthobranchia * Philippines * Batangas * 30 ft (9 m).

872 - *Phyllidia madangensis* * Phyllidiidae * Nudibranchia * Opisthobranchia * Papua New Guinea * Madang * lagoon * 80 ft (24 m).

873 - *Phyllidia ocellata* * Phyllidiidae * Nudibranchia * Opisthobranchia * Palau * Idim's Corner * 50 ft (15 m). This, and the following two pictures, shows color variation in what is known as *P. ocellata*. These color variations could be due to diet, or perhaps the nudibranchs actually represent separate species.

874 - *Phyllidia* cf. *ocellata* * Phyllidiidae * Nudibranchia * Opisthobranchia * Papua New Guinea * Albatross Channel * 100 ft (30 m).

870 - *Phyllidia carlsonhoffi* * Fiji

871 - *Phyllidia elegans* * Philippines

872 - *Phyllidia madangensis* * Papua New Guinea

874 - *Phyllidia* cf. *ocellata* * Papua New Guinea

873 - *Phyllidia ocellata* * Palau

875 - *Phyllidia* cf. *ocellata* * Federated States of Micronesia

876 - ***Phyllidia pustulosa*** * Philippines

877 - ***Phyllidia tula*** * Indonesia

878 - ***Phyllidia varicosa*** * Philippines

879 - ***Phyllidia* sp.** * Papua New Guinea

880 - ***Phyllidia* sp.** * Papua New Guinea

881 - ***Phyllidia* sp.** * Papua New Guinea

882 - ***Phyllidia* sp.** * Philippines

883 - ***Phyllidiella* sp.*** Papua New Guinea

884 - *Phyllidiopsis shirenae* * Papua New Guinea

885 - *Nembrotha cristata* * Indonesia

886 - *Nembrotha cristata* * Papua New Guinea

887 - *Nembrotha kubaryana* * Indonesia

875 - *Phyllidia* cf. *ocellata* * Phyllidiidae * Nudibranchia * Opisthobranchia * Federated States of Micronesia * Chuuk Atoll * lagoon reef * 100 ft (30 m).

876 - *Phyllidia pustulosa* * Phyllidiidae * Nudibranchia * Opisthobranchia * Philippines * Pamalican Island * 30 ft (9 m). This species is abundant in areas of live coral in Hawaii. There are well over fifty described species of *Phyllidia* and many mor that are undescribed. Many of these species superficially resemble each other. Most *Phyllidia* feed on sponges and they have few, if any, oredators. When removed from the water most have a characteristic, pungent odor.

877 - *Phyllidia tula* * Phyllidiidae * Nudibranchia * Opisthobranchia * Indonesia * Biak Island * reef * 50 ft (15 m).

878 - *Phyllidia varicosa* * Phyllidiidae * Nudibranchia * Opisthobranchia * Philippines * Pamalican Island * 40 ft (12 m). This species is common in Hawaii and the Indo-west Pacific.

879 - *Phyllidia* sp. * Phyllidiidae * Nudibranchia * Opisthobranchia * Papua New Guinea * Kavieng * 70 ft (21 m).

880 - *Phyllidia* sp. * Phyllidiidae * Nudibranchia * Opisthobranchia * Papua New Guinea * Madang * reef * 100 ft (30 m).

881 - *Phyllidia* sp. * Phyllidiidae * Nudibranchia * Opisthobranchia * Papua New Guinea * Madang * lagoon patch reef * 30 ft (9 m).

882 - *Phyllidia* sp. * Phyllidiidae * Nudibranchia * Opisthobranchia * Philippines * Cebu * Mactan Island * reef * 33 ft (10 m).

883 - *Phyllidiella* sp. * Phyllidiidae * Nudibranchia * Opisthobranchia * Papua New Guinea * Madang * Barracuda Point * 50 ft (15 m).

884 - *Phyllidiopsis shirenae* * Phyllidiidae * Nudibranchia * Opisthobranchia * Papua New Guinea * Madang * barrier reef * 100 ft (30 m).

885 - *Nembrotha cristata* * Polyceridae * Nudibranchia * Opisthobranchia * Indonesia * Manado * algal flat * 3 ft (1 m). This, and the next photo, shows different color forms of the same species.

886 - *Nembrotha cristata* * Polyceridae * Nudibranchia * Opisthobranchia * Papua New Guinea * New Britain * Kimbe Bay * 60 ft (18 m).

887 - *Nembrotha kubaryana* * Polyceridae * Nudibranchia * Opisthobranchia * Indonesia * Manado * fringing reef * 50 ft (15 m).

888 - *Nembrotha lineolata* * Polyceridae * Nudibranchia * Opisthobranchia * Papua New Guinea * Madang * barrier reef * 60 ft (18 m). This nudibranch is on the sponge *Gelliodes fimbriata*.

889 - *Nembrotha purpureolineata* * Polyceridae * Nudibranchia * Opisthobranchia * Indonesia * Manado * vertical wall * 110 (33 m).

890 - *Nembrotha* sp. * Polyceridae * Nudibranchia * Opisthobranchia * Philippines * Batangas * Pulang Buli * 20 ft (6 m).

891 - *Nembrotha* sp. * Polyceridae * Nudibranchia * Opisthobranchia * Indonesia * Manado * 130 ft (36 m). This pair of nudibranchs was not disturbed for the photograph. It is common to find a pair or several individuals of a species very close to one another, probably for ease in mating. Since nudibranchs are hermaphrodites, these are not male-female pairs. The spectacular color pattern of this pair is probably warning coloration intended to make them conspicuous as a bad bargain to any potential predators.

888 - *Nembrotha lineolata* * Papua New Guinea

889 - *Nembrotha purpureolineata* * Indonesia

890 - *Nembrotha* sp. * Philippines

891 - *Nembrotha* sp. * Indonesia

892 - *Tambja morosa* * Federated States of Micronesia

893 - *Notodoris minor* * Papua New Guinea

894 - *Notodoris minor* eggs * Papua New Guinea

895 - *Notodoris* sp. * Philippines

896 - *Notodoris* sp. * Papua New Guinea

897 - *Melibe fimbriata* * Philippines

898 - *Bornella anguilla* * Marshall Islands

899 - *Armina sp.* * Papua New Guinea

892 - *Tambja morosa* * Polyceridae * Nudibranchia * Opisthobranchia * Federated States of Micronesia * Chuuk * lagoon * 85 ft (25 m).

893 - *Notodoris minor* * Aegiridae * Nudibranchia * Opisthobranchia * Papua New Guinea * Madang * lagoon * 20 ft (6 m). These yellow nudibranchs are known in the aquarium trade as "banana slugs" because of their yellow color. They can be locally quite abundant.

894 - *Notodoris minor* eggs * Aegiridae * Nudibranchia * Opisthobranchia * Papua New Guinea * Dyaul Island * 20 t (6 m). Like their parents, the eggs of *Notodoris minor* are yellow. The color makes them very distinctive on the reef and is probably intended to warn potential predators of their distasteful nature. It is common for nudibranch eggs to resemble the color of their parent species.

895 - *Notodoris minor* * Aegiridae * Nudibranchia * Opisthobranchia * Papua New Guinea * Madang * lagoon * 30 ft (9 m).

896 - *Notodoris* sp. * Aegiridae * Nudibranchia * Opisthobranchia * Philippines * Puerto Princessa * lagoon * 50 ft (15 m). This species of *Notodoris* is not as brightly colored, but still has the characteristic shape of the genus.

897 - *Melibe fimbriata* * Tethydidae * Nudibranchia * Opisthobranchia * Philippines * Cebu * Mactan Island * 6 ft (2 m). This nudibranch has some remarkable feeding behavior. It has an extensible oral hood, which can be seen in the photograph appearing like a translucent bubble, which it uses as a throw net. It covers and engulfs small crustaceans with the oral hood and then ingests them.

898 - *Bornella anguilla* * Bornellidae * Nudibranchia * Opisthobranchia* Marshall Island * Enewetak * lagoon pinnacle * 35 ft (11 m). This nudibranch is nocturnally active, hiding in caves or under rocks during the day. It feeds on hydroids and when disturbed shifts to swimming behavior. It swims in the same manner as an eel, what is called anguilliform swimming (hence the specific name), and appears to be a fish, rather than a mollusc! The species is known from South Africa to the Marshall Islands, but is not known to occur in Hawaii (where its close relative *B. adamsii* occurs).

899 - *Armina sp.* * Arminidae * Nudibranchia * Opisthobranchia * Papua New Guinea * Manam Island * reef slope * 90 ft (27 m). Members of *Armina* have long, flat tapering bodies and are more commonly found in temperate areas than the tropics.

900 - *Armina* sp. * Arminidae * Nudibranchia * Opisthobranchia * Palau * Ngerdewais * mud bottom * 60 ft (18 m). This colorful species was found on a silty muddy bottom in a bay in Palau. These bays are rich in unique marine life that is never found on reefs in clear water.

901 - *Flabellina exoptata* * Flabellinidae * Nudibranchia * Opisthobranchia * Marshall Islands * Kwajalein Atoll * Roi Namur Pass * 40 ft (12 m). This nudibranch is probably a hydroid feeder like most other aeolids. When they eat cnidarians, somehow they manage to avoid triggering the stinging cells (nematocysts) of the cnidarian and then concentrate these undischarged nematocysts in the cerata along the back (which are actually outpocketings of the gut). Here they will discharge if the nudibranch is touched.

902 - *Cuthona* cf. *sibogae* * Tergepedidae * Nudibranchia * Opisthobranchia * Indonesia * Manado * 40 ft (12m).

903 - *Phyllodesmium briareus* * Glaucidae * Nudibranchia * Opisthobranchia * Papua New Guinea * Madang * 50 ft (15 m).

904 - *Phyllodesmium longicirra* * Glaucidae * Nudibranchia * Opisthobranchia * Papua New Guinea * barrier reef * 36 ft (11 m). This nudibranch was observed feeding on the soft coral, *Sarcophyton*.

905 - *Pteraeolidia ianthina* * Glaucidae * Nudibranchia * Opisthobranchia * Marshall Islands * Medren Pinnacle * 40 ft

900 - ***Armina* sp.** * Palau

901 - ***Flabellina exoptata*** * Marshall Islands

902 - ***Cuthona* cf. *sibogae*** * Indonesia

903 - ***Phyllodesmium briareus*** * Papua New Guinea

904 - ***Phyllodesmium longicirra*** * Papua New Guinea

905 - ***Pteraeolidia ianthina*** * Marshall Islands

906 - ***Pteraeolidia* sp.** * Philippines

907 - ***Arca ventricosa*** * Papua New Guinea

908 - *Lithophaga zittelliana* * Federated States of Micronesia

909 - *Lithophaga zittelliana* * Federated States of Micronesia

910 - **mussel** * Palau

911 - *Atrina pectinata* * Indonesia

(12 m). This species is the largest aeolid in Hawaii, but it is still relatively small, only a few inches long and very slender. It is widespread in the Indo-Pacific. It eats hydroids, including the widespread *Halichordyle disticha*, and stores the nematocysts for future use. This species also contains zooxanthellae, which may account for many of the color differences found among individuals.

906 - *Pteraeolidia* sp. * Glaucidae * Nudibranchia * Opisthobranchia * Philippines * Batangas * Pulang Buli * 60 ft (18 m).

907 - *Arca ventricosa* * Arcidae * Bivalvia * Papua New Guinea * West New Britain * Kimbe Bay * 40 ft (12 m). This bivalve occurs on coral heads where it forms depressions in the coral head into which the clam pulls if threatened or disturbed.

908 - *Lithophaga zittelliana* * Mytilidae * Bivalvia * Federated States of Micronesia * Chuuk * Dublon Island * 20 ft (6 m). There are more than a dozen *L. zittelliana* present in this photograph. Their presence is revealed by the openings of their siphons on the surface of the coral head. This mussel burrows into coral heads, producing tunnels which greatly weaken the coral and, combined with other burrowers, can cause the eventual death of the coral.

909 - *Lithophaga zittelliana* * Mytilidae * Bivalvia * Federated States of Micronesia * Chuuk * Éten Island * 40 ft (12 m). This photograph shows the animal and shell of *L. zittelliana* in an accidentally broken open coral head. The mussel is normally found deeply buried in the coral head with only the dark openings of siphons visible externally. The siphons, with their dark ends, are long and the shell remains in the tunnel-like burrow.

910 - mussel * Mytilidae * Bivalvia * Palau * Jellyfish lake * 3 ft (1 m). Mussels are not particularly common organisms on coral reefs in the Pacific, but in some of the marine lakes of Palau they are abundant; the conditions in the lakes differ enough from the outside ocean to allow them to flourish.

911 - *Atrina pectinata* * Pinnidae * Bivalvia * Indonesia * Banka Island * 10 ft (3 m). It occurs in areas of sand and mud in and around reefs, particularly in inshore areas rich in food for this filter-feeding bivalve.

912 - *Atrina vexillum* * Pinnidae * Bivalvia * Federated States of Micronesia * Chuuk * Dublon Island * 10 ft (3 m). This pen shell reaches nearly 19 inches (480 mm) in length. The mantle of this pen shell is clearly visible inside the opening of the valves.

913 - *Pinctada margaritifera* * Pteriidae * Bivalvia * Federated States of Micronesia * Chuuk * lagoon reef * 20 ft (6 m). The black-lip pearl oyster is the major pearl oyster of the Pacific and is found from the western Indian Ocean to the western Pacific.

914 - *Pinctada maxima* * Pteriidae * Bivalvia * Papua New Guinea * Madang * lagoon * 66 ft (20 m). This is another species which contains pearls, but is not nearly as common as the black-lip pearl oyster. The shell of *P. maxima* was and still is highly valued in the coastal and highland areas of Papua New Guinea where it is used for personal ornamentation.

915 - *Pteria penguin* * Pteriidae * Bivalvia * Federated States of Micronesia * Chuuk * lagoon * 90 ft (27 m). The winged oyster is found on gorgonians, sea fans and black corals and can reach sizes larger than a hand. It is often used in local handicrafts.

916 - *Pteria* sp. * Pteriidae * Bivalvia * Federated States of Micronesia * Chuuk * lagoon reef * 80 ft (24 m). This species also occurs on gorgonians, but is much smaller than *P. penguin*.

917 - *Pedum spondyloideum* * Pectinidae * Bivalvia * Palau * Rock Island * 10 ft (3 m). This scallop is found deeply embedded in coral heads with only the opening of the valves visible.

918 -*Spondylus* sp. * Spondylidae * Bivalvia * Federated States of Micronesia * Chuuk * Shinkoku Maru * 85 ft (25 m). The spiny oysters are known for their heavy valves with long, strong spines on the outer surface.

912 - ***Atrina vexillum*** * Federated States of Micronesia

913 - ***Pinctada margaritifera*** * Federated States of Micronesia

914 - ***Pinctada maxima*** * Papua New Guinea

915 - ***Pteria penguin*** * Federated States of Micronesia

916 - ***Pteria* sp.** * Federated States of Micronesia

917 - ***Pedum spondyloideum*** * Palau

918 - ***Spondylus* sp.** * Federated States of Micronesia

919 - ***Spondylus* sp.** * Indonesia

920- ***Spondylus* sp.** * Papua New Guinea

921 - ***Isognomon* sp.** * Papua New Guinea

922 - ***Malleus malleus*** * Papua New Guinea

919 -*Spondylus* sp. * Spondylidae * Bivalvia * Indonesia * Biak Island * reef * 40 ft (12 m).

920 -*Spondylus* sp. * Spondylidae * Bivalvia * Papua New Guinea * Dyaul Island * 66 ft (20 m). It is not uncommon for a bright orange thin encrusting sponge to grow on the outer shell of some *Spondylus*.

921 -*Isognomon* sp. * Isognomonidae * Bivalvia * Papua New Guinea * Manam Island * 33 ft (10 m).

922 - *Malleus malleus* * Malleidae * Bivalvia * Papua New Guinea * Manam Island * 10 ft (3 m). This is commonly known as the hammer oyster.

923 - *Lima* sp. * Limidae * Bivalvia * Hong Kong * Cape d'Aguilar * 20 ft (6 m). The file shells have long fingers of mantle extending out. They are capable of jetting away if disturbed.

924 - *Hyotissa hyotis* * Ostreidae * Bivalvia * Federated States of Micronesia * Chuuk * Shinkoku Maru * 66 ft (20 m). The ribs of *hyotis* sp.do not form a zig-zag pattern that is as uniform as in another species, *Lopha cristagalli*.

925 - *Hyotissa* sp. * Ostreidae * Bivalvia * Papua New Guinea * New Ireland * Kalili Plantation * 50 ft (15 m).

926 - *Lopha cristagalli* * Ostreidae * Bivalvia * Federated States of Micronesia * Chuuk * Fujikawa Maru * 40 ft (12 m). The ribs on *Lopha cristagalli* are very angular, with a regualr zig-zag pattern.

927 - *Lopha frons Ostreidae * Bivalvia * Hong Kong * Shek Ngau Chau * 33 ft (10 m).** This species has a velvety black mantle with a striking thin white edge. It was found on the under surfaces of huge blocks of rock on an offshore island in Mirs Bay, Hong Kong. We have not yet been able to identify it, but it is almost certainly a member of the oysters (Ostreidae).

928 - *Alectryonella plicata* * Ostreidae * Bivalvia * Philippines * Pamalican Island * intertidal. This bivalve is found around the intertidal level on rock . It has one valve firmly cemented to rock and the other free so it is extremely resistant to wave action.

929 - *Chama lazarus* * Chamidae * Bivalvia * Federated States of Micronesia * Chuuk * lagoon * 77 ft (23 m). These rock oysters form beautiful frilly fronds.

930 - *Chama* sp. * Chamidae * Bivalvia * Palau * lagoon reef * 40 ft (12 m). This rock oyster has a very heavily calicfied shell and does not have any projections.

931 - Unidentified bivalve * Galeommatidae * Bivalvia * Papua New Guinea * Madang * lagoon * 10 ft (3 m). This unusual small bivalve was found at the entrance to a mantis shrimp (stomatopod) burrow. The family is poorly known but typically they are commensal with a variety of invertebrates. This individual cannot even be placed into a genus.

932 - *Hippopus hippopus* * Tridacnidae * Bivalvia * Palau * Koror * 15 ft (5 m). The tissue of giant clams in the genus *Hippopus* does not extend over the edges of the shell. The razor sharp edges of the valves close so tightly that any flesh caught between the valves would be severely cut. This species can reach sixteen inches in length, and occurs in the western Pacific. There are two species in the genus, *H. hippopus* and *H. porcellanus*, the latter which is restricted to the Philippines and nearby areas..

933 - *Tridacna terove* * Tridacnidae * Bivalvia * Fiji * Kaimbu Island * 20 ft (6 m). This species of giant clam is known only from Fiji and Tonga. It is a medium-sized giant clam with a rather plain shell. The mantle is brownish-gray with small protuberances.

934 - *Tridacna crocea* * Tridacnidae * Bivalvia * Philippines * Pamalican Island * 13 ft (4 m). This is the smallest giant clam species, growing only to about six inches in length. It bores deeply into coral boulders and reef so that only the upper edge of the shell can be seen. Its mantle is brightly colored, often with iridescent blues and greens. It can be found in high densities in some areas, and is often intertidal. It occurs in the western Pacific.

935 - *Tridacna maxima* * Tridacnidae * Bivalvia * Palau * German Channel * 10 ft (3 m). This giant clam can be easily confused with *T. cro-*

923 - ***Lima* sp.** * Hong Kong

924 - ***Hyotissa hyotis*** * Federated States of Micronesia

925 - ***Hyotissa* sp.** * Papua New Guinea

926 - ***Lopha cristagalli*** * Federated States of Micronesia

927 - ***Lopha frons*** * Hong Kong

928 - ***Alectryonella plicata*** * Philippines

929 - ***Chama lazarus*** * Federated States of Micronesia

930 - ***Chama* sp.** * Palau

931 - Galeommatidae * Papua New Guinea

932 - *Hippopus hippopus* * Palau

933 - *Tridacna terove* * Fiji

934 - *Tridacna crocea* * Philippines

cea and *T. squamosa*. It has numerous close, set scutes on the shell valves, and most often bores into the reef, though not as deeply as *T. crocea*, so that the scutes are still visible. It reaches about fifteen inches in length. The mantle is variable in pattern and often brightly colored, though not usually iridescent. It has the widest distribution of all the giant clams, from east Africa to Polynesia.

936 - *Tridacna squamosa* * Tridacnidae * Bivalvia * Marshall Islands * Enewetak * 10 ft (3 m). *T. squamosa* can grow slightly larger than *T. maxima*, up to twenty inches. Its shell is covered with numerous well-spaced scutes, which distinguishes it from *T. maxima*. The mantle is also variable in pattern, and brightly colored. It is difficult to distinguish these two species by mantle color alone. *T. squamosa* is most often in the open, or nestled among loose coral, but it does not normally bore into the reef. It is also widely distributed throughout the Indo-Pacific.

937 - *Tridacna derasa* * Tridacnidae * Bivalvia * Fiji * Kaimbu Island * 10 ft (3 m). This is the second largest species of giant clam, growing up to two feet. Adult shells do not have protruding scutes, and are more elongate and narrow than *T. gigas*. The mantle is often striped or spotted, with brilliant colors.

938 - *Tridacna gigas* * Tridacnidae * Bivalvia * Marshall Islands * Enewetak Atoll * 10 ft (3 m). If any bivalve can be considered the embodiment of the tropical Pacific, it is the giant clam. The largest of all bivalves, it can reach weights (animal and shell) of over five hundred pounds. The shell makes up much of this weight. They can reach over three feet in length. *T. gigas* generally have a brown mantle with numerous iridescent blue-green circles. The shell valves are more distinctly ribbed than *T. derasa*, with no scutes. The tissue (mantle) is so thick that a large individual cannot close its shell valves tightly. The adductor muscle of these clams (which holds the two valves together) is highly prized in the Orient and illegal poaching of the species has occurred throughout the Pacific. This species has become extinct, or at least very rare, in many islands in the Pacific, and is on the CITES 'threatened' species list. Its natural distribution is the western Pacific.

939 - *Periglypta "clathrata"* * Veneridae * Bivalvia * Papua New Guinea * Kavieng * 40 ft (12 m).

940 - *Nautilus pompilius* * Nautilidae* Nautiloidea* Cephalopoda * Papua New Guinea * New Ireland * Albatross Channel * surface. Although the species of *Nautilus* usually occur at depths below safe diving, these endlessly fascinating creatures are regularly captured using traps at depths of 500-660 feet (150-200 m) and brought to the surface to be photographed by divers. They survive this upward trip in good condition and are later released at the drop off to return to the depths. They cannot tolerate water warmer than about 77° F (25° C) for long periods, but in areas where the water is relatively cool at diving depths, such as New Caledonia, they can actually be visited in their natural habitat by divers. *Nautilus pompilius* is believed to have the most widespread distributon, but several others species occur in the Pacific. *N. macropthalmus* occurs in Australia and New Caledonia. *Nautilus belauensis* is the largest species and is known only from the Palau Islands. The rarest and most enigmatic, *Nautilus scrobiculatus,* is known from the Bismarck Archipelago and was not seen alive until the last decade.

941 - *Metasepia pfefferi* *Sepiolidae * Sepioidea * Cephalopoda * Indonesia * Manado * 20 ft (6 m). This amazing little cuttlefish hardly looks like a cephalopod at all, but appears to be a sunken leaf or piece of algae. Like most cephalopods, it can change its color and appearance instantly, to be less conspicuous. This species is known from the Great Barrier Reef and Indonesia.

942 - *Sepia* sp.* Sepiidae* Sepioidea * Cephalopoda * Papua New Guinea * Madang * lagoon * 15 ft (5 m).

943 - *Sepia* sp. * Sepiidae * Sepioidea * Cephalopoda * Papua New Guinea * Port Moresby * Lion Island * 33 ft (10 m).

944 - *Sepia latimanus* * Sepiidae * Sepioidea * Cephalopoda * Papua New Guinea * Bagabag Island * 33 ft (10 m). Most cuttlefish are able to alter their color and texture, this species is no exception. *Sepia latimanus* is common throughout much of the region. It is easily regognized when displaying the yellow color pattern.

935 - ***Tridacna maxima**** Palau

936 - ***Tridacna squamosa**** Marshall Islands

938 - ***Tridacna gigas*** * Marshall Islands

937 - ***Tridacna derasa*** * Fiji

939 - ***Periglypta "clathrata"*** * Papua New Guinea

940 - ***Nautilus pompilius*** * Papua New Guinea

941 - ***Metasepia pfefferi*** * Indonesia

942 - ***Sepia* sp.** * Papua New Guinea

943 - ***Sepia* sp.** * Papua New Guinea

944 - ***Sepia latimanus*** * Papua New Guinea

945 - ***Sepia latimanus*** * Indonesia

945 - *Sepia latimanus* * Sepiidae * Sepioidea * Cephalopoda * Indonesia * Ruang Island * 33 ft (10 m).

946 - *Euprymna* sp. * Sepiolidae * Sepioidea * Cephalopoda * Palau * Lighthouse Reef * 10 ft (3 m). This species buries in the sand and covers itself completely, then jumps out to catch prey. It may belong in the genus *Idiosepious.*

947 - *Sepioteuthis lessoniana* * Teuthoidea * Cephalopoda * Indonesia * Manado * night * 60 ft (18 m). This is the common reef squid and it occurs throughout the region. Similar species occur throughout the tropics.

948 - *Sepioteuthis* eggs * Teuthoidea * Cephalopoda * Indonesia * Manado * 20 ft (6 m). These eggs cases each contain 5-6 developing squid. The shape of the egg case is characteristic of the genus.

949 - *Hapalochlaena lunulata* * Octopoda * Cephalopoda* Indonesia * Manado 20 ft (6 m. This octopus is commonly known as the blue-ringed octopus. The rings, however, are not always evident because of the octopus's ability to change colors rapidly. It is a small octopus, hiding under rocks and dead coral during the day, and seen out in the open at night. It feeds on crustaceans. It has an extremely venomous bite, which has been fatal to humans.

950 - *Octopus macropus* * Octopodidae * Octopoda * Cephalopoda * Bahrain * reef * night * 55 ft (17 m). The distinct white spots on this octopus make it one of the easier species to identify.

951 - *Octopus* sp. * Octopodidae * Octopoda * Cephalopoda * Federated States of Micronesia * Chuuk * Onang Island * 30 ft (9m). This octopus is usually observed at night when it swims near boat lights and docks. It is an oceanic species that occurs worldwide in the tropics.

952 - *Octopus* sp. * Octopodidae * Octopoda * Cephalopoda * Bahrain * 30 ft (9 m).

953 - *Octopus cyanea* * Octopodidae * Octopoda * Cephalopoda * Palau * 30 ft (9 m). This octopus is common on and around coral reefs throughout much of the region.

Above- *Octopus* sp. * Octopodidae * Octopoda * Cephalopoda * Papua New Guinea * Kavieng * Albatross Channel * night * 20 ft (6 m). This small unidentified *Octopus* was found on an open sand bottom at night. It immediately hid, very effectively, by pushing itself into the bottom. Despite the color differences, it blends into the bottom very well.

946 - ***Euprymna* sp.** * Palau

947 - ***Sepioteuthis lessoniana*** * Indonesia

948 - ***Sepioteuthis*** eggs * Indonesia

949 - ***Hapalochlaena lunulata*** * Indonesia

950 - ***Octopus macropus*** * Bahrain

951 - ***Octopus* sp.*** Federated States of Micronesia

952 - ***Octopus luten.*** * Philippines

953 - ***Octopus* cyanea** * Palau

Crustaceans

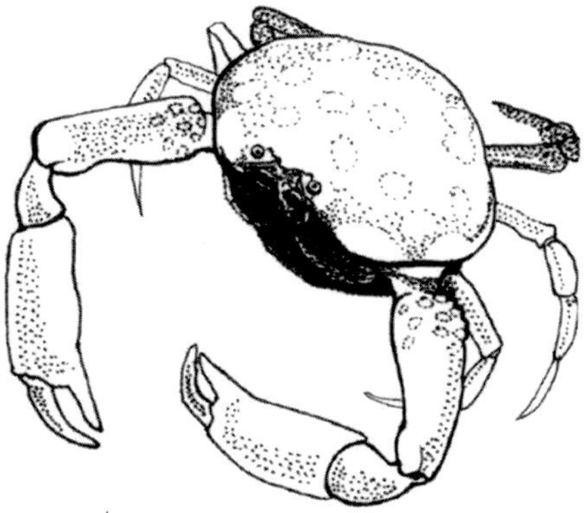

The Arthropods comprise the largest phylum of living creatures, with roughly one million species. Of these, the subphylum Crustacea is the only sizable group (perhaps 45,000 species) occurring in shallow tropical waters, a very small part of the entire Phylum. Crustaceans have an exoskeleton, with a carapace covering the central body of the animal, jointed appendages, two pairs antennae and compound eyes. Some authorities believe that the crustaceans form their own Phylum. Even if they do not, they still comprise one of the four main lines of Arthropod evolution. A few groups of marine arthropods, such as the pycnogonids, exist outside the Crustacea, (Chelicerata) but are included here.

The crustaceans are arranged in a complicated taxonomic hierarchy which is beyond the scope of this book. The different layers of subclasses, infraorders and other divisions all bring order to this diverse group. The majority of species considered here belong to the Class Malacostraca (lobsters, shrimp, crabs, etc.).

Most of the larger crustaceans found in the shallow Pacific tropics have planktonic larvae, which is one reason for the wide geographic distribution of most species. In crustaceans such as spiny lobsters, the planktonic larval stage can last as long as 6 months, ample time for currents to carry larvae thousands of miles. In other crustaceans planktonic larvae are short-lived or or absent, and these species have more restricted distributions.

The barnacles, members of the class Maxillopoda, subclass Cirripedia, superficially resemble molluscs, but are actually crustaceans. Sessile organisms, barnacles filter feed using their "legs" (cirri) to actively or passively strain food items from passing water. Barnacles are renown as fouling organisms on ship hulls and buoys, but many others occur as commensals or parasites. There are three orders of barnacles: the acorn and goose barnacles (Thoracica), the burrowing barnacles (Acrothoracica) and parasitic (on crabs) barnacles (Rhizocephala). A number of the sponges illustrated in this book contain barnacles buried in the tissue of the sponge so that only the opening of the barnacle is exposed. Virtually unnoticed until the sponge is broken open, these barnacles are poorly known. Some of the acorn barnacles, *Acasta* sp., also grow in sponges.

Opposite- The caridean shrimp *Rhynchocinetes uritai* is a common inhabitant of reef caves throughout the Indo-west Pacific region. The shrimp can number in the tens or hundreds in some caves, and are often found with other cave-dwelling shrimps. This photo was taken near Nama Island, Chuuk State, in the Federated States of Micronesia.

Top- The crab *Platypodia ceylonica*, photographed here at night, inhabits crevices and other sheltered areas of the reef. Center- *Dasycaris* sp. is one of many species of Palaemonid shrimps which live as commensals with other reef organisms. This small species lives on wire-like black coral. Bottom- This hermit crab occupying a small *Natica* shell is typical of the many small species which may occur in vast numbers in areas of reefs and seagrass.

The stomatopods or mantis shrimp (order Stomatopoda, class Maxillopoda) are highly successful predators living in cavities of coral and rock or within smooth-walled burrows found in sandy bottoms. There are about 400 species worldwide. Stomatopods are noted for their raptorial claws, which are capable of lightning fast strikes against potential prey. The larger species are capable of breaking open human skin, earning them the common name "thumbspliters".

The sand dwelling stomatopods, such as the colorful *Odontodactylus scyllarus*, can reach considerable size, as much as two feet in length. There are fisheries for these larger species. They are prepared and eaten like shrimp or small lobsters. Their burrows are large in diameter and usually have a mucous flap around their rim, which may help to make the circular opening less conspicuous to prey which they ambush. Prey are stabbed or clubbed with the raptorial chelae, then held against the mouth parts and shredded. The stalked compound eyes of stomatopods can be rotated independent of each other.

The order Decapoda, or decapods, as their name implies, have 10 legs arranged in five pairs. With some 10,000 described species, they have an incredible variety of forms and life habitats. The order contains the most conspicuous crustaceans, but many others are small and cryptic. Most are marine, but a limited number are fresh water or terrestrial. The class is divided into two orders, and several suborders, but many useful divisions of the decapods occur at the infraorder level (below suborder), which separate the penaeid shrimps, caridean shrimps, coral shrimps, ghost shrimps, spiny lobsters, hermit crabs and related crabs, and the true crabs.

The penaeid shrimps are commercially important, being common in shallow mud, sand, seagrass beds and reef flats. They are difficult to identify due to a large number of similar species. They, and the closely related sergestid shrimps, are the only decapods which shed their eggs free into the water. All other groups carry the eggs on the abdomen until hatching.

The carideans include most other decapods commonly called "shrimp" and number over 1,000 species worldwide. In our region these shrimps are found in seagrass beds, on reef flats, as commensals (see box), as cleaners and in all areas of the reef. These include the alpheids (snapping shrimps), the palaemonids (commensal shrimps) and various reef shrimps.

The coral shrimps, stenopodids, are cleaners, removing ectoparasites from fishes, and are found on reefs throughout the tropics. Their long white antennae make them unmistakable and they often occur in male-female pairs. They typically dwell in dark crevices and some species seldom venture where there is any appreciable light.

Above left- This small crab, *Quadrella* sp. lives on the octocoral *Siphonogorgia* sp. The crabs usually occur in pairs. Top right- The horseshoe crabs have survived for at least 400 million years and are often refered to as living fossils. They are the only living representatives of the Arthropod class Merostomata (not Crustacea). There are several horseshoe crab species, usually placed in the genus *Limulus,* they live on sandy or mud bottoms in temperate and tropical seas. Below right- This female reef shrimp, *Stenopus zanzibaricus* has many bluish-colored eggs on her abdomen. The genus *Stenopus* occurs throughout the tropics.

The ghost shrimps, or thalassinids, are extremely common in reef, seagrass and mangrove areas, but are essentially invisible. They live in complex burrow systems underlying virtually all sediment bottoms. Their presence is made apparent only by the conical mounds of sediment expelled from their burrows which dominate many sediment bottoms. In reef areas members of *Callianassa* predominate, while in mangrove areas *Thalassina* is found.

The spiny lobsters of the infraorder Palinura, a different infraorder than that of the Maine lobster *Homarus*, are the basis of important fisheries worldwide in the tropics and some temperate areas. They are members of the family Palinuridae, but in the original description of the genus, the generic name was misprinted as *Panulirus*, instead of the intended *Palinurus*, and the misprinted name, despite the intentions of the author, is now permanently attached to the genus. Also in the infraorder are the slipper lobsters, which have their second antennae modified in paddle-like structures. Lobsters are most active at night, some of the spiny lobsters come onto atoll reef flats at night from deeper water to feed. Spiny lobsters typically are cave-dwelling. Female lobsters carrying eggs beneath the abdomen are known as "berried".

The hermit crabs are not true crabs, but are more closely related to lobsters and squat lobsters. They live in the shells of gastropod molluscs, although a few other types of objects can be used. A few species spend the major part of the their lives on land. Among these is the largest hermit crab, the coconut crab *Birgus latro* which does not utilize a gastropod shell, except as a small juvenile. Like all land crabs, these land hermit eggs hatch and undergo their larval development in the sea. They return to land as juveniles and do not venture far from the ocean.

The squat lobsters, member of the Galatheids, are generally small commensals, found with crinoids, sea pens and others. They are often colored to closely resemble their host organism and can be found only by careful examination of the host.

The true crabs come in many varieties, depending on their living habits. Some are swift, predatory swimmers, the insides of their claws (chelae) lined with sharp teeth and their last legs modified as swimming paddles. Most slower species clamber over the bottom, scurrying on their legs, searching for food. Many crabs are intertidal, living on sandy beaches, mud flats and rocky shores. Those found on intertidal rocks are gen-

Commensal Crustaceans

Commensalism is the association of two organisms in which the smaller (the commensal) obtains some benefit from the larger without causing significant harmand in some cases both participants benefit. The two may share food, sometimes the commensal is carried about on the surface of the host (phoresy) or commensals may live inside the burrow of the larger (inquilinism). Crustacean commensals fit this range of relationships nicely.

Commensal crustaceans are found among the palaemonid shrimps, porcellanid crabs, squat lobsters (galatheids) and true crabs. Numerous examples from these groups are included in the following section. Other crustaceans, particularly members of *Stenopus, Periclimenes*, and *Lysmata*, are cleaners, they remove ectoparasites from the bodies of fishes which solicit this service.

Probably the most conspicuous crustaceans commensals are those that occur with sea anemones. Small porcellanid crabs of the genus *Neopetrolisthes* occur on the oral disk of *Stichodactylus* anemones, and scurry to shelter beneath the margin when threatened. Nearly transparent shrimps of *Periclimenes* occur on the oral disk, their presence revealed only by their small spots of bright colors. Certain species of hermit crabs in the genus *Dardanus (*upper left*)* place and carry sea anemones, *Calliactis* spp. and *Adamsia* spp. on their shells. These anemones are only found with hermit crabs.

With careful observation and search a diver can locate a further wealth of commensals, many of which make great photographic subjects. Among cnidarians, the jellyfish (upper right), soft coral, gorgonians, black corals and many fleshy stony corals have commensal crustaceans. One group of porcellanid crabs is found at the base of tubes of the cerianthid anemones. Among echinoderms, commensals are found with all types, usually on the oral disk, other areas of the body surface and even in the openings of the gut. Commensals are found on molluscs, which include large opisthobranchs. Some commensals are specific to one type of host, while others seems adaptable to a number of different species. For example, *Periclimenes soror* is found with both starfishes and sea cucumbers, while *Periclimens imperator* occurs with echinoderms and opisthobranchs.

While commensals generally do not harm their hosts, some crustaceans are actually beneficial to their hosts. The crabs of *Trapezia,* (lower left) and some snapping shrimps (alpheids), live deep among the branches of the corals *Pocillopora*. If a crown-of-thorns starfish approaches the coral colony, the crustaceans move to meet it. They pinch the delicate tube feet of the starfish and cause it to move rapidly away from their home coral. The crabs may be looking out for their own best interests, but the benefit to the coral is certainly apparent. Other alpheids live with burrow-dwelling gobies, the alpheid maintaining the burrow while the goby serves as the sentinel which provides mutual benefit (lower right).

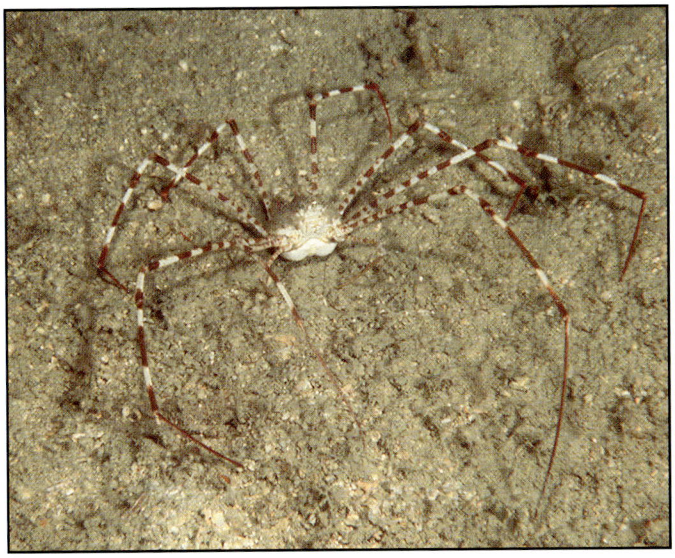

Left- This beautiful Spider crab, *Phalangipus* sp., from the Philippines, lives on muddy bottoms near patch reefs, in the Philippines. Middle- The urchin shrimp, *Stegopontonia commensalis* lives on the spines of *Astropyga radiata* and several other sea urchins. This individual was photographed at night in the Philippines. Right- A commensal shrimp in the genus *Periclimenes* lives among the tentacles of a sea anemone in Madang, Papua New Guinea. Species of this genus are found in associations with a wide variety of invertebrates.

erally fast and agile, scurrying over the rocks both above and below the water. The mud dwellers, typified by the fiddler crabs *Uca*, live in burrows they dig in the mud. Some slow species have developed various means of camouflage or deception. Some are cryptic, closely resembling a mixed algal bottom, where they hide. Other actually employ pieces of algae and invertebrates, attaching these materials to their exoskeleton (decorator crabs), to aid in their cryptic endeavors. A few species attach or hold, with their modified last legs, pieces of sponge on the top of the carapace to discourage predators with the distasteful sponge.

Finally, there are some other orders of non-decapod, generally small crustacea which are abundant on reefs, but not very apparent to human visitors. These include the mysid shrimps, amphipods, isopods and copepods. Mysids occur in dense schools in and around crevices and caves, looking more like baby fishes than crustaceans. Some copepods have similar habits, forming dense swarms of individuals even smaller than the mysids in the water above and near crevices of reefs. Amphipods are occasionally found on reefs, living on the bottom in groups. Isopods are most evident on reefs as parasites of fishes.

Above- This photograph shows an advanced larval stage of a slipper lobster in the process of settling on a reef after its life of several months in the plankton. The transparent nature of this and most crustacean larvae is a benefit for living in the open water of the planktonic environment.

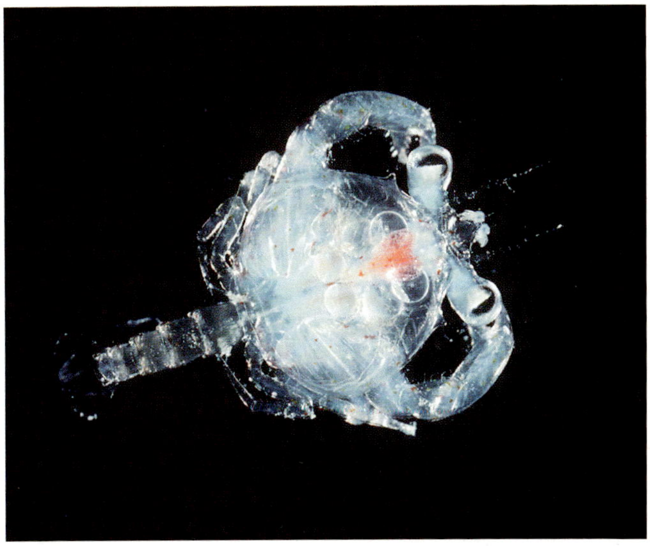

Above- This is the megalopa or last larval stage of a crab. The larva swims in the plankton, it is about one half inch in length. After it settles to the bottom, it may walk for a short time until it molts and completes the transformation into a juvenile crab. Such transparent plankton stages quickly produce pigment after taking up residence on the bottom.

954- Unidentified pycnogonid * Hawaii

955- Unidentified pycnogonid * Hawaii

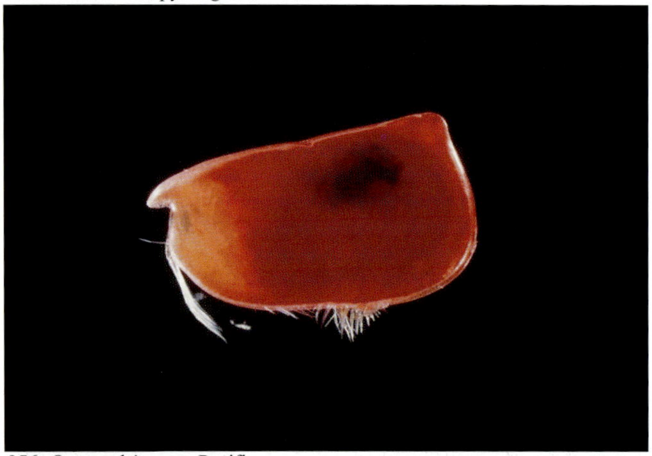
956- Ostracod * open Pacific

957- ***Lironeca*** **sp.*** Hawaii

954 - Unidentified pycnogonid * Pycnogonida * Chelicerata * Hawaii * Puako * on sponge * 20 ft (6 m). Many pycnogonids are small, like these in the photograph, and are tricky to spot. This individual was found living on a sponge.

955 - Unidentified pycnogonid * Pycnogonida * Chelicerata * Hawaii * Kewalo * 30 ft (9 m). This pycnogonid was found on the undersurface of a rock. Part of a brachiopod shell is visible in the upper corner.

956 - Unidentified Ostracod * Ostracoda * Crustacea * Open Pacific Clam-like in appearance, ostracods have a bivalved carapace that is hinged at the top. Most ostracods are less than one half inch in length. This one was photographed over deep water at night. Some species of ostracods are brightly bioluminescent.

957 - *Lironeca* sp.* Isopoda * Crustacea * Hawaii * Puako * night * 30 ft (9 m). Some of the more apparent isopods in shallow water are those found as external parasites on fishes. This one is on the pectoral fin of a porcupine fish *Diodon.*

958 - Unidentified isopod * Isopoda * Crustacea * Papua New Guinea * Madang * barrier reef * 20 ft (6 m). This is another parasitic isopod, in this case on the head of a soldierfish of the genus *Myripristis*.

959 - Unidentified amphipod * Amphipoda * Crustacea * Federated States of Micronesia * Chuuk * reef * 60 ft (18 m). These isopods are tiny, less than 1/4 of an inch long, but they are easily seen by careful observers. They usually occur on sponges. these amphipods actually appear to be red in color at depth, using a strobe they photograph as green in color.

960 - *Caprella* sp. * Caprellidae * Amphipoda * Crustacea * Philippines * Batangas * Pulang Buli Island * 30 ft (9 m). Commonly known as skeleton shrimp, these small amphipods are found clinging to algae and other material growing on the reef and seagrass beds. They are highly modified to cling to their host substrata.

961 - *Lepas* sp. * Cirripedia * Crustacea * Federated States of Micronesia * Chuuk * Pizion Reef * 100 ft (30 m).

962 - Unidentified barnacle * Cirripedia * Crustacea * Federated States of Micronesia * Chuuk * Pizion Reef * 100 ft (30 m). These gooseneck barnacles are growing on a black coral on the outer reef dropoff.

963 - *Megabalanus* sp. * Cirripedia * Crustacea * Federated States of Micronesia * Chuuk * Pis Moen Channel * 20 ft (6 m). This is a large barnacle for tropical waters, typically found growing on *Millepora platyphilla*.

964 - Unidentified barnacle * Cirripedia * Crustacea * Philippines * Batangas * reef * 20 ft (6 m).

965 - Unidentified barnacle * Cirripedia * Crustacea * Coral barnacle * Federated States of Micronesia * Chuuk * Anaw Wall * 40 ft (12 m).

966 - Unidentified barnacle * Cirripedia * Crustacea * Coral barnacle * Federated States of Micronesia * Chuuk * barrier reef * 30 ft (9 m).

967 - *Heteromysis* sp. * Mysidacea * Crustacea * Federated States of Micronesia * Chuuk * lagoon reef * 50 ft (15 m). Most shallow water mysids are found swimming near overhangs on patch reefs. They may occur in large numbers and supeficially resemble schools of larval fish. The ones in this photograph are less than a quarter of an inch in length.

958- Unidentified isopod * Papua New Guinea

959- Unidentified amphipod * Federated States of Micronesia

960- ***Caprella* sp.** * Philippines

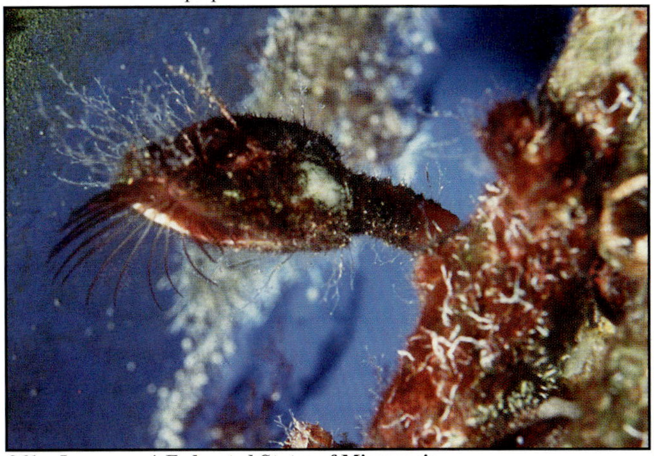

961- ***Lepas* sp.** *** Federated States of Micronesia**

962- Unidentified Barnacle * Federated States of Micronesia

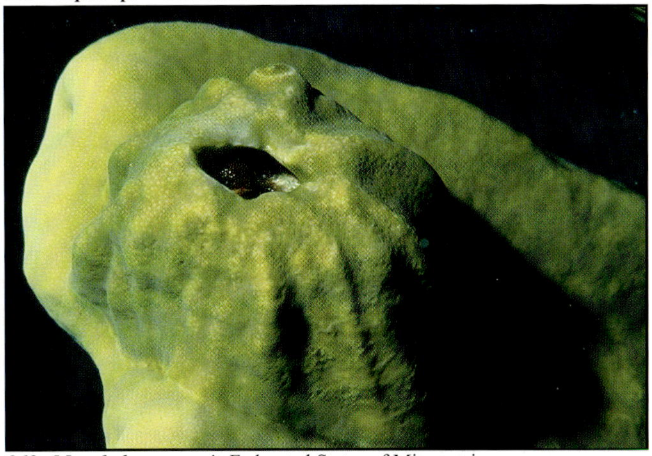

963- ***Megabalanus* sp.** * Federated States of Micronesia

964- Unidentified barnacle * Philippines

965- Unideintified barnacle * Federated States of Micronesia

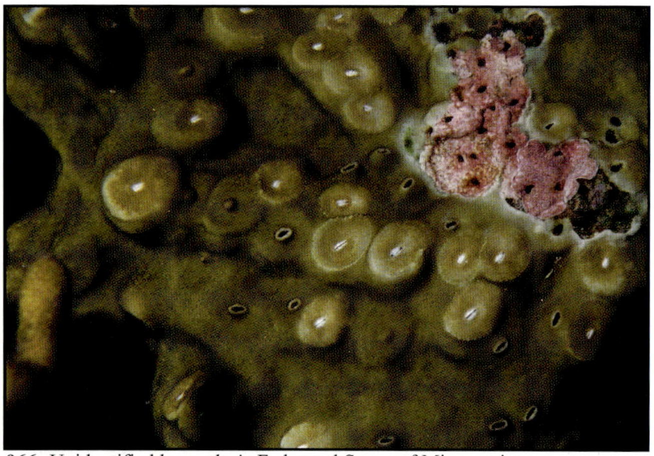

966- Unidentified barnacle * Federated States of Micronesia

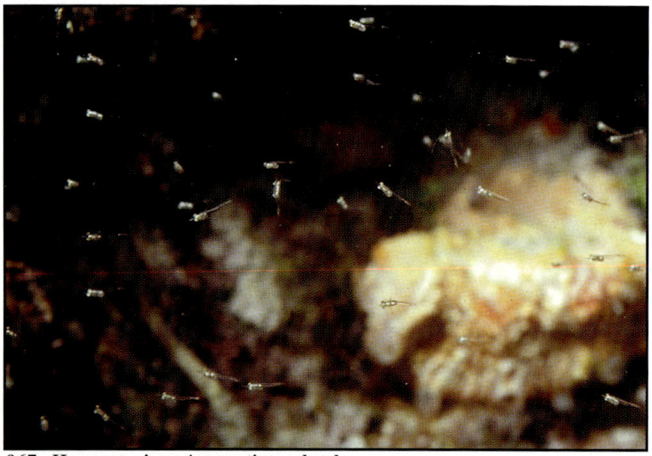

967- ***Heteromysis* sp.* mention schools**

968- ***Euchaeta* sp.** * Hawaii

969- ***Lysosquilla* sp.** * Philippines

968 - Euchaeta sp. * Copepoda * Crustacea * Hawaii * Open Pacific. Copepods can be found, with careful examination, in the water column. Most species are less than a quarter of a inch long. Copepods are primary consumers of phytoplankton and are in turn an important food source for other carniverous zooplankton.

969 - *Lysiosquilla* sp. * Squillidae * Stomatopoda * Crustacea * Philippines * Pamalican Island * 20 ft (6 m).

970 - *Gonodactylus* sp. * Gonodactylidae * Stomatopoda * Crustacea * Hawaii * Oahu * Coconut Island * 30 ft (9 m).

971 - Unidentified stomatopod * Squillidae * Stomatopoda * Crustacea * Hawaii * Puako * 33 ft (10 m).

972 - *Thalassina anomala* * Thalassinidae * Anomura * Decapoda * Philippines * Cebu * Mactan Island * mangroves * intertidal. This ghost shrimp lives in mangrove areas where it produces large mud mounds. It is very similar in appearance to the callianassids (ghost shrimps) which occur in nearly all sediment bottoms around reefs. Callianassids are responsible for producing the "volcano" mounds found at depths below forty feet on broad expanses of sediment. The ghost shrimp live in burrow systems deep in the sediment and produce the mounds by pumping water and sediment out of their burrows via vertical tubes which occur at the center of each volcano.

973 - *Ranina ranina* * Raninidae * Anomura * Decapoda * Indonesia * Manado * Bangka Island * 33 ft (10 m). It may first appear there is nothing in this photograph, but sand. However there are two compound eyes, yellow in color, on white stalks sticking out of the sand. This is *Ranina ranina*, one of the mole crabs. Most mole crabs (*Emerita)* live on sandy beaches, in areas where waves wash up and down the beach. They filter food from the water rushing back to the sea after a wave has broken on the beach. Some others, like the pictured species, live on the reef where they are usually buried in sand and seldom seen.

974 - *Birgus latro* * Coenobitidae * Anomura * Decapoda * Fiji * Kaimbu Island * land. This is the coconut crab, the largest hermit crab, which lives its juvenile and adult life on land. Its larval life is spent in the sea, the females release their fully developed eggs by wading into the sea. It is a delicacy in many areas of the western Pacific and is heavily exploited.

975 - *Aniculus maximus* * Diogenidae * Anomura * Decapoda * Papua New Guinea * Madang * fringing reef * 33 ft (10 m). This distinctive hermit crab grows quite large. It can be recognized by its hairy yellow legs. The species is known from the Indo-west Pacific, including Hawaii.

976 - *Dardanus guttatus* * Diogenidae * Anomura * Decapoda * Federated States of Micronesia * Chuuk * lagoon reef * 40 ft (12 m). The white spotted legs with blue "knees" instantly identify this common hermit crab. It is known throughout the Indo-west Pacific.

977 - *Dardanus megistos* * Diogenidae * Anomura * Decapoda * Fiji * Kaimbu Island * fringing reef * 20 ft (6 m). This is the largest species of its family, reaching 12 inches (30 cm) in length on the Great Barrier Reef. Large individuals are often found in trumpet triton shells.

978 - *Dardanus pedunctulatus* * Diogenidae * Anomura * Decapoda * Indonesia * Manado * fringing reef * 50 ft (15 m). This species of hermit crab has *Calliactis* sea anemones on its shell. *Dardanus deformis* from the Great Barrier Reef also has the anemones, including a second small species of anemone, *Sagartiomorphe paguri,* inside the opening of the shell.

979 - *Paguritta* sp. * Paguridae * Anomura * Decapoda * Marshall Islands * Enewetak Atoll * Medren Island * 40 ft (12 m).

970- ***Gonodactylus* sp.** * Hawaii

971- Unidentified stomatopod * Hawaii

972- ***Thalassina anomala*** * Philippines

973- ***Ranina ranina*** * Indonesia

974- ***Birgus latro*** * Fiji

975- ***Aniculus maximus*** * Papua New Guinea

976- ***Dardanus guttatus*** * Federated States of Micronesia

977- ***Dardanus megistos*** * Fiji

978- ***Dardanus pedunctulatus*** * Indonesia

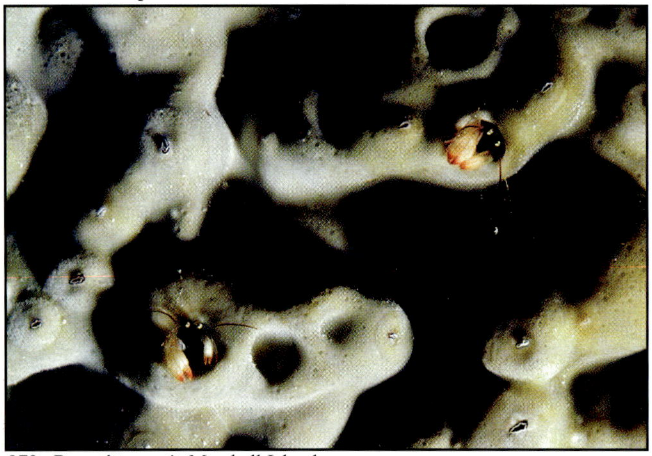

979- ***Paguritta* sp.** * Marshall Islands

980- ***Allogalathea elegans*** * Marshall Islands

981- ***Galatheid on sea pen*** * Philippines

These unusual hermit crabs do not occur in mollusc shells, but live in tubes in *Millepora* sp. fire coral. They are tiny and filter food from the water using their feathery antennae. *Paguritta harmsi* occurs on the Great Barrier Reef where it lives in dead serpulid worms tubes in corals.

980 - *Allogalathea elegans* * Galatheidae * Anomura * Decapoda * Marshall Islands * Enewetak Atoll * Cement ship reef * 30 ft (9 m). This brightly colored galatheid is a commensal on crinoids, often occuring as a male and female pair. In the Marshall Islands it is known to occur on *Comanthus bennetti* and *Comanthina schlegeli*. The galatheid is believed distributed from the Red Sea to the western Pacific, but there are some questions regarding its taxomony.

981 - Unidentified galatheid * Galatheidae * Anomura * Decapoda * Philippines * Pamalican Island * reef * 40 ft (12 m). This galatheid is a commensal on sea pens, but its identification is not yet known.

982 - *Neopetrolisthes maculatus* * Porcellanidae * Anomura * Indonesia * Manado * fringing reef * 30 ft (9 m). The species of *Neopetrolisthes* live among the tentacles of sea anemones. There are at least three species known to occur with sea anemones, particularly *Stichodactyla haddoni*. The photographed species is reported to occur in the Indian Ocean and western Pacific. The second species, *N.ohshimai*, (following) occurs in the western Pacific and a third species, *N. alobatus*, is known only from areas near east Africa.

983 - *Neopetrolisthes oshimai* * Porcellanidae * Anomura * Decapoda * Philippines * Cebu * Mactan Island * 10 ft (3 m). This crab also lives with sea anemones, and appears similar to the previous species. It has irregular patches of larger spots rather than the tiny spots of *N. maculatus*. Its distribution is reported to be the western Pacific, but the exact limits of it and *N. maculatus* are uncertain.

984 - *Porcellana picta* * Porcellanidae * Galatheoidea * Anomura * Decapoda * Hong Kong * Cape d'Aguilar * 66 ft (20 m). This species lives with sea pens. Porcellanid crabs have the last pair of their five pairs of legs modified for cleaning the body. The third maxillipeds are modified to filter feeding structures which they wave through the water catching small plankton or organic particles which they convey to the mouth. They are also capable of feeding directly, using the pincers.

985 - Unidentified porcellanid * Porcellanidae * Anomura * Decapoda * Indonesia * Manado * inshore bay * night * 3 ft (1 m). This small porcellanid crab lives at the base of cerianthid anemones. It can also enter the top of the cerianthid tube if threatened.

986 - *Dromidiopsis edwardsi* * Dromiidae * Decapoda * Indonesia * Biak Island * near dock * night * 30 ft (9 m). This crab uses sponges for camouflage, holding the sponge on its back with a pair of modified legs.

987 - Unidentified Dromiid crab* Dromiidae * Decapoda * Indonesia * Biak Island * dock * 33 ft (10 m).

988 - *Calappa calappa* * Calappidae * Decapoda * Palau * lagoon * sand bottom * night * 10 ft (3 m). The species of *Calappa* are known as "shame -faced" crabs because their claws are held in front of the head making it appear they are hiding their face behind their claws. In actuality they are modified to enable them to deal with their preferred prey, gastropod molluscs. Their claws, somewhat reminiscent of can openers, are different on each arm, but together are highly adapted to hold and open up gastropod shells. These crabs also can dig quickly into sand, burying themselves in only a few seconds in order to hide. There are a number of species in the region, but they all share the basic calappid morphology.

989 - Unidentified crab * Calappidae * Decapoda * Palau * Lighthouse Reef * seagrass bed * night * 10 ft (3 m). We do not

982- ***Neopetrolisthes maculatus*** * Indonesia

983- ***Neopetrolisthes oshimai*** * Philippines

984- ***Porcellana picta*** * Indonesia

985- Unidentified porcellanid

986-***Dromidiopsis edwardsi*** * Indonesia

987- Dromeid crab with sponges * Indonesia

988- ***Calappa calappa*** * Palau

989- Unidentified crab * Palau

990- *Unidentified crab* * Palau

991- *Camposia retusa* * Palau

992- *Oncinopus* sp.* Palau

993- Schizophyrs * Palau

yet have an indentification for this crab which is known in the aquarium trade as the "frog -faced" crab. They occur in sea grassbeds in shallow water in Palau anb vary greatly in color.

990 - Unidentified crab * Calappidae * Decapoda * Palau * Lighthouse Reef * seagrass bed * night * 10 ft (3 m).

991 - *Camposia retusa* * Majidae * Decapoda * Palau * Lighthouse Reef * night * 3 ft (1 m). These spider crabs are masters of camouflage, covering themselves with sponges, algae and other benthic organisms. Many times it is impossible to detect their presence unless they move. This species is widespread in the region. This species is known from the Indian Ocean to the western Pacific north to Japan.

992 - *Oncinopus* sp. * Majidae * Decapoda * Palau * Mutremdiu Wall * 30 ft (9 m). The small spider crabs of this genus have curved tips to the legs and the claws are also curved. The photographed individual is on a sponge.

993 - *Schizophyrs* sp. * Majidae * Decapoda * Palau * cave to marine lake * 27 ft (6 m). This is a typical spider crab, possessing a round thick body, pointed carapace, thin legs and claws.

994 - *Xenocarcinus* sp. * Majidae * Decapoda * Palau * Mutremdiu Wall * 66 ft (20 m). This spider crab lives on red gorgonians where it blends in well. This species is found in Micronesia and the Great Barrier Reef.

995 - *Xenocarcinus conicus* * Majidae * Papua New Guinea * Madang * barrier reef * night * 50 ft (15 m). This spider crab lives on antipatharian (black) corals among the polyps on the branches. This individual is well camouflaged on the photographed species of *Antipathes*.

996 - *Parthenope validus* - Parthenopidae * Decapoda * Hong Kong * Cape d'Aguilar * 66 ft (20 m). The genus *Parthenope* has several species which are most often found on muddy bottoms, particularly at night, and rarely are found on reefs. This species is known from the western Pacific, including China, Japan, Samoa, Australia and Indonesia.

997 - *Charybdis* sp. * Portunidae * Decapoda * Hong Kong * Cape d'Aguilar * 66 ft (20 m). The Portunids are the swimming crabs, so named because they have a pair of legs adapted for swimming. They have claws designed for ripping and tearing animal prey and generally they are scavengers. The illustrated genus, *Charybdis*, is typical of portunids. *Portunus* is another major genus of the family.

998 - *Lissocarcinus orbicularis* * Portunidae * Decapoda * Hawaii * Puako reef * 40 ft (12 m). This crab is a commensal with holothurians, occurring in the anus of sea cucumbers.

999 - Unidentified portunid * Portunidae * Decapoda * Federated States of Micronesia * Chuuk Atoll * lagoon * 33 ft (10 m). This is a typical portunid crab with its last leg modified to a swimmerette. The crabs typically swim sideways when in a hurry to escape.

1000 - *Atergatis floridus* * Xanthidae * Decapoda * Bahrain * reef * 10 ft (3 m). The xanthids are one of the largest groups of crabs and in many respects are the typical crabs with which people are most familiar. They have carapaces which are wider than long, strong crushing claws and a varied diet. This species is reported to be poisonous to eat, the poison similar to the tetradotoxin of puffer fishes. *Atergatis floridus* is found from the Arabian Gulf to Hawaii.

1001 - *Atergatis intergerrimus* * Xanthidae * Decapoda * Palau * Lighthouse Reef * night * 13 ft (4 m). This photograph shows the shape of a typical xanthid crab carapace. This species is known from the western Pacific (Palau, Japan, Philippines) to east Africa.

994- ***Xenocarcinus* sp.** * Palau

995- ***Xenocarcinus conicus*** * Papua New Guinea

996- ***Parthenope validus*** * Hong Kong

997- ***Charybdis* sp.** * Hong Kong

998- ***Lissocarcinus orbicularis*** * Hawaii

999- Unidentified portunid * Federated States of Micronesia

1000- ***Atergatis floridus*** * Bahrain

1001- ***Atergatis intergerrimus*** * Palau

1002- *Carpilius maculatus* * Papua New Guinea

1003- *Cymo* sp. * Federated States of Micronesia

1004- *Etisus dentatus* * Marshall Islands

1005- *Liagore rubromaculatus* * Hong Kong

1002 - *Carpilius maculatus* * Xanthidae * Papua New Guinea * Madang * Pig Island * night * 20 ft (6 m). This species is named for the dark spots on the carapace. The genus is found circumtropically while this species is known from Hawaii to Japan and Australia, then westward to the Red Sea and east Africa. This species has been reported to be poisonous to eat.

1003 - *Cymo* sp. * Xanthidae * Decapoda * Federated States of Micronesia * Chuuk Atoll * Otta Island * 30 ft (9 m). This is one of the gall crabs, the crab is barely visible within a gall formed by *Acropora* coral. Eventually the coral gall grows to the point the crab is entrapped (and protected) in the coral skeleton. The males, which are smaller than the females, are able to crawl out of the gall to visit the female and fertilize eggs.

1004 - *Etisus dentatus* * Xanthidae * Decapoda * Marshall Islands * Kwajalein Atoll * west reef * night * 60 ft (18 m). *Etisus* is another genus with members who are "typical" xanthids. This species is known from Hawaii and Tahiti to the Red Sea and east Africa.

1005 - *Liagore rubromaculatus* * Xanthidae * Decapoda * Hong Kong * Cape d'Aguilar * 50 ft (15 m). This attractive small xanthid is known from Hawaii and Japan to east Africa and the Red Sea.

1006 - *Lybia tessellata* * Xanthidae * Decapoda * Marshall Islands * Kwajalein Atoll * western barrier reef * oceanside * 50 ft (15 m). This small crab has one of those amazing relationships found on the reef in that it maintains small sea anemones on its claws for defense. It is found from Hawaii and Japan to the Red Sea.

1007 - *Quadrella maculosa* * Xanthidae * Decapoda * Indonesia * Biak Island * fringing reef * on black coral * 40 ft (12 m). In this photograph the small commensal black coral crab raises its claws to ward off the human photographer from approaching its antipatharian home, *Antipathes abies*. Usually a pair of these crabs resides on a single antipatharian.

1008 - *Trapezia rufopunctata* * Xanthidae * Decapoda * Federated States of Micronesia * Nama Island * in *Pocillopora* coral * 30 ft (9 m). The species of *Trapezia* are usually associated with corals or other invertebrates. The species shown here is one that helps to prevent crown-of-thorns starfish from eating their corals. If the starfish approaches the coral, the crab pinches the tube feet of the starfish, which prompts the starfish to move on to another coral which is not protected before it can feed. It is found throughout the western Pacific, including Hawaii.

1009 - *Trapezia* sp. * Xanthidae * Decapoda * Palau * Mutremdiu Wall * 33 ft (10 m). *Trapezia* is considered by some authorities to be a separate family, the Trapeziidae, rather than of part of Xanthidae. The relatively large claws of this small crab can be clearly seen in the photograph. In this case the crab resides in a coral of the genus *Seratiopora.*

1010 - *Zozymus aeneus* * Xanthidae * Decapoda * Indonesia * Manado * fringing reef * 40 ft (12 m). This is a fairly common crab which has a very attractive color pattern and sculpturing on its carapace and claws. It is reported to be poisonous to eat. It is known from Hawaii to the Red Sea and east Africa.

1011 - Unidentified crab * Xanthidae * Decapoda * Federated States of Micronesia * Chuuk * Fujikawa Maru * 40 ft (12 m).

1012 - Unidentified crab * Xanthidae * Decapoda * Federated States of Micronesia * Chuuk * Fujikawa Maru * night * 50 ft (15 m). This species is a typical herbivorous xanthid with flattened "fingers" on its claws useful in picking up algae to convey it to the mouth.
1013 - Unidentified crab * Xanthidae * Decapoda * Palau * Lighthouse Reef * night * 10 ft (3 m). In many xanthids the fingers of the claw are black. This species has the opposite situation with the fingers being lighter than the rest of the claw.

1006- *Lybia tessellata* * Marshall Islands

1007- *Quadrella maculosa* * Indonesia

1008- *Trapezia rufopunctata* * Federated States of Micronesia

1009- *Trapezia* sp. * Palau

1010- *Zozymus aeneus* * Indonesia

1011- Unidentified crab * Federated States of Micronesia

1012- Unidentified crab * Federated States of Micronesia

1013- Unidentified crab * Palau

1014- ***Grapsus*** **sp.** * Palau

1015- ***Percnon*** **sp.** * Palau

1016- ***Ocypode cerathopthalma*** * Marshall Islands

1017- ***Uca*** **sp.** * Philippines

1014 - ***Grapsus*** **sp. * Grapsidae * Decapoda * Palau * Rock Islands * intertidal.** The grapsid crabs are commonly known as the "Sally Lightfoot" crabs as they are very fast and agile, scampering around the intertidal zone in rocky splash areas.

1015 - ***Percnon*** **sp. * Grapsidae * Decapoda * Palau * Rock Islands * intertidal.** The species of *Percnon* are found around the intertidal in rocky areas.

1016 - ***Ocypode cerathopthalma*** *** Ocypodidae * Decapoda * Marshall Islands * Enewetak Atoll * beach.** These crabs are known as ghost crabs. They are most often found on beaches and sandy intertidal areas where they have burrows into which they can disappear, hence the name ghost crabs.

1017 - ***Uca*** **sp. * Ocypodidae * Decapoda * Philippines * Cebu * Mactan Island * intertidal.** Fiddler crabs, named because of the large claw of the male, live on mud flats and very shallow subtidal areas. Typically they have burrows in the mud and emerge at low tide to feed on the mud.

1018 - ***Dorippe granulata*** *** Oxystomatidae * Decapoda * Hong Kong * Cape d'Aguilar * 66 ft (20 m).** This small crab carries the anemone *Carcinactis ichikawai* on its back. The third and fourth legs are small and used for holding the anemone on the top of the carapace.

1019 - ***Aethra scruposa*** *** Decapoda * Parthenopidae * Papua New Guinea * West New Britain * 10 ft (3 m).** This is quite a strange looking crab, found in seagrass beds and rubble areas. It is known from Japan, New Caledonia, Papua New Guinea to east Africa.

1020 - ***Matuta lunaris*** *** Decapoda * Calappidae * Philipines * Cebu * Mactan Island * seagrass bed * 6 ft (2 m).** This species is widespread, from China, Japan and Australia to the Red Sea.

1021 - Penaeid shrimp * Penaeidae * Penaeidea * Philippines * Pamalican Island * night * 3 ft (1 m). The penaeid shrimps are the commercial shrimps of the world, being caught in large number by mechanized trawlers. Penaeids are also found on reefs and seagrass beds where they form a group of crustaceans which are easily recognized as members of that group, but not easily identified to species. The taxonomy of penaeids is difficult, and the commercially important species have been worked out the best. The smaller reef-dwelling species are not as well known.

1022 - ***Penaeus latisulcataus*** *** Penaeidae * Penaeidea * Bahrain * sea grass bed * 3 ft (1 m).** This photograph shows a number of the features of penaeid shrimps; the pointed serrated rostrum, the small walking legs on the thorax, the swimmerets on the abdomen and the long antennae.

1023 - Penaeid shrimp* Penaeidae * Penaeidea * Papua New Guinea * Madang * lagoon reef * 33 ft (10 m). Another reef - dwelling penaeid on the bottom at night. During the day these shrimp remain hidden as they would be easy target for many predators.

1024 - ***Stenopus hispidus*** *** Stenopodidae * Stenopodidea * Federated States of Micronesia * Chuuk * Northeast Pass * night * 40 ft (12 m).** Known as the banded coral shrimp, this species is a cleaner, removing ectoparasites from fishes. They often occur in pairs, this duo residing in a sponge. The species is found worldwide in the tropics.

1025 - ***Stenopus pyrsonotus*** *** Stenopodidae * Stenopodidea * Papua New Guinea * West New Britain * Farmers Reef * night * 33 ft (10 m).** Despite being a different species, this coral shrimp has the "look" which immediately identifies it as a member of *Stenopus*. There are several species in the genus not included here.

1026 - ***Hymenocera picta*** *** Gnathophyllidae * Caridea * Palau * Lighthouse Reef * seagrass bed * 6 ft (2 m).** These bizarre crea-

1018- *Dorippe granulata* * Hong Kong

1019- *Aethra scruposa* * Papua New Guinea

1020- *Matuta lunaris* * Philipines

1021- Penaeid shrimp * Philippines

1022- *Penaeus latisulcataus* * Bahrain

1023- Penaeid shrimp * Philippines

1024- *Stenopus hispidus* * Federated States of Micronesia

1025- *Stenopus pyrsonotus* * Papua New Guinea

1026- ***Hymenocera picta*** * Palau

1027- ***Lysmata amboinensis*** * Papua New Guinea

1028- ***Parhippolyte* cf. *uveae*** * Fiji

1029- ***Saron marmoratus* male** * Federated States of Micronesia

tures are known as "paddle shrimp" because their massively enlarged second walking legs which are quite flattened and they grow to about 2 inches (5 cm). There is some debate as to how many species are represented by these shrimps. The Indian Ocean species is known as *H. elegans*. Only a single species is known from the Pacific, however, and this occurs from Hawaii throughout the tropical Pacific. They have amazing behavior in that they are voracious predators of starfish, particularly species of *Nardoa* and *Linckia*, many times larger than them. They are believed to anesthetize the starfish, then turn the starfish onto its "back" (aboral surface) and attack through the openings on the arms (ambulacral grooves).

1027 - *Lysmata amboinensis* * Hippolytidae * Caridea * Papua New Guinea * Madang * barrier reef * 40 ft (12 m). A cleaner shrimp, it lives in caves where the white antennae, stripe on the back and tiny pincers stand out. It is often found in the same caves with the little hingebeak shrimp, *Rhynchocinetes uritai*. The members of *Lysmata* are mostly cleaner shrimps and there are a number of other species in the region which might be encountered. *Lysmata debelius* is nearly all bright red with a few white spots on the thorax, while *L. multicissa* is clear with thin red lines down the body.

1028 - *Parhippolyte* cf. *uveae* * Hippolytidae * Caridea * Fiji * Kaimbu Island * brackish pond * 3 ft (1 m). This striking shrimp in Hawaii is found in deep caves and lava tubes on the reef. In other areas, such as Fiji, it is known from brackish ponds.

1029 - *Saron marmoratus* male * Federated States of Micronesia * Nama Island * reef cave * night * 30 ft (9 m). The extremely long arms identify a mature male *S. marmoratus*, a species which is nocturnally active on reefs. During the day, these shrimps hide in crevices and caves.

1030 - *Saron* sp. * Federated States of Micronesia * Chuuk * fringing reef * night * 3 ft (1 m). An unidentified species of *Saron* shelters in the reef at night.

1031 - *Saron* sp. * Hippolytidae * Caridea * Federated States of Micronesia * Nama Island * reef cave * night * 30 ft (9 m). This photograph shows the complicated color patterns and ornamentation of the genus *Saron*. This genus has the upturned rostrum and frilly fringes on body plates.

1032 - *Thor amboinensis* * Hippolytidae * Caridea * Philippines Despite its small size, *Thor amboinensis* is found worldwide in the tropics occurring with sea anemones.

1033 - *Rhynchocinetes* sp. *Rhynchocinetidae * Caridea * Federated States of Micronesia * Chuuk * night * 20 ft (6 m). Shrimps of the genus *Rhynchocinetes* are the epitome of a nocturnal organism. They were believed to be rare until people started night diving. Then they were revealed to be quite common in areas with well developed caves.

1034 - *Rhynchocinetes hiatti* * Rhynchocinetidae * Caridea * Federated States of Micronesia * Nama Island * reef cave * night * 20 ft (6 m). This shrimp is reasonably well known and occurs over a wide range. The white tip of the rostrum stands out in the dark areas these shrimp inhabit.

1035 - *Rhynchocinetes rugulosa* * Rhynchocinetidae * Caridea * Hong Kong * Breaker Reef * 33 ft (10 m). This species is distinguished by the dark marking on the top of the bend in the abdomen, which is missing in the similar *R. uritai*. This species occurs in groups in small caves, such as that shown in the photograph, where a number of *Diadema* sea urchins also occurred.

1036 - *Rhynchocinetes uritai* * Rhynchocinetidae * Caridea * Philippines * Pamalican Island * fringing reef * 50 ft (15 m). This species occurs in groups of up to several dozen shrimps in caves and

1030- *Saron* sp. * Federated States of Micronesia

1031- *Saron* sp. * Federated States of Micronesia

1032- *Thor amboinensis* * Philippines

1033- *Rhynchocinetes* sp. * Federated States of Micronesia

1034- *Rhynchocinetes hiatti* * Federated States of Micronesia

1035- *Rhynchocinetes rugulosa* * Hong Kong

1036- *Rhynchocinetes uritai* * Philippines

1037- *Rhynchocinetes* sp. * Palau

crevices. It is often found in the company of *Lysmata amboinensis*, a cleaner shrimp.

1037 - *Rhynchocinetes* sp. * Rhynchocinetidae * Caridea * Palau * German Channel * night * 20 ft (6 m). This unidentified species of *Rhynchocinetes* is in the open next to a crown-of-thorns starfish.

1038 - *Rhynchocinetes* sp. * Rhynchocinetidae * Caridea * Philippines * Cebu * Mactan Island * cave * 78 ft (23 m). This attractively banded, unidentified species was photographed in an extensive submarine cave at Mactan Island. The cave is so dark that the shrimps, normally inactive during the day, were out moving around in the middle of the day.

1039 - *Periclimenes amboinensis* * Palaemonidae * Caridea * Marshall Islands * Enewetak Atoll * Cement Ship Reef * 40 ft (12 m). This shrimp is a commensal on the crinoid *Comanthus bennetti* and the shrimp is colored to match the color variety of the crinoid upon which it is living.

1040 - *Periclimenes amboinensis* * Palaemonidae * Caridea * Marshall Islands * Enewetak Atoll * Deep Channel * 40 ft (12 m). In this photograph the commensal shrimp is on a different color variety of the same species of crinoid as the previous photograph and the shrimp matches this different color variety.

1041 - *Periclimenes* cf. *brevicarpalis* * Palaemonidae * Caridea * Marshall Islands * Enewetak Atoll * lagoon * 33 ft (10 m). This commensal shrimp is found on sea anemones throughout much of the western Pacific. It is generally transparent with a series of colored spots. The photograph shows two shrimp, the large male and the smaller, less colored female.

1042 - *Periclimenes brevicarpalis* * Palaemonidae * Caridea * Palau * Lighthouse Channel * 6 ft (2 m). This shrimp was photographed on a sea anemone in an inshore environment. The shrimp is not as transparent as in the previous photograph. This may be a response to the different, less clear environment where it was living.

1043 - *Periclimenes holthuisi* * Palaemonidae * Caridea * Federated States of Micronesia * Nama Island * reef cave * 33 ft (10 m). This commensal is living with a fleshy coral, either a species of *Symphyllia* or *Catalaphyllia*.

1044 - *Periclimenes imperator* female * Palaemonidae * Caridea * Papua New Guinea * Kalihi Harbor * 20 ft (6 m). This is one of the best known of the commensal *Periclimenes*, being found on opisthobranchs, including *Pleurobranchus forskali*, and the spanish dancer nudibranch, *Hexabranchus sanguiensis*. The male and female differ in color and generally only a single pair are found per mollusc.

1045 - *Periclimenes imperator* * Palaemonidae * Caridea * Papua New Guinea * Port Moresby * barrier reef * 50 ft (15 m). This photograph shows a large male *P. imperator* on *Pleurobranchus grandis*. The shrimp crawls all over the surface of the opisthobranch and the exact nature of their relationship is not well known.

1046 - *Periclimenes kororensis* * Palaemonidae * Caridea * Palau * Mutremdiu Wall * 30 ft (9 m). This commensal shrimp is distinctive with reddish orange and white markings on the forward portion of the body. It occurs with *Heliopora actiniformis*, living among the long tentacles.

1047 - *Periclimenes soror* * Palaemonidae * Caridea * Papua New Guinea * Madang * Barracuda Point * 66 ft (20 m). This individual is on a sea cucumber, *Boschadsia argus*, and is also a commensal on asteroids. It is very similar in appearance to *P. imperator*, but differs in color pattern.

1038- *Rhynchocinetes* sp. * Philippines

1039- *Periclimenes amboinensis* * Marshall Islands

1040- *Periclimenes amboinensis* * Marshall Islands

1041- *Periclimenes* cf. *brevicarpalis* * Marshall Islands

1042- *Periclimenes brevicarpalis* * Palau

1043- *Periclimenes holthuisi* * Federated States of Micronesia

1044- *Periclimenes imperator* * Papua New Guinea

1045- *Periclimenes imperator* * Papua New Guinea

1046- *Periclimenes kororensis* * Palau

1047- *Periclimenes soror* * Papua New Guinea

1048- *Periclimenes tenuis* * Marshall Islands

1049- *Periclimenes tenuipes* * Palau

1050- ***Periclimenes sp.*** * Federated States of Micronesia

1051- ***Periclimenes sp.*** * Papua New Guinea

1052- ***Periclimenes*** **sp.** * Indonesia

1053- ***Alpheus*** **sp.** * Philippines

1048 - *Periclimenes tenuis* * Palaemonidae * Caridea * Marshall Islands * Enewetak Atoll * Cement Ship Reef * 40 ft (12 m). This shrimp is a commensal on the crinoid *Comanthus bennetti* with as many as 15-20 shrimp per crinoid. Interestingly this same species of crinoid harbors other commensal shrimp, including *P. amboinensis*, plus alpheids. This species is known from the Red Sea to the central Pacific.

1049 - *Periclimenes tenuipes* * Palaemonidae * Caridea * Palau * barrier reef * 50 ft (15 m). This species has long arms which make it distinctive. It occurs with corals.

1050 - *Periclimenes sp.* * Palaemonidae * Caridea * Federated States of Micronesia * Chuuk * Northeast Pass * 50 ft (15 m). These unidentified species of *Periclimenes* occur with a sea anemone. There is still a great deal to be learned about the occurrence and relationships between commensal crustaceans and host organisms of the tropical Pacific.

1051 - *Periclimenes sp.* * Palaemonidae * Caridea * Papua New Guinea * Madang * Barracuda Point * 50 ft (15 m). This shrimp occurs with black corals, in this case a species of *Antipathes*.

1052 - *Periclimenes* sp. * Paleomonidae * Caridea * Indonesia * Manado * fringing reef * 30 ft (9 m). This unidentified species of shrimp is commensal with the anemone *Actinostephanus haeckeli*.

1053- *Alpheus* sp. * Alpheidae * Caridea * Philippines * Cebu * Sand bottom* 60 ft (20 m). This symbiotic shrimp shares the burrow of the goby *Amblyeleotris* sp.

1054- Unidentified Alpheid * Federated States of Micronesia * Chuuk * South Pass Pinnacle * 100 ft (30 m). This unidentified alpheid was found on a species of sponge of the genus *Diacarnus*. The large claw is clearly visible.

1055 - *Alpheus* sp. burrows * Hawaii * Kaneohe Bay * 15 ft (5 m). While the alpheids themselves are not easily visible, their handiwork, in the form of these extensive grooves systems in some stony corals (in this case *Porites lobata)* is hard to miss. These grooves develop over a long time and are probably more the result of the coral not growing where the grooves occur while the surrounding coral has continued to grow.

1056 - *Alpheus djiboutiensis* * Alpheidae * Caridea * Marshall Islands * Enewetak Atoll * 20 ft (6 m). This snapping shrimp lives in a commensal relationship with a gobiid fish of the genus *Cryptocentrus*. They live together in a burrow, the fish sitting guard at the burrow entrance most of the time, while the alpheid maintains the burrow, constantly moving sand out of it. This activity results in the common name "bulldozer shrimp" for these alpheids.

1057 - Unidentified alpheid * Alpheidea * Caridea * Marshall Islands * Enewetak Atoll * sand flat * 20 ft (6 m). This is a different species of goby and different species of alpheid from the previous photograph. It appears that each alpheid is specific for each goby, an arrangement which has evolved over a long time.

1058 - *Synalpheus carinatus* * Alpheidae * Caridea * Marshall Islands * Enewetak Atoll * lagoon reef * 50 ft (15 m). This alpheid is commensal on the crinoid *Comanthina schlegeli*.

1059 - Unidentified alpheid * Alpheidae * Caridea * Federated States of Micronesia * Chuuk * lagoon reef * 50 ft (15 m). This unidentified species occurs with the sponge *Clathria basilana*, a common tubular sponge in Micronesia. The color of the alpheid certainly matches that of the inside of this sponge.

1060 - *Justitia longimanus* * Palinuridae * Palinura * Hawaii * cave * 66 ft (20 m). This strange little lobster has distinctive long

1054- Unidentified Alpheid * Federated States of Micronesia

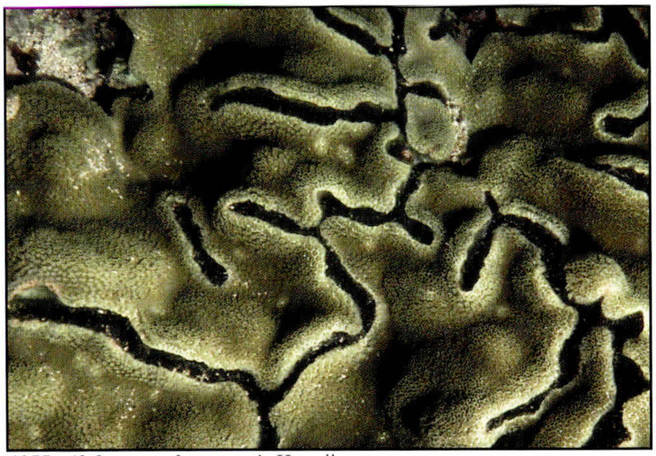

1055- *Alpheus* sp. burrows * Hawaii

1056- *Alpheus djiboutiensis* * Marshall Islands

1057- Unidentified alpheid * Marshall Islands

1058- *Synalpheus carinatus* * Marshall Islands

1059- Unidentified alpheid * Federated States of Micronesia

1060- *Justitia longimanus* * Hawaii

1061- *Panulirus marginatus* * Hawaii

1062- ***Panulirus pencillatus*** * Hawaii

1063- ***Panulirus versicolor*** * Papua New Guinea

1064- ***Panulirus versicolor*** * Palau

1065- ***Palinurella wienecki*** * Hawaii

claws with curved tips. It is generally found in deep water reef areas in caves during the day. At night it emerges to feed. The species occurs circumtropically.

1061 - *Panulirus marginatus* * Palinuridae * Palinura * Hawaii * reef cave * 50 ft (15 m). This spiny lobster is known only from the Hawaiian Islands, where it supports a modest fishery.

1062 - *Panulirus pencillatus* * Palinuridae * Palinura * Hawaii * night * 50 ft (15 m). This photo also has another spiny lobster *Panulirus marginatus* in the photograph. *P. pencillatus* is found from the Red Sea and Madagascar across the Indian and Pacific Oceans to the eastern Pacific, including some areas along the Pacific coast of Mexico. It supports important fisheries throughout its range.

1063 - *Panulirus versicolor* * Palinuridae * Palinura * Papua New Guinea * Madang * lagoon reef * 15 ft (5 m). This is a very easily recognized species. It ranges from the east coast of Africa and the Red Sea through Micronesia and Melanesia, but is not known from Hawaii. This is typically a species of reef areas as it likes clear water.

1064 - *Panulirus versicolor* * Palinuridae * Palinura * Palau * Lighthouse Reef channel * 10 ft (3 m). This is the juvenile color form of *P. versicolor* with brilliant white antennae.

1065 - *Palinurella wienecki* * Synaxidae * Palinura * Hawaii * Puako * night * 80 ft (24 m). This uncommonly seen species has a broad Indo-Pacific distribution.

1066 - *Enoplometopus* sp. * Nephropidae * Astacidea * Indonesia * Lembeh Strait * night * 33 ft (10 m). Members of this genus are generally known as reef lobsters, although they are not closely related to the more familiar spiny lobsters of tropical waters. They tend to live in deep reef areas in caves and may only be seen at night.

1067 - *Enoplometopus occidentalis* * Nephropidae * Astacidea * Hawaii * Puako * night * 70 ft (21 m). There are several species in the Indo-Pacific area, including the one shown here.

1068 - *Arctides regalis* * Scyllaridae * Palinura * Hawaii * 66 ft (20 m). This is the smallest species of slipper lobster in Hawaii, reaching only about 6 inches (15 cm) in length. It tends to be found in deeper water in Hawaii.

1069 - *Parribacus antarcticus* * Scyllaridae * Palinura * Federated States of Micronesia * Chuuk * lagoon reef * 50 ft (15 m). This species grows to about 8 inches (20 cm) in length.

1070 - *Scyllarides haanii* * Scyllaridae * Palinura * Hawaii * 50 ft (15 m).

1071 - *Scyllarides tumidus* * Scyllaridae * Palinura * Hawaii * Oahu * night * 60 ft (18 m). This species lives in caves and only emerges at night.

1072 - *Scyllarides* sp. * Scyllaridae * Palinura * Philippines * reef slope * 33 ft (10 m). This small slipper lobster may be a juvenile, it is about one inch long.

1073- *Thenus orientalis* * Scyllaridae * Palinura * Bahrain * reef * 33 ft (10 m). This slipper lobster is found from east Africa to the Philippines, Indonesia, Ryukyus and tropical Australia.

1066- ***Enoplometopus* sp.** * Indonesia

1067- ***Enoplometopus occidentalis*** * Hawaii

1068- ***Arctides regalis*** * Hawaii

1069- ***Parribacus antarcticus*** * Federated States of Micronesia

1070- ***Scyllarides haanii*** * Hawaii

1071- ***Scyllarides tumidus*** * Hawaii

1072- *Scyllarides* sp. Philipines

1073- ***Thenus orientalis*** * Bahrain

Phyla Ectoprocta (Bryozoa), Phoronida, Brachiopoda and Kamptozoa

Lophophorates

The first three phyla in this section are superficially quite different. However, because they share a common feeding structure, the lophophore, they are grouped together. The lophophore is a ciliated tentacular crown surrounding the mouth. They have other similarities: their general body plan, a U-shaped gut, a transient reproductive system, and outer casings, including tubes, with compartments or shells. The fourth phylum, the Kamptozoa, is a closely related group.

Phylum Phoronida- phoronids

Phoronids are found only in marine waters, living in chitinous tubes which they secrete. There are only two genera (*Phoronis* and *Phoronopsis*) with about 15 species. The lophophore functions in feeding, respiration and protection. Tentacular ciliary bands filter particulates from the water and deliver them to the mouth. The gut is U-shaped, with the mouth at the base of the lophophore and anus just outside the lophophore. Each tentacle of the lophophore contains a coelonic extension. Phoronids have a free-swimming larvae, called an actinotroch, which usually has a lengthy existence in the plankton. In our region, phoronids are most often seen as associates of various other invertebrates such as sponges and tube anemones.

Phylum Ectoprocta- bryozoans or moss animals

Members of the Phylum Ectoprocta, usually called bryozoans from the outmoded Phylum name Bryozoa, are sessile colonial animals which encrust on rocks and various living organisms and resemble algae, hence their common name "moss animals". The colonies are composed of zooids, in essence replicated individuals. Like hydroid polyps and individuals of other colonial animals the zooids of many bryozoans are polymorphic (different in form and function). Autozooids occur in a horny or calcified exoskeleton, sometimes with a small door (operculum) which can cover the opening where the lophophore is extended. Zooids specialized for feeding, the autozooids, have a U-shaped gut with the mouth inside the lophophore, similar to that of the phoronids, and the anus opening outside the lophophore. Other types of zooids include avicularia, which have the operculum modified into a jaw;

Left- This bright red bryozoan is not yet identified, but underwater it is hard to distinguish from a calcareous red algae. Only when carefully examined is the zooid structure of the fine flexible branches evident. The species is found on drop offs and sloping reefs in the Philippines.

Above- This bryozoan, *Caulibugula intermis* seems to like areas with strong currents. In Palau it is found on the bottom and sides of deep channels between the lagoon and ocean. The flower-like structures bear the clonial zooids.

and marine species. The marine members have traditionally been considered as comprising the Class Gymnolaemata. Overall the bryozoans are one of the most poorly known groups of marine invertebrates on tropical Pacific reefs. Inner reef flats on the Great Barrier Reef, although not an ideal habitat for bryozoans, are known to support at least 80 species, while a similar number of species is known from Enewetak Atoll in the Marshall Islands. Studies from Chuuk Atoll indicate that perhaps as many as 300-400 species might occur in that environmentally diverse area. Many species, even from shallow water, remain undescribed. Among known species many appear to have broad geographic ranges, often from Hawaii to the western Indian Ocean within the tropical and subtropical belts.

and vibracula, with the operculum modified into a bristle.

Sessile colonies of zooids are produced by asexual budding. The structure and repetitive nature of the zooids quickly allow determination of whether the organism in question is a bryozoan or not. Some bryozoans are encrusting on mangrove roots or rocks. Some species can be heavily calcified, finely branched, often white in color, and could easily be confused with stylasterine corals (*Stylasterina*). Bryozoans are often surprisingly common on reefs.

The phylum Ectoprocta has has both freshwater

Above- The ciliated tentacles of the zooids of this bryozoan are easily seen protruding out of the opening of the exoskeleton. The exoskeletons of bryozoans can be somewhat calcified, in some species to the point where the resemble small coral colonies.

Recently a potential anti-cancer compound called bryostatin was isolated from the bryozoan *Bugula neretina* and is presently undergoing clinical trials. There may well be other compounds of medicinal value in the bryozoans.

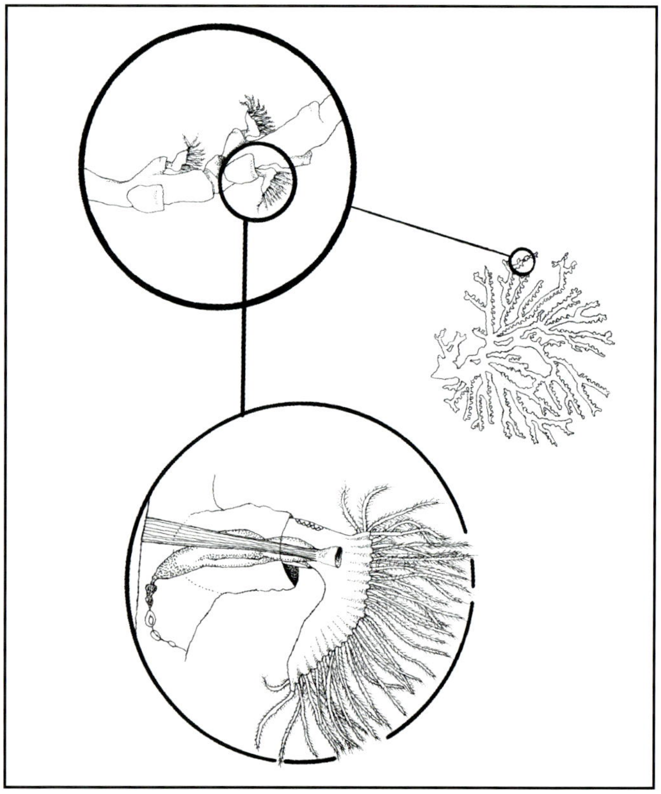

Above- Bryozoans are colonial animals, consisting of many small zooids, each encased in a chitinous cup, and then organized into delicate colonies. Each zooid has a lophophore with tentacles which protrude when the animal is feeding.

Phylum Brachiopoda- lamp shells or brachiopods.

The brachiopods, commonly called lamp shells, consist of only about 300 living species, but over 12,000 extinct species dating back 600 million years are known from the fossil record. During that time brachiopods have changed little; they are truly living fossils. Brachiopods superficially resemble bivalve molluscs in having a bivalved calcareous shell, but they are actually quite different. The mantle cavity has coiled arms, the brachia, that bear the lophophore which is used in collecting suspended food particles. The body is organized similarly to the phoronids, hinting at a common ancestry. It has been suggested competition from bivalve molluscs after the Paleozoic led to the gradual decline of the brachiopods.

Brachiopods are solitary and live in benthic marine environments. There are two basic types. In the articulate brachiopods, the body is enclosed in hinged, dorsoventrally-oriented calcium carbonate valves (shells), which contrast with the laterally oriented shells of bivalve molluscs. They attach to the substratum by the pedicle, a fleshy foot, (although some lack it) and normally sit ventral side up. Most occur in caves or in sheltered areas beneath boulders. In the second group, the inarticulate brachiopods, the valves are unhinged and composed of calcium phosphate, plus chitin and protein. Most inarticulates occur in sand and mud and the pedicle is adapted for burrowing and anchoring in soft substrata.

One species, *Lingula reevei*, can be particularly common. It lives in vertical sand burrows in shallow water; its presence is evident by a three (two incurrent, one excurrent) siphonal openings visible on the sand surface. It is widespread in the Pacific, including Hawaii, Indonesia, and the Philippines.

Above- - The small brown polychaete-like animal on the lower left side of the cerianthid tube is a phoronid, *Phoronis* sp. Phoronids are one of the few organisms that live on the tube of the cerianthid anemones. This species is sometimes found living in soft sediment as well.

Phylum Kamptozoa (Entoprocta)

The final group in this section, the Kamptozoa or Entoprocts, are not lophophorates, but are similar in appearance and are included with the lophophorates here. The entoprocts differ from the ectoprocts (bryozoans) in having their anus within the ring of tentacles (hence their name). They are tiny, less than a quarter of an inch high, and they are most often found attached to sponges, rocks and plants. There are perhaps 100 species with an unknown number occurring in the Indo-Pacific region.

1073 - *Bugula* sp. * Indonesia

1074 - *Caulibugula intermis* * Palau

1075 - *Retiflustra cornea* * Papua New Guinea

1076 - *Membranipora savartii* * Papua New Guinea

1073- *Bugula* sp. * Bugulidae * Cheilostomata * Indonesia * Bangka Island * 20 ft (6 m). This is a typical "hairy" bryozoan that is not easily noticed by divers. The branching structure of the chains of zooids can be seen on close examination.

1074- *Caulibugula intermis* * Bugulidae * Cheilostomata * Palau * Airai Channel * channel bottom * 115 ft (35 m). This bryozoan looks like a tiny stalked crinoid. It occurs in various areas of Palau, particularly in the bottom of deep tidal channels, where it is attached to rock. When the tidal currents are running in the areas where this species lives, it seems as though they will be swept away by the force of the water. Evidently this species likes such areas as it can be very abundant.

1075- *Retiflustra cornea* * Flustridae * Cheilostomata * Papua New Guinea * Port Moresby * Motupore Island * mud bottom * 60 ft (18 m). This species lives on mud bottoms in inshore areas of Papua New Guinea. It forms a dome-shaped network held above the sediment by a number of "legs".

1076- *Membranipora savartii* * Membraniporidae * Cheilostomata * Papua New Guinea * Kavieng * Albatross Channel wall * 100 ft (30 m). This family of bryozoans forms a lightly calcified skeleton which makes it seem more like a delicate coral or coralline algae. These also superficially resemble the tubes of the serpulid polychaete worm *Filograna implexa*. The network structure of the bryozoan is apparent on close examination.

1077- *Membranipora savartii* * Membraniporidae * Cheilostomata * Papua New Guinea * Manam Island * 66 ft (20 m). This delicate bryozoan superficially resembles a finely branched coral or coralline algae.

1078- *Membranopora* sp. * Membraniporidae * Cheilostomata * Indonesia * Biak Island * reef * 66 ft (20 m). This colony of *Membranopora* is growing on a black coral colony, a common occurrence in this and many other species of bryozoans. They are also common fouling organisms, growing on man-made objects.

1079- *Serripetraliella* sp. * Petraliellidae * Cheilostomata * Palau * Koror * Rock Islands * 3 ft (1 m). This species grows as a thin calcified crust on rocky areas with abundant shade. The photographed specimens were found on a small section of limestone rock lining a shallow channel in the Rock Islands of Palau. This species also occurs in Papua New Guinea.

1080- *Iodictyum* sp. * Phidoliporidae * Cheilostomata * Federated States of Micronesia * Chuuk * Northeast Pass * 66 ft (20 m). The delicate lacy fans of this species are surprisingly tough and strong. They also reach a size of several inches across. They occur in caves and crevices on reefs. Although this species appears similar to *Triphyllozoon*, they are in separate families. Both groups, though, occupy the same types of habitats and the similarities are probably due to convergent evolution.

1081- *Reteporella* sp. * Phidoliporidae * Cheilostomata * Papua New Guinea * Hansa Bay * 50 ft (15 m). This is believed to be an undescribed species which occurs as tiny white fans. In the miniature world of these bryozoans, tiny red and yellow gorgonians of the genus *Acabaria* mix with them to form an inch high forest in a reef cave.

1082- *Schizoporella serialis* * Schizoporellidae * Cheilostomata * Federated States of Micronesia * Chuuk * boat hull * 6 ft (2 m). A common fouling bryozoan, occuring on boat hulls as a thin calcareous crust.

1083- *Stylopoma* sp. * Schizoporellidae * Cheilostomata * Papua New Guinea * Madang Channel * 10 ft (3 m). This forms a thick crust on mangrove roots and other objects with many layers of dead bryozoan skeleton. Superficially it resembles a small polyped coral or coralline algae and could be easily confused.

1077 - *Membranipora savartii* * Papua New Guinea

1078 - *Membranopora* sp. * Indonesia

1079 - *Serripetraliella* sp. * Palau

1080 - *Iodictyum* sp. * Federated States of Micronesia

1081 - *Reteporella* sp. * Papua New Guinea

1082 - *Schizoporella serialis* * Federated States of Micronesia

1083 - *Stylopoma* sp. * Papua New Guinea

1084 - *Catenicella* sp. * Papua New Guinea

1085 - *Celleporaria sibogae* * Palau

1086 - *Scrupocellaria ferox* * Federated States of Micronesia

1087 - *Triphyllozoon trifoliatum* * Papua New Guinea

1088 - *Triphyllozoon* sp. * Papua New Guinea

1084- *Catenicella* sp. * Scrupocellariidae * Cheilostomata * Papua New Guinea * Port Moresby * barrier reef * 60 ft (18 m). This is an undescribed species we have found in Papua New Guinea and Palau. It is a lovely bryozoan, a golden brown tuft of the softest sort of branches. It is most abundant along deep reef dropoffs and varies greatly in density along any given section of wall.

1085- *Celleporaria sibogae* * Scrupocellariidae * Cheilostomata * Palau * Denges Pass * 66 ft (20 m). This is an unusual bryozoan we have seen in Chuuk, Palau, Papua New Guinea, the Philippines, and Indonesia. It occurs as a hard calcareous crust with many layers of dead bryozoan. Tiny white commensal hydroids protrude from the zooids and are retracted if the bryozoan is touched or disturbed.

1086- *Scrupocellaria ferox* * Scrupocellariidae * Cheilostomata * Federated States of Micronesia * Chuuk * lagoon reef * 75 ft (23 m). This is a typical "moss animal" bryozoan and is common in lagoonal areas of Chuuk and Palau.

1087- *Triphyllozoon trifoliatum* * Sertellidae * Cheilostomata * Papua New Guinea * Madang * fringing reef * cave * 66 ft (20 m). The delicate white skeletons of this species are found in caves and crevices in moderate depths. They vary greatly in size, but can be up to several inches across. To the uninitiated these highly contoured and sculptured colonies would appear to be some strange coral or other calcareous organism, not a lowly bryozoan.

1088- *Triphyllozoon* sp. * Sertellidae * Cheilostomata * Papua New Guinea * Kavieng * Albatross Channel * cave * 100 ft (30 m). This species appears similar to the previous, but is actually different. The colony here has the zooids extended, the lophophores visible in the photograph.

1089- Genus species unknown * Federated States of Micronesia * Mortlock Islands * reef face * 20 ft (6 m). This species has long branches radiating out in whorls from its base and is a dark gray in color. There is a group of *Tubastrea* corals to one side of the photograph.

1090- Genus species unknown * Papua New Guinea * Eastern Fields * reef wall * 100 ft (30 m). The lophophores are clearly visible on this branching form.

1091- *Zoobotryon* sp. * Vesiculariidae * Ctenostomata * Papua New Guinea * Kavieng * Albatross Channel * 3 ft (1 m). This bryozoan looks more like an alga than a bryozoan. It has flexible clear branches with inconspicuous zooids scattered along them. We call this the "noodle bryozoan" for obvious reason. The taxonomy of the genus is poorly known, but the genus is found circumtropically.

1092- Genus species unknown * Federated States of Micronesia * Chuuk * Fujikawa * 66 ft (20 m).

1093- Genus species unknown * Gymnolaemata * Philippines * Palawan * Honda Bay * sediment * 66 ft (20 m). This bryozoan occurs as small clumps on sediment bottoms. The branches are calcified but brittle and the colonies are quite fragile.

1094- Unidentified Phoronid- * Phoronida * Palau * Lighthouse Reef * 20 ft (6 m). This unidentified phoronid occurs as a dark tiny bush a few inches high, little is known of its occurrence or biology.

1095- Genus species unknown * Brachiopoda * Federated States of Micronesia * Mortlock Islands * cave * 20 ft (6 m). Brachiopds are "living fossils". This once diverse and abundant group is now limited to a few species, many of which live on the walls of caves on the reef.

1096- *Lingula reevi* * Philippines * Cebu * Mactan Island * sediment bottom * 3 ft (1 m). This inarticulate brachiopod is found in mud and sand over a broad geographic range. The photographed specimens were dug up from a mud flat in the Philippines where they were very common. In nature, only the openings of the siphons, which look like fine slits, are visible on the surface of the sediment and the animal can dig down quickly if disturbed.

1089 - Genus species unknown * Federated States of Micronesia

1090 - Genus species unknown * Papua New Guinea

1091 - *Zoobotryon* sp. * Papua New Guinea

1092 - Unidentified bryozoan * Federated States of Micronesia

1093 - Unidentified bryozoan * Philippines

1094 - ***Phoronis* or *Phoronopsis*** * Philippines

1095 - Brachiopod * Federated States of Micronesia

1096 - *Lingulua reevi* * Philippines

Echinoderms

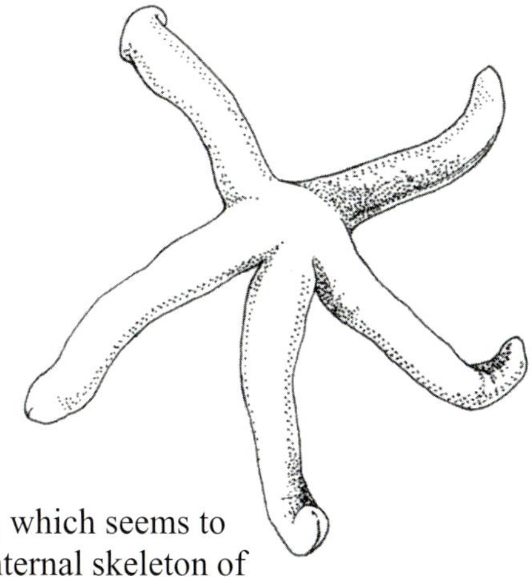

All adult Echinoderms display a five-part radial symmetry, which seems to serve their sedentary benthic lifestyle well. They also possess an internal skeleton of calcium carbonate plates, called ossicles, and a water vascular system unique to echinoderms. The water vascular system consists of a series of canals which radiate throughout the body and terminate in structures called tube feet. The tube feet penetrate the body wall and often have a tiny suction or adhesive cup at their tip. Body fluid, with an ionic composition close to seawater, circulates through the water vascular system for hydraulic expansion or contraction of the tube feet. Tube feet serve echinoderms in a variety of ways, including adhesion, locomotion, feeding, and respiration.

The echinoderms are exclusively marine and are widely distributed in benthic habitats from the intertidal zone to the deep sea. The phylum name, Echinodermata, means "spiny skin", in reference to their characteristically tough, spiny exterior. Living representatives comprise at least five classes: the Crinoidea (feather stars and sea lilies), the Asteroidea (sea stars), the Echinoidea (sea urchins and sand dollars), the Holothuroidea (sea cucumbers) and the Ophiuroidea (brittle stars and basket stars). They number about 6000 living species and have a fossil record extending back over 500 million years.

Above- The adult pin-cushion starfish *Culcita novaguineae* is brilliantly colored and a prominent echinoderm on Indo-Pacific coral reefs. A large *Culcita* is about the size of a volleyball.

Crinoids are abundant on most tropical Pacific reefs. Most, if not all, are believed to be filter feeders using their arms, which number from 5 to 200 depending on the species, to capture food particles and plankton from the water. Most cling to the reef surface with modified arms called cirri. Others lack cirri, and rely on their longer arms to hold them in place. Some crinoids climb to exposed positions on gorgonians and reef structure at night in areas

Opposite- The crinoid, *Oxycomanthus bennetti* is common on corals and reef pinnacles near the reef crest in about 6 to 30 feet of water where it is exposed to strong water flow. The species exhibits a wide variety of colors and patterns including black, green, yellow, orange, peach and grey.

of currents (rheophilic) to more effectively capture plankton by filter feeding. A number of crinoids are capable swimmers, moving the arms in a graceful manner when dislodged from the bottom. Crinoid species and genera are difficult to identify, as many show radical structure and color variation. There are a significant number of undescribed species and genera .

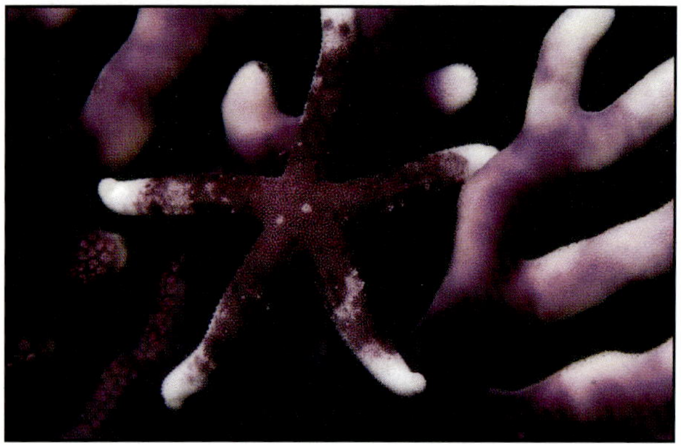

Top- Crinoids often perch on seafans to achieve better access to nutrients carried by currents passing reef points. Center- Among the organisms that live on crinoids (shrimps, crabs, worms) is the clingfish. The clingfish lives on the oral disk of some large crinoids. Below- An unidentified starfish lives with a stylasterine coral (Hydrozoa).

Above- Two brightly colored starfish, *Gomphia gomphia* on the left and *Celerina* sp. on the right, are grazing the surface of an orange sponge, *Agelas* sp.

The sea stars or starfish are the group where the pentagonal symmetry of echinoderms is most readily visible. Although most starfish have five arms, the number and length of arms can vary among sea stars, in some to the point the arms are essentially non-existent in adults. All crawl using the tube feet which arise from the ambulacral grooves on the lower (oral) surface of each arm and converge on the mouth. They are typically predators or detritus feeders. Many of the species occurring on reefs are quite colorful.

Brittlestars typically have only five arms with a discrete central disk. In many species the arms are easily broken off, hence the common name. Some brittlestars can move surprisingly fast, "walking" with their arms across the bottom. Some species are luminescent. They are generally detritus feeders or predators of small organisms. A few are able to capture active prey such as small fishes by trapping them beneath the central disk. Most brittlestars are cryptic, hiding among rocks and crevices during the day and emerging at night. Some extend only their arms, leaving the central disk protected within a hole or crevice. Some of the brittlestars have extremely sharp spines along their arms and are often difficult to see on a gorgonian or sponge. Uncautious grabbing of sponges and gorgonians by divers sometimes results in penetration of these spines resulting in a burning, irritating sensation which lasts for some time.

Sea urchins are found in many different habitats. The regular urchins have a spherical test with stiff, often sharp spines while the so-called irregular urchins, the sand dollars and sea bisquits are flattened or elongate with short, relatively fine spines which are used for digging and locomotion. Many sea urchins are difficult echinoderms to handle. The long, sharp and brittle spines of species of *Diadema* easily penetrate and break

Above- This small sea cucumber, *Cucumaria miniata* lives loosely attached to and on top of other reef organisms. In some areas of the Philippines and Indonesia thousands of these sea cucumbers occur along large sections of the reef.

off in human skin. A mild venom in the spines adds to the misery of the victim. Some other sea urchins have venomous pedicellariae, small beak-like structures on their outer surface which can inflict possibly fatal wounds. Certain sea urchins, particularly those living inshore, will cover themselves with bits of seagrass or algae, for unknown reasons. It has been suggested such behavior protects the urchin from sunlight or from predators, but is still a matter of conjecture.

The Holothuroids, commonly known as sea cucumbers or beche-de-mer, are the only echinoderms which lie on their "sides". They often have the lower surface modified with abundant tube feet for attachment or creeping and there are often color differences between the upper and lower surfaces. Some species are capable of exuding sticky, distasteful, white tubules (cuvierian tubules) if disturbed. These are believed to deter predators. A related but more violent defensive mechanism is evisceration, in which internal organs are expelled by rupture through the body if attacked or disturbed. In either case, the holothurian is capable of regenerating the lost structures and survives.

Holothurians bear 5 to 30 oral tube feet modified as tentacles for feeding. Most species are deposit feeders, however, some of the most remarkable holothuroids are those which have large, highly branched tentacles which are used for filter feeding. Some of these, such as *Neothyonidium magnum*, can be easily confused with sea anemones, as the body of the holothuroid is buried in sediment while the tentacles protrude vertically from the bottom, the array of branches looking like the tentacles of an anemone. These species do not pass sand through their digestive tracts, like more "typical" holothurians. Rather, the tentacles are bent down and inserted into the mouth one at a time and food is removed. The tentacle is raised

Above- These colorful brittlestars are living on a gorgonian from Bahrain in the Arabian Sea. All individuals in the photograph are probably color varieties of the same species.

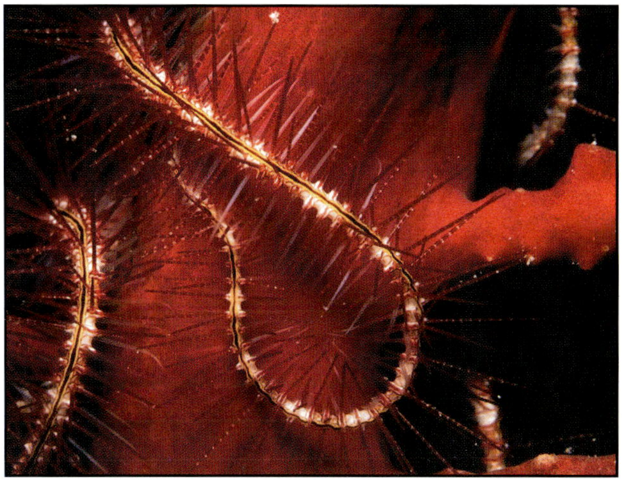

Above- This photograph shows the detail of a brittlestar arm with long spines. The spines have venom, are very sharp and inflict a painful wound if the brittlestar is handled.

back into position and another is then brought to the mouth. This fascinating process is easily observed since the holothurian is undisturbed by the presence of divers, unless actually touched.

There are a number of fisheries based on echinoderms, principally sea urchins and sea cucumbers. When ripe, the gonads of some sea urchins are highly valued as food items in the Orient. Sea cucumbers are taken for trepang, the dried body wall, which

Above- These sea urchins *Asthenosoma varium*, are extremely venomous. This species is variable in color and although beautiful to look at, it should never be handled.

Above- This sea cucumber is getting ready to release sperm into the water. Males and females release gametes in response to environmental cues. Eggs, fertilized externally, develop into planktonic larvae.

Above- This sand-star, in this picture, *Luidia* sp. has captured and is eating another starfish *Linkia laevigata*.

is also popular in the Orient. The valued species are large with thick body walls and lack sticky cuverian tubules. These include *Thelenota ananas, Actinopyga* spp. and *Holothuria nobilis*. The holothurians are cleaned and dried before being shipped.

Overall the echinoderms are very important in coral reef ecology. The sea urchins are major grazers of the reef surface; they crop algae growing on reef substrate with their five-toothed feeding structure (Aristotle's lantern). Without their grazing and that of such herbivorous fish as parrotfishes, reefs would be far different places for algae would certainly dominate the bottom. The echinoderms that feed on organic material in the sand pass a great amount of material through their gut and by doing so help turn over reef sediment. Many holothurians feed this way and are constantly ingesting sand and passing pelletized excreta back onto the reef. Echinoderms can also be voracious predators on molluscs and other invertebrates. The crown-of-thorns starfish, *Acanthaster planci*, has destroyed immense areas of stony corals and altered the basic structure of many Indo-Pacific reefs. Some echinoderms, crinoids for example, engage in filter feeding and particulate trapping, using a variety of techniques to capture plankton or other material from the water

Night is prime time for the echinoderms. Brittlestars leave their daytime hiding places to scavenge over the bottom at night. Basket stars, highly modified brittlestars, also appear and climb on to promontories on the reef where they spread their arms to form a basket for catching plankton from passing currents. Many crinoids behave similarly, perching in the open and spreading their arms to catch plankton only at night.

Many small reef organisms have developed intimate commensal relationships with echinodermss. Certain shrimps (particularly the Pontoniid shrimps) occur on crinoids, holothurians and sea stars. The pearlfishes, of the family Carapidae, spend the day inside the body cavity of some holothurians, emerging from the anus at night to feed.

The majority of echinoderms have separate sexes, but hermaphrodites occur among the asteroids, holothurians and ophiuroids. Many species have external fertilization which produces planktonic larvae, but some brood their eggs, never releasing free-swimming larvae. The planktonic larval stages of echinoderms consist of a number of distinct stages which bear little resemblance to their parents. Larvae are normally produced in vast numbers, but most perish in the plankton. When conditions favor increased survival of larvae, the result can be an overproduction of juveniles and adults, producing, for example, the "plagues" of the crown-of-thorns starfish. Most echinoderms are capable of regenerating lost arms and some, such as the sea stars of the genus *Linckia* grow back from a single detached arm.

Above- The tube feet of the crown-of-thorns starfish protrude from the ambulacral groove on the bottom of each arm.

The crown-of-thorns starfish, *Acanthaster planci*, (above) is probably the single most important echinodermin influencing the nature of Pacific reef communities. This large species reaches 20 inches (50 cm) in diameter. The long spines on the arms and disk are sharp, stiff and venomous, easily penetrating the skin of a diver who accidently steps on or is thrown by a wave against this starfish. It may be the only venomous starfish in the world. Long a part of the normal community on Pacific coral reefs, the crown-of-thorns is a predator of stony corals. It feeds by everting its stomach out of its mouth in a thin sheet that covers all or part of a coral colony and digesting the living coral tissue from the skeleton. After several hours, the starfish moves away leaving a portion of dead coral which is starkly white. This feeding site soon becomes darkened with algae, but remains apparent for many days. In low numbers the crown-of-thorns is a predator which helps to keep the balance between stony corals and the many other organisms competing for space on the reef. In the 1960's or early 1970's, though, conditions changed on some Pacific reefs which allowed the population of starfish to explode to "plague" levels. The starfish devastated the coral communities of many reefs and entire islands. The "plague" levels of *Acanthaster* receded in later years, but the damage has remained on many reefs which have not regenerated to any significant degree. Large populations, although not at previous "plague" levels, still remain on many reefs in Micronesia and elsewhere. Whether these are equilibrium levels or not is a matter of conjecture.

How these waves of starfish abundance occur is still largely a mystery. A single individual can produce as many as 65 million eggs in a spawning season. Small changes in the survival of larvae can have a magnifying effect on the number of juveniles surviving. The most widely accepted theory suggests that increased phytoplankton production, perhaps from increased "fertilization" of waters from runoff from agricultural land or other human activities, resulted in increased larval survival with resultant increases in juvenile and adult populations. It has also been suggested that waves of starfish abundance are natural fluctuations, not the result of human activities, and have occurred many times in the past.

The effect of the crown-of-thorns on reefs goes far beyond the simple predation on coral. When large amounts of coral are killed, algae colonize the bare coral skeletons. The many organisms which live in and around coral colonies have lost their proper habitat and become rare or disappear from a reef. Herbivorous fishes and other invertebrates feed upon the benthic algae, their scraping and biting eroding the reef surface. Some organisms increase in abundance, but many others are reduced, and the overall result is a decrease in the diversity of organisms living on a given reef. If all the reefs in an area are affected, then entire species may disappear where they previously occurred in abundance.

There are few predators of the crown-of-thorns. The trumpet triton shell is known to eat the starfish, but populations of this gastropod are never very high and have potentially been decreased by shell collecting. The small paddle shrimp, *Hymenocera picta,* will kill and eat *Acanthaster* if starving, but much prefers other starfish prey such as *Linckia.* Only a single fish, the humphead wrasse, *Cheilinus undulatus,* is known to eat *Acanthaster,* but again populations of that large tasty fish are reduced in many areas, eliminating predation as an effective biological control on the starfish.

Where juvenile crown-of-thorns occur was a mystery for a number of years. They were finally found to be sheltering under rocks and rubble during daytime and were not normally visible. This illustrates how often we are ignorant of even the most basic facts about the organisms of the sea.

The crown-of-thorns occurs from the east coast of Africa and the Red Sea across the tropical Indian and Pacific Oceans to the west coast of the Americas, from the Sea of Cortez south to Panama. It is not found in the Galapagos. Some authorities consider eastern Pacific populations to represent a separate species.

1097 - *Comanthina schlegelii* * Palau

1098 - *Comanthus alternans* * Indonesia

1099 - *Comanthus parvicirrus* * Papua New Guinea

1100 - *Comanthina schlegelii* * Federated States of Micronesia

1097 - *Comanthina schlegelii* * Comasteridae * Crinoidea * Palau * barrier reef * 33 ft (10 m). This crinoid is a large bushy species that typically sits on exposed pinnacles, although its oral disk may be tucked in a crevice. It has small cirri, claw-like "feet" on the bottom of the oral disk, used for clinging onto the reef, but it also uses its arms, as shown in the photograph, to hang on to the reef. It is often found in the passages between lagoon and ocean.

1098 - *Comanthus alternans* * Comasteridae * Crinoidea * Indonesia * Biak Island * fringing reef * 40 ft (12 m).

1099 - *Comanthus parvicirrus* * Comasteridae * Crinoidea * Papua New Guinea * Madang * barrier reef * 50 ft (15 m). This individual has the central disk situated in a crevice in the reef face from which the arms are extended. In Palau it is highly variable in color, in other areas it is typically brown with blue pinnules. It is a common crinoid in Palau and appears to be most abundant in areas exposed to current flow. It is known from the Indian Ocean to the Marshall Islands.

1100 - *Comanthina schlegelii* * Comasteridae * Crinoidea * Federated States of Micronesia * Chuuk Atoll * Northeast Pass * 33 ft (10 m). As mentioned above, this crinoid is, like many others, in that it is quite variable in color. At Enewetak Atoll the species is polychromatic with different colors on oral disk, cirri, arms and pinnules, the colors varying from white to black, orange and yellow. Different suites of color may be found at other locations. This usually makes it difficult to identify crinoids based on color alone and similarly limits the ability to make identification from photographs without an actual specimen. This crinoid is known from the Maldives and Sri Lanka in the Indian Ocean to the Marshall Islands.

1101 - *Comanthus mirabilis* *Comasteridae * Crinoidea * Papua New Guinea * Madang * barrier reef * 40 ft (12 m). This specimen is clinging to a sponge of the genus *Callyspongia* which may be a productive location for a filter-feeding organisms. Many crinoids will climb after sunset to an exposed location on the reef, often up a gorgonian, black coral or sponge, to reach a position where they can feed more effectively than if it had to remain close to the reef.

1102 - *Comanthus suavia* * Comasteridae * Crinoidea * Federated States of Micronesia * Chuuk Atoll * barrier reef * 66 ft (20 m).

1103 - *Comaster gracilis* * Comasteridae * Crinoidea * Marshall Islands * Kwajalein Atoll * barrier reef * 30 ft (9 m). This species is known from the Maldives to the Marshall Islands. It has a few color varieties. During the day it extends its arms from a protected crevice, as is seen in the photograph, but its behavior at night is not known.

1104 - *Comaster multifidus* * Comasteridae * Crinoidea * Marshall Islands * Enewetak Atoll * leeward barrier reef * 70 ft (21 m). This species has a few color varieties and is reported to be cryptic during the day. The species may emerge at night. Known from Singapore and northeast Australia to eastern Micronesia and the Philippines.

1105 - *Comaster multibrachiatus* * Comasteridae * Crinoidea * Indonesia * Manado * fringing reef * 33 ft (10 m). This species is known from the Andaman Islands to the Philippines.

1106 - *Comatella "maculata"* * Comasteridae * Crinoidea * Papua New Guinea * Port Moresby * barrier reef * 50 ft (15 m). This crinoid is generally red in color, both arms and oral disk. It is cryptic during the day, lying curled up beneath coral heads or among coral branches. It is known from the Indian Ocean to Palau and the far western Pacific. It is probably a juvenile form of the next species.

1107 - *Comatella stelligera/nigra* * Comasteridae * Crinoidea * Papua New Guinea * Madang * fringing reef * 84 ft (25 m). The

1101 - *Comanthus mirabilis* * Papua New Guinea

1102 - *Comanthus suavia* * Federated States of Micronesia

1103 - *Comaster gracilis* * Marshall Islands

1104 - *Comaster multifidus* * Marshall Islands

1105 - *Comaster multibrachiatus* * Indonesia

1106 - *Comatella maculata* * Papua New Guinea

1107 - *Comatella stelligera* * Papua New Guinea

1108 - *Comissia pectinifer* * Indonesia

1109 - ***Oxycomantus bennetti*** * Papua New Guinea

1110 - ***Colobometra perspinosa*** * Papua New Guinea

1111 - ***Cenometra bella*** * Indonesia

exact identity of the photographed individual cannot be determined. It is resting among *Anacropora* coral branches. Known from Indonesia, Australia, Papua New Guinea, Philippines and Palau, where it is not common.

1108 - *Comissia pectinifer*a* Comasteridae * Crinoidea * Indonesia * Biak Island * fringing reef * 50 ft (15 m). This crinoid has long arms, which it extends up from the bottom leaving the disk protected. Its occurs in the East Indies, Micronesia and Australia.

1109 - *Oxycomantus bennetti* * Comasteridae * Crinoidea * Papua New Guinea * Madang * barrier reef * 33 ft (10 m). This species is often placed in the genus *Comanthus*. It is a very common crinoid which is active and exposed both day and night. Hence their appearance as shown in the photograph is the rule. There are several color varieties. This species likes currents and it is very common in passes and other areas with good water flow. It hangs on with its cirri to coral, rocks or gorgonians. It can occur singly or in aggregations and is known from the eastern Indian Ocean to the Marshall Islands.

1110 - *Colobometra perspinosa* * Colobometridae * Crinoidea * Papua New Guinea * Madang * barrier reef * 66 ft (20 m). This species lives on gorgonians and antipatharians, as is shown in the photograph, usually in reef waters below 66 ft (20 m). It has ten arms and a number of color varieties. It is known from Papua New Guinea, Australia, Lord Howe Island and Fiji.

1111 - *Cenometra bella* * Colobometridae * Crinoidea * Indonesia * Biak Island * fringing reef * 66 ft (20 m). This species is usually found on gorgonians and antipatharians. The photograph shows an individual holding on to a whip black coral of the genus *Cirrhipathes* with its cirri. Crinoids with such cirri are able to climb out into open water where filter feeding is better than close to the reef. This species is also capable of active swimming. It has several color varieties and generally between 19 and 39 arms. It is known from the western Pacific, including Indonesia, Vietnam, the Philippines and Marshall Islands.

1112 - *Oligometra serripinna* * Colobometridae * Crinoidea * Papua New Guinea * Madang * Cape Croiselles * 3 ft (1 m). This small crinoid has only ten arms and it is almost always found on gorgonian fans or wire corals such as *Cirrhipathes* sp. It is known from the Red Sea and east Africa to the Philippines, Palau and New Caledonia, but is not known from the Marshall Islands. In Palau it is not common, but can sometimes be found in large numbers elsewhere in the western Pacific.

1113 - *Himerometra robustipinna* * Himerometridae * Crinoidea * Indonesia * Manado * fringing reef * 60 ft (18 m). This species sits in the open on corals near the reef crest, both day and night. The deep red color is typical. It is known from Sri Lanka, Australia, the Philippines and Indonesia.

1114 - *Liparometra regalis* * Mariametridae * Crinoidea * Marshall Islands * Kwajalein Atoll * 60 ft (18 m). The photograph shows clearly how crinoids hold on with their cirri. This species was not previously known from the Marshall Islands, the geographic distribution is poorly known.

1115 - *Stephanometra indica* * Mariametridae * Crinoidea * Federated States of Micronesia * Chuuk Atoll * Yanagi Island * night * 6 ft (2 m). This is a common crinoid on some inshore reefs in Chuuk, climbing up corals and gorgonians at night to feed. During the day they hide among the coral branches. In the photograph one individual can be clearly seen hanging on to the coral with its cirri. This species does not get very large, only about one foot across. It occurs over a broad area of the region, but the exact limits are not known.

1116 - Genus species unknown * Mariametridae * Crinoidea * Federated States of Micronesia * Nama * reef face * night * 30 ft (9 m). This unidentified species is clearly seen holding on to the reef with its cirri.

1117 - Genus species unknown* Mariametridae * Crinoidea * Papua New Guinea * Madang * barrier reef * 60 ft (18 m).

1118 - Genus species unknown* Mariametridae * Crinoidea * Papua New Guinea * West New Britain * Kimbe Bay * night * 33 ft (10 m).

1119 - *Luidia* cf. *avicularia* * Luidiidae * Asteroidea * Philippines * Cebu * Mactan Island * 3 ft (1 m).

1120 - *Archaster typicus* * Archasteridae * Asteroidea * Indonesia * Manado * sand bottom * 6 ft (2 m). This species is very common on sandy bottoms inshore. The starfish is often slightly buried in sand and only the general impression in the surface of the sand

1112 - ***Oligometra sirripinna*** * Papua New Guinea

1113 - ***Himerometra robustipinna*** * Indonesia

1114 - ***Liparometra regalis*** * Marshall Islands

1115 - ***Stephanometra indica*** * Federated States of Micronesia

1116 - **Unidentified crinoid** * Federated States of Micronesia

1117 - **Unidentified crinoid** * Papua New Guinea

1118 - **Genus species unknown*** Papua New Guinea

1119 - ***Luidia cf. avicularia*** * Philippines

1120 - *Archaster typicus* * Indonesia

1121 - *Choriaster granulatus* * Palau

1122 - *Halityle regularis* * Palau

may be visible. It has numerous gastropod parasites. *A. typicus* is known from southeast Asia, Melanesia, Australia and western Polynesia, but not from eastern Polynesia. In the Indian Ocean it is replaced by *Archaster angulatus*, a similar species.

1121 - *Choriaster granulatus* * Oreasteridae * Asteroidea * Palau * Lighthouse Reef * 10 ft (3 m). This is a large and unmistakable starfish common in the Indian Ocean and western Pacific Ocean.

1122 - *Halityle regularis* * Oreasteridae * Asteroidea * Palau * Lighthouse Reef Channel * 3 ft (1 m). The Pentagon starfish is found in the Indian and western Pacific Oceans.

1123 - *Pentaster obtusatus* * Oreasteridae * Asteroidea * Philippines * Cebu * Mactan Island * 15 ft (5 m). This and several following species are members of the family Oreasteridae, all of which have a characteristic appearance with heavy tapering arms and the aboral surface rising to a peak. They are generally found in seagrass beds in fairly shallow water.

1124 - *Pentaster regulus* * Oreasteridae * Asteroidea *Philippines * Cebu * Mactan Island * 10 ft (3 m). This species is found on sea grass and sediment bottoms. On the Great Barrier Reef this species is found on sandy bottoms.

1125 - *Pentaceraster* cf. *multispinus* * Oreasteridae * Asteroidea * Philippines * Cebu * Mactan Island * 15 ft (5 m).

1126 - *Protoreaster nodosus* * Oreasteridae * Asteroidea * Indonesia * Manado * seagrass bed * 6 ft (2 m). This species is quite variable in color and ornamentation. It is very common in many areas in the western Pacific and eastern Indian Oceans.

1127 - *Protoreaster nodosus* * Oreasteridae * Asteroidea * Indonesia * Manado * sea grass bed * 4 ft (1.5 m). This is a color variant of the previous species from the same area.

1128 - *Culcita novaeguineae* * Oreasteridae * Asteroidea * Marshall Islands * Enewetak Atoll * barrier reef * 66 ft (20 m). This unusual echinoderm, often called the pin-cushion star, looks more like a basketball than a starfish. Common on many types of reefs, it is often a coral predator, roughly similar to the crown-of-thorns starfish. However, its ability to tackle the larger branching corals is limited by its shape and lack of flexibility. It can kill only small colonies, two to three inches in diameter and can only partially eat larger colonies. A number of color varieties exist, even in the same location. It occurs throughout the western Pacific. In the Indian Ocean it is replaced by *Culcita schmideliana*.

1129 - *Culcita novaeguineae* * Oreasteridae * Asteroidea * Marshall Islands * Kwajalein Atoll * barrier reef * 20 ft (6 m). The juveniles of *C. novaeguineae* look more like typical starfish, with definite arm-like extensions. As they grow, they gradually lose the rudimentary arms and become nearly spherical.

1123 - *Pentaster obtusatus* * Philippines

1124 - *Pentaster regulus* * Philippines

1125 - ***Pentaceraster* cf. *multispinus*** * Philippines

1126 - ***Protoreaster nodosus*** * Indonesia

1127 - ***Protoreaster nodosus*** * Indonesia

1128 - ***Culcita novaeguineae*** * Marshall Islands

1129 - ***Culcita novaeguineae*** * Marshall Islands

1130 - ***Celerina heffernani*** * Papua New Guinea

1131 - ***Celerina heffernani*** * Papua New Guinea

1132 - ***Fromia indica*** * Indonesia

1133 - ***Fromia* cf. *milleporella*** * Marshall Islands

1134 - ***Fromia monilis*** * Philippines

1135 - ***Gomophia egeria*** * Papua New Guinea

1136 - ***Gomophia gomophia*** * Papua New Guinea

1137 - ***Gomophia watsoni*** * Palau

1130 - *Celerina heffernani* * Ophidiasteridae * Asteroidea * Papua New Guinea * Madang * Pig Island * 27 ft (8 m). This species is a common reef-dwelling asteroid in the China Sea, Melanesia and wesern Polynesia.

1131 - *Celerina heffernani* * Ophidiasteridae * Asteroidea * Papua New Guinea * West New Britain * Kimbe Bay * 40 ft (12 m). This photograph shows a color variation of the previous species.

1132 - *Fromia indica* * Ophidiasteridae * Asteroidea * Indonesia * Manado * fringing reef * 20 ft (6 m). This species is common from the Maldives to Polynesia.

1133 - *Fromia* cf. *milleporella* * Ophidiasteridae * Asteroidea * Marshall Islands * Enewetak Atoll * reef flat quarry * 3 ft (1 m).

1134 - *Fromia monilis* * Ophidiasteridae * Asteroidea * Philippines * Pamalican Island * 20 ft (6 m). This species is common in the western Pacific, but not known from Hawaii.

1135 - *Gomophia egeriae* * Ophidiasteridae * Asteroidea * Papua New Guinea * West New Britain * Kimbe Bay * 33 ft (10 m)

1136 - *Gomophia gomophia* * Ophidiasteridae * Asteroidea * Papua New Guinea * West New Britain * Kimbe Bay * 50 ft (15 m).

1137 - *Gomophia watsoni* * Ophidiasteridae * Asteroidea * Palau * Mutremdiu Wall * 60 ft (18 m). This species is known from New Caledonia and eastern Australia.

1138 - *Leiaster leachi* * Ophidiasteridae * Asteroidea * Papua New Guinea * Madang * reef * 50 ft (15 m). This species is common in the Indian Ocean and western Pacific Ocean.

1139 - *Linckia guildingi* * Ophidiasteridae * Asteroidea * Papua New Guinea * Madang * barrier reef * 10 ft (3 m). The various species of *Linckia* are generally found on reefs and are an important part of that fauna. With five long arms their distinctive appearance separates them from other genera. This species has a circumtropical distribution. In New Caledonia it is generally found in lagoons, but not on outer barrier reefs. It is duller and has thinner arms than the similar *L. laevigata*.

1140 - *Linckia guildingi* * Ophidiasteridae * Asteroidea * Papua New Guinea * Dyaul Island * fringing reef * 20 ft (6 m). This is the coloration of juvenile *L. guildingi*. This individual has some of the arms damaged, a common occurrence in the long armed species of *Linckia*. The arms are regenerating, testament to the ability of asteroids to regrow damaged or lost arms.

1141 - *Linckia laevigata* * Ophidiasteridae * Asteroidea * Chuuk Atoll * Northeast Pass * 40 ft (12 m). This blue starfish is easy to recognize and is the most commonly seen member of the genus. It is found on reefs flats and other shallow areas. It is often parasitized

1138 - *Leiaster leachi* * Papua New Guinea

1139 - *Linckia guildingi* * Papua New Guinea

1140 - *Linckia guildingi* * Papua New Guinea

1141 - *Linckia laevigata* * Federated States of Micronesia

1142 - *Linckia laevigata* * Indonesia

1143 - *Linckia multifora* * Philippines

1144 - *Linckia multifora* * Federated States of Micronesia

1145 - *Linckia* sp. * Philippines

by small gastropod molluscs of the genera *Stylifer* and *Thyca*, which can be found by examining the underside and arms of the starfish (see Mollusc section for photograph). It is common throughout the Indo-Pacific, particularly in areas with wave action.

1142 - *Linckia laevigata* * Ophidiasteridae * Asteroidea * Indonesia * Manado * fringing reef * 20 ft (6 m). This is a color variant of *L. laevigata*. The blue form, shown previously, is most common, but some variants do exist.

1143 - *Linckia multifora* * Ophidiasteridae * Asteroidea * Philippines * Cebu * Mactan Island * fringing reef * 20 ft (6 m). This starfish is very small compared to the species above. It is another species with wide color variation. It is very common in the western Pacific and Indian Oceans.

1144 - *Linckia multifora* * Ophidiasteridae * Asteroidea * Federated States of Micronesia * Chuuk Atoll * Pizion Reef * 10 ft (3 m).

1145 - *Linckia* sp. * Ophidiasteridae * Asteroidea * Philippines * Cebu * Mactan Island * fringing reef * 6 ft (2 m). This and the following photograph are specimens of *Linckia* which cannot be assigned to a particular species.

1146 - *Linckia* sp. * Ophidiasteridae * Asteroidea * Philippines * Cebu * Mactan Island * fringing reef * 10 ft (3 m).

1147 - *Nardoa galatheae* * Ophidiasteridae * Asteroidea * Philippines * Batangas * Pulang Buli Island * 30 ft (9 m).

1148 - *Nardoa novaecaledoniae* * Ophidiasteridae * Asteroidea * Federated States of Micronesia * Chuuk Atoll * Northeast Pass * 40 ft (12 m). This species is common in the western Pacific Ocean, but is not known from Hawaii.

1149 - *Nardoa tuberculata* * Ophidiasteridae * Asteroidea * Indonesia * Manado * seagrass bed * 3 ft (1 m).

1150 - *Nardoa tuberculata* * Ophidiasteridae * Asteroidea * Papua New Guinea * Madang * Riwo mangroves * 6 ft (2 m). This is the "pauciforis" form of *N. tuberculata*, a color form previously described as *Nardoa pauciforis* (but later synonymized with *N. tuberculata*). This photographed individual is living on mud in a mangrove swamp area.

1151 - *Neoferdina cumingi* * Ophidiasteridae * Asteroidea * Papua New Guinea * Kavieng * fringing reef * 30 ft (9 m). This species is known from northern Australia to western Polynesia.

1152 - *Mithrodia clavigera* * Mithrodiidae * Asteroidea * Papua New Guinea * Madang * Pig Island lagoon * 30 ft (9 m). This species is known from the western Pacific and Indian Oceans.

1153 - *Thromidia catalai* * Mithrodiidae * Asteroidea * Papua New Guinea * Madang * barrier reef * 50 ft (15 m). This is a very

1146 - *Linckia* sp. * Philippines

1147 - *Nardoa galatheae* * Philippines

1148 - *Nardoa novaecaledoniae* * Federated States of Micronesia

1149 - *Nardoa tuberculata* * Indonesia

1150 - *Nardoa tuberculata* * Papua New Guinea

1151 - ***Neoferdina cumingi*** * Papua New Guinea

1152 - ***Mithrodia clavigera*** * Papua New Guinea

1153 - ***Thromidia catalai*** * Papua New Guinea

1154 - ***Acanthaster planci*** * Palau

1155 - ***Acanthaster planci*** * Federated States of Micronesia

1156 - ***Echinaster callosus*** * Federated States of Micronesia

1157 - ***Echinaster luzonicus*** * Palau

1158 - ***Echinaster luzonicus*** * Palau

1159 - ***Echinaster luzonicus*** * Papua New Guinea

1160 - ***Astroboa nuda*** * Papua New Guinea

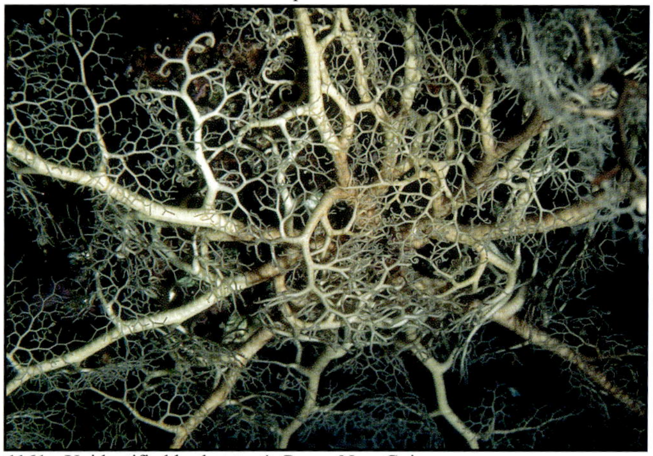

1161 - Unidentified basketstar * Papua New Guinea

1162 - Astroboa granulatus * Indonesia

1163 - ***Ophiothela dana**** Papua New Guinea

large starfish, the arms a foot (30 cm) or more in length, and looks more like an overstuffed toy than a living organism. It is known from Guam and the Bonin Islands to New Caledonia and Hawaii.

1154 - *Acanthaster planci* * Acanthasteridae * Asteroidea * Palau * Mutremdiu Wall * 20 ft (6 m). The crown-of-thorns starfish is found on reefs. The illustrated individual is quite large. They are very common in the western Pacific and Indian Oceans. See page 245 in the Section Introduction for more information.

1155 - *Acanthaster planci* * Acanthasteridae * Asteroidea * Federated States of Micronesia * Nama Island * 15 ft (5 m). This is a juvenile *A. planci*, only an inch or so across. It is eating its way across a stony coral.

1156 - *Echinaster callosus* * Echinasteridae * Asteroidae * Chuuk Atoll * patch reef * 30 ft (9 m). This is a very distinctive asteroid which is found in the western Pacific and Indian Oceans. It varies somewhat in color, but in all cases has the nobby appearance seen in the photograph.

1157 - *Echinaster luzonicus* * Echinasteridae * Asteroidae * Palau * Lighthouse Reef * 20 ft (6 m). This species usually has six arms and is somewhat variable in color. This species often feels slimy to the touch. It is common in the tropical Pacific, but is not found in Hawaii. Some other species of *Echinaster* are similar and a definitive identification is often difficult.

1158 - *Echinaster luzonicus* * Echinasteridae * Asteroidea * Palau * Denges Channel * 30 ft (9 m). This is a color variant of this common starfish.

1159 - *Echinaster luzonicus* * Echinasteridae * Asteroidea * Papua New Guinea * Manam Island * 66 ft (20 m). This individual was found on a black volcanic sand slope and is fairly light in color.

1160 - *Astroboa nuda* * Gorgonocephalidae * Ophiuroidea * Papua New Guinea * Madang * barrier reef * night * 40 ft (12 m). This large and common basketstar occurs throughout the region.

1161 - Unidentified basketstar * Gorgonocephalidea * Ophiuroidea * Papua New Guinea * Eastern Fields * outer reef * 50 ft (15 m). The basketstars differ from other ophiuroids in having branching arms. There are two families, the Gorgonocephalidae (which lack spines on the dorsal surface of the disk) and the Euryalidae (which have dorsal spines).

1162 - *Astroboa granulatus* * Gorgonocephalidae * Ophiuroidea * Indonesia * Manado * fringing reef * 30 ft (9 m). This is the normal appearance of a basketstar during the day, rolled up into a tight ball. It is impossible to identify this basketstar from the photograph when it is rolled up tightly during the day.

1163 - *Ophiothela danae* * Ophiotrichidae * Ophiouroidea * Papua New Guinea * Madang * barrier reef * 20 ft (6 m). These

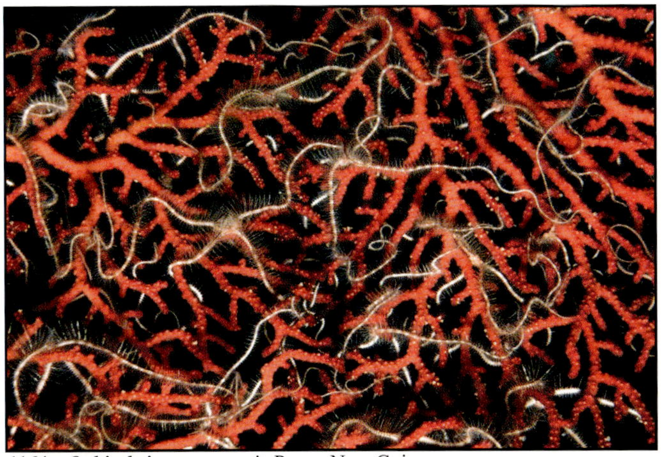

1164 - *Ophiothrix purpurea* * Papua New Guinea

1165 - *Ophiothrix* sp. * Papua New Guinea

1166 - *Ophiothrix* sp. * Solomon Islands

1167 - *Ophiothrix* sp. * Marshall Islands

1168 - *Macrophiothrix* sp. * Indonesia

1169 - *Ophionereis porrecta* * Hawaii

1170 - *Ophionereis* sp. * Marshall Islands

1171 - *Ophiarachna incrassata* * Indonesia

1172 - *Ophioarachna incrassata* * Indonesia

1173 - *Ophioarachna incrassata* * Palau

1174 - *Ophiolepis cincta* * Marshall Islands

brittlestars are found tightly clinging to gorgonians and soft corals, their arms wound tightly around the branches of the cnidarians. Color is quite variable in this species and it is known thoughout the tropical Indo-west Pacific.

1164 - *Ophiothrix purpurea* * Ophiotrichidae * Ophiouroidea * Papua New Guinea * Madang * barrier reef * 20 ft (6 m).

1165 - *Ophiothrix* sp. * Ophiotrichidae * Ophiouroidea * Papua New Guinea * Madang * Rasch Passage * 50 ft (15 m). This brittlestar is on a *Dendronephthya* soft coral.

1166 - *Ophiothrix* sp. * Ophiotrichidae * Ophiouroidea * Solomon Islands * Savo Island * 30 ft (9 m).

1167 - *Ophiothrix* sp. * Ophiotrichidae * Ophiouroidea * Marshall Islands * Enewetak Atoll * Medren Island * 30 ft (9 m).

1168 - *Macrophiothrix* sp. * Ophiotrichidae * Ophiouroidea * Indonesia * Manado * reef flat * 26 ft (2 m).

1169 - *Ophionereis porrecta* * Ophionereidae * Ophiuroidea * Hawaii * Hauula * 10 ft (3 m).

1170 - *Ophioneris* sp. * Ophionereidae * Ophiuroidea * Marshall Islands * Enewetak Atoll * lagoon pinnacle * 40 ft (12 m).

1171 - *Ophiarachna incrassata* * Ophiodermatidae * Ophiuroidea * Indonesia * Talisei * 3 ft (1 m). This is usually the largest brittlestar found in our area. It is reportedly capable of eating small fishes which take shelter beneath the disk. The brittlestar then rotates the disk to close off the spaces between the arms and trap them.

1172 - *Ophioarachna incrassata* * Ophiodermatidae * Ophiuroidea * Indonesia * Manado * seagrass bed * 3 ft (1 m).

1173 - *Ophioarachna incrassata* * Ophiodermatidae * Ophiuroidea * Palau * Lighthouse Reef * 3 ft (1 m). There is some color variation in this brittlestar, as indicated between this and the previous photographs.

1174 - *Ophiolepis cincta* * Ophiuroidae * Ophiuroidea * Marshall Islands * Kwajalein Atoll * offshore reef * 30 ft (9 m).

1175 - *Ophiolepis superba* * Ophiuridae * Ophiuroidea * Indonesia * Manado * sea grass bed * 3 ft (1 m). This is a distinctive brittlestar, hard to confuse. It has strongly marked, relatively short orange and black arms with a distinctive black pentagonal marking on the disk with an orange spot in the center.

1176 - *Ophiomastix* sp. * Ophiocomidae * Ophiuroidea * Indonesia * Manado * fringing reef * 10 ft (3 m).

1177 - *Ophiomastix* sp. * Ophiocomidae * Ophiuroidea * Philippines * Pamalican Island * 30 ft (9 m).

1175 - *Ophiolepis superba* * Indonesia

1176 - *Ophiomastix* sp. * Indonesia

1177 - *Ophiomastix* sp. * Phillipines

1178 - *Ophioarthrum pictum* * Philippines

1179 - *Ophioarthrum* sp. * Marshall Islands

1180 - *Ophiomyxa* sp. * Marshall Islands

1181 - *Ophiomyxa* sp. * Marshall Islands

1182 - *Ophiactis* sp. * Hawaii

1183 - Unidentified ophiuroid * Marshall Islands

1184 - Unidentified ophiuroid * Philippines

1185 - ***Eucidaris metularia*** * Papua New Guinea

1186 - ***Eucidaris* cf. *metularia*** * Indonesia

1187 - ***Phyllacanthus* cf. *imperialis*** * Indonesia

1188 - ***Asthenosoma ijimai*** * Palau

1189 - ***Asthenosoma varium*** * Philippines

1178 - *Ophiarthrum pictum* * Ophiocomidae * Ophiuroidea * Philippines * Cebu * Mactan Island * algal bed * 6 ft (2 m).

1179 - *Ophiarthrum* sp. * Ophiocomidae * Ophiuroidea * Marshall Islands * Kwajalein Atoll * offshore reef * 20 ft (6 m).

1180 - *Ophiomyxa* sp. * Ophiomyxidae * Ophiuroidea * Marshall Islands * Kwajalein Atoll * lagoon pinnacle * 40 ft (12 m).

1181 - *Ophiomyxa* sp. * Ophiomyxidae * Ophiuroidea * Marshall Islands * Kwajalein Atoll * lagoon pinnacle * 40 ft (12 m).

1182 - *Ophiactis* sp. * Ophiactidae * Ophiuroidea * Hawaii * Mak Reef * 30 ft (9 m).

1183 - Unidentified ophiuroid * Ophiuroidea * Marshall Islands * Enewetak Atoll * lagoon pinnacle * 40 ft (12 m).

1184 - Unidentified ophiuroid * Ophiuroidea * Philippines * Cebu * Mactan Island * 20 ft (6 m).

1185 - *Eucidaris metularia* * Cidaridae * Echinoidea * Papua New Guinea * Madang * barrier reef * 20 ft (6 m). This species is widespread in the Pacific and Indian Oceans, including Hawaii.

1186 - *Eucidaris* cf. *metularia* * Cidaridae * Echinoidea * Indonesia * Manado * fringing reef * 30 ft (9 m).

1187 - *Phyllacanthus* cf. *imperialis* * Cidaridae * Echinoidea * Indonesia * Manado * fringing reef * night * 30 ft (9 m). This species is known from the Red Sea and Madagascar to Australia.

1188 - *Asthenosoma ijimai* * Echinothuridae * Echinoidea * Palau * Malakal Harbor * 10 ft (3 m).

1189 - *Asthenosoma varium* * Echinothuridae * Echinoidea * Philippines * Batangas * 40 ft (12 m). Known from the Philippines, Indonesia and Japan to the Gulf of Suez.

1190 - *Asthenosoma* sp. * Echinothuridae * Echinoidea * Philippines * Batangas * 84 ft (25 m).

1191 - *Astropyga radiata* * Diadematidae * Echinoidea * Philippines * Cebu * Mactan Island * seagrass bed * 6 ft (2 m). This urchin is found throughout the entire Indo-Pacific region. Small fishes are often found sheltering around the spines of this urchin, including *Lutjanus sebae* and *Apogon* sp.

1192 - *Astropyga radiata* * Diadematidae * Echinoidea * Indonesia * Manado * Biaro Island * 84 ft (25 m). This urchin was found in dense clumps of several hundred individuals in a sheltered sandy basin lined with mangroves. This species varies in coloration, as is indicated between this and the previous photograph.

1190 - *Asthenosoma* sp. * Philippines

1191 - *Astropyga radiata* * Philippines

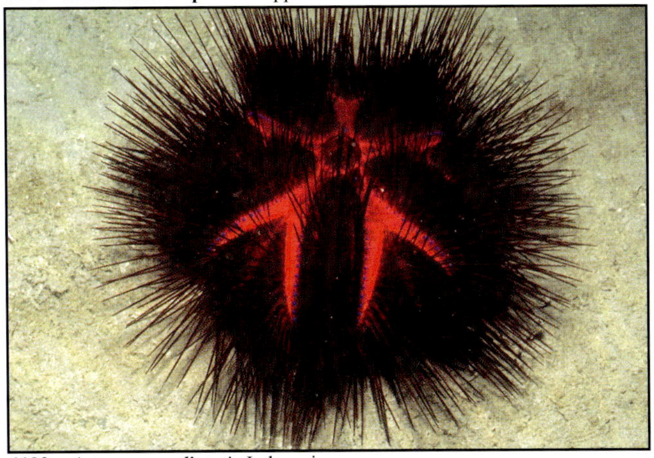

1192 - *Astropyga radiata* * Indonesia

1193 - *Diadema savignyi* * Papua New Guinea

1194 - *Diadema setosum* * Philippines

1195 - *Echinothrix calamaris* * Indonesia

1196 - *Mespilia globulus* * Papua New Guina

1197 - *Temnopleurus toreumaticus* * Bahrain

1198 - **Unidentified echinoid** * Indonesia

1199 - ***Pseudoboletia maculata*** * Papua New Guinea

1200 - ***Toxopneustes pileolus*** * Palau

1201 - ***Toxopneustes pileolus*** * Philippines

1193 - *Diadema savignyi* * Diadematidae * Echinoidea * Papua New Guinea * Port Moresby * Bootless Bay * 50 ft (15 m).

1194 - *Diadema setosum* * Diadematidae * Echinoidea * Philippines * Cebu * Mactan Island * 6 ft (2 m). This urchin is found from the western Indian Ocean to Australia, Tahiti, Fiji and the Bonin Islands.

1195 - *Echinothrix calamaris* * Diadematidae * Echinoidea * Indonesia * Manado * seagrass beds * 3 ft (1 m). This species is known from Polynesia to the Red Sea and east Africa.

1196 - *Mespilia globulus* * Temnopleuridae * Echinoidea * Papua New Guinea * West New Britain * fringing reef * 66 ft (20 m). This is commonly called the royal urchin. It is known from the Philippines, Palau, New Guinea, the Loyalty Islands and Australia.

1197 - *Temnopleurus toreumaticus* * Temnopleuridae * Echinoidea * Bahrain * 10 ft (3 m). This species is known from East Africa and Madagascar east to South Pacific Islands, including New Caledonia.

1198 - Unidentified sea urchin * Temnopleuridae * Echinoidea * Indonesia * Manado * seagrass bed * 3 ft (1 m).

1199 - *Pseudoboletia maculata* * Toxopneustidae * Echinoidea * Papua New Guinea * West New Britain * Kimbe Bay * seagrass bed * 6 ft (2 m).

1200 - *Toxopneustes pileolus* * Toxopneustidae * Echinoidea * Palau * Lighthouse Reef * 10 ft (3 m). This species is well known for its venomous pedicellaria, small structures which project out from the test over most of the surface. It is found in much of the western Pacific, including Samoa, Fiji, Palau, New Caledonia and Enewetak Atoll, then westward across the Indian Ocean to the African coast. It seems to be very rare or absent in some areas, such as Chuuk.

1201 - *Toxopneustes pileolus* * Toxopneustidae * Echinoidea * Philippines * Batangas * Pulang Buli Island * 10 ft (3 m). This photograph shows the great many circular venomous pedicellaria waiting to touch something! The stings from this urchin can be quite painful and dangerous, so it should be treated with caution at all times.

1202 - *Tripneustes gratilla* * Toxopneustidae * Echinoidea * Fiji * Kaimbu Island * 10 ft (3 m). This species is found throughout the Indo-Pacific region, including Hawaii. It has a number of color forms, as comparison between this and the next photo will indicate, and lives in seagrass beds.

1203 - *Tripneustes gratilla* * Toxopneustidae * Echinoidea * Palau * Lighthouse Reef * seagrass bed * 3 ft (1 m).

1204 - *Colobocentrotus atratus* * Colobocentrodidae * Echinoidea * Mariana Islands * Saipan * the Grotto * intertidal. This strange little urchin is adapted for holding on to wave swept rocks in the intertidal zone. It is found over a wide area to Hawaii in wave swept areas only.

1205 - *Echinometra mathaei* * Echinometridae * Echinoidea * Indonesia * Manado * seagrass bed * 3 ft (1 m). This species is found in the tropical and subtropical Indo-west Pacific. It often is found on rocky areas where the urchins, over time, wear holes or grooves into the rock.

1206 - *Echinostrephus* sp. * Echinometridae * Echinoidea * Papua New Guinea * Laing Island * 10 ft (3 m). This species excavates depressions in rock that protect the urchin.

1207 - *Anthocidaris crissipinina* * Echinometridae * Echinoidea * Hong Kong * Breaker Reef * 20 ft (6 m). This urchin usually has seagrass blades or algae held over its body to protect it from sunlight.

1208 - *Heterocentrotus mammillatus* * Echinometridae * Echinoidea * Marshall Islands * Enewetak Atoll * night * 10 ft

1202 - *Tripneustes gratilla* * Fiji

1203 - *Tripneustes gratilla* * Palau

1204 - *Colobocentrodus atratus* * Mariana Islands

1205 - *Echinometra mathaei* * Indonesia

1206 - *Echinostrephus* sp. * Papua New Guinea

1207 - *Anthocidaris crissipinina* * Hong Kong

1208 - *Heterocentrodus mammillatus* * Marshall Islands

1209 - *Heterocentrotus trigonarius* * Marshall Islands

1210 ***Heterocentrotus trigonarius*** * Papua New Guinea

1211 - *Clypeaster humilus* * Bahrain

1212 - *Clypeaster reticulatus* * Philippines

1213 - *Clypeaster* sp. * Philippines

(3 m). During the day these urchins hide in cracks and crevices in shallow water in wave swept areas on outer reefs. At night they leave their shelter holes and graze over the reef. They are known for their stout spines which remain intact after the urchin dies, often washing up on beaches or being found in reef sediments. The species is known from Hawaii through the tropical Pacific and Indian Oceans to the coast of Africa. Their spines are often used in handicrafts.

1209 - *Heterocentrotus trigonarius* * Echinometridae * Echinoidea * Marshall Islands * Enewetak Atoll * offshore reef * 10 ft (3 m). This species seems similar to *H. mammilatus*, but differs in a number of morphological characters. The two species often occur together in the same location. It occurs in the western Pacific from Polynesia through the Indian Ocean, but is not known from Hawaii.

1210 - *Heterocentrotus trigonarius* * Echinometridae * Echinoidea * Marshall Islands * Kwajalein Atoll * leeward reef * 10 ft (3 m). These urchins have three types of spines which are all visible in this photograph; long triangular spines on top to deter predators, shorter and flatter spines underneath to clamp into crevices of the reef and flattened platelike spines to protect the body. Urchins of the genus *Heterocentrotus* have commensal caridean shrimp *Athanas dorsalis*.

1211 - *Clypeaster humilus* * Clypeasteridae * Echinoidea * Bahrain * sand bottom * 6 ft (2 m). This and the following species are commonly known as sand dollars due to their flattened, often circular test. They are inhabitants of sandy bottoms and can be quite abundant in the right environment. Its distribution ranges from East Africa and Madagascar to the Philippines and Australia.

1212 - *Clypeaster reticulatus* * Clypeasteridae * Echinoidea * Philippines * Cebu * Mactan Island * 6 ft (2 m). This species has long spines for a sand dollar. These creatures live on sandy bottoms.

1213 - *Clypeaster* sp. * Clypeasteridae * Echinoidea * Philippines * Cebu * Mactan Island * 6 ft (2 m). This unidentified species from the Philippines is similar to *C. humilus*, but differs in the shape of the test and the grooves on the aboral surface.

1214 - *Peronella lesueuri* * Laganidae * Echinoidea * Papua New Guinea * Madang * Kranket Island lagoon * 3 ft (1 m).

1215 - *Laganum laganum* * Laganidae * Echinoidea * Indonesia * Manado * 3 ft (1 m).

1216 - *Maretia planulata* * Spatangidae * Echinoidea * Marshall Islands * Enewetak Atoll * back reef sand slope * 20 ft (6 m). The following species are known as heart urchins. They are somewhat unusual for echinoderms in that their tests are bilaterally symmetrical. They have a "head" end and can move and dig surprisingly fast using their spines. This species is found in the tropical and subtropical Indian and western Pacific Oceans. It can occasionally be very common and buries in sand during the day.

1217 - *Maretia* sp. * Spatangidae *Echinoidea * Philippines * Cebu * Mactan Island * 6 ft (2 m). *M. planulata*, like most of the spatangoids, remains buried in the sand during the day and emerges at night to feed. The markings on the test vary in different areas.

1218 - Genus species undetermined * Spatangidae * Echinoidea *Palau * Lighthouse Reef * 6 ft (2 m). This species could not be identified, but it appears close to *Maretia*. It has a greater number of long spines, however, and certainly represents another species.

1219 - *Metalia* sp. * Spatangidae * Echinoidea * Philippines * Cebu * Mactan Island * 6 ft (2 m). Some spatangoids have short spines which makes them easy to pick up. It is amazing how efficient these creatures are at digging into the bottom when released on a sandy area.

1220 - *Brissus latecarinatus* * Brissidae * Echinoidea * Indonesia * Manado * 10 ft (3 m). This is a large species of heart urchin, reaching about 5 inches (12 cm) in length. The test of dead heart urchins are often found while diving or on the beach and from these

1214 - *Peronella lesueuri* * Papua New Guinea

1215 - *Laganum laganum* * Indonesia

1216 - *Maretia planulata* * Marshall Islands

1217 - *Maretia* sp. * Philippines

1218 - Unidentified irregular urchin * Palau

1219 - *Metalia* sp * Philippines

1220 - *Brissus latecarinatus* * Indonesia

1221 - *Actinopyga mauritiana* * Mariana Islands

1222 - ***Actinopyga miliaris*** * Federated States of Micronesia

1223 - ***Actinopyga palauensis*** * Federated States of Micronesia

1224 - ***Actinopyga* sp.** * Papua New Guinea

1225 - ***Bohadschia argus*** * Federated States of Micronesia

it is easy to appreciate the design of the animal. This species is known from the Indian and western Pacific Oceans.

1221 - *Actinopyga mauritiana* * Holothuridae * Aspidochirotida * Mariana Islands * Saipan * fringing reef * 20 ft (6 m). This species is quite variable in color (brown, green or mottled) and shape, often with the body very hard. Sometimes it is difficult to even realize they are sea cucumbers. They are found in reef environments, often in areas with strong wave action, firmly attached to rocks. It is found throughout the Indo-west Pacific region.

1222 - *Actinopyga miliaris* * Holothuridae * Aspidochirotida * Federated States of Micronesia * Chuuk Atoll * Weno * west reef * 50 ft (15 m). The short, fat body of this black sea cucumber is hard to confuse. It has small papillae covering most of the body. It is found throughout the tropical Indo-west Pacific.

1223 - *Actinopyga palauensis* * Holothuridae * Aspidochirotida * Federated States of Micronesia * Chuuk Atoll * Eten Island * 40 ft (12 m). This dark sea cucumber is common in lagoon areas of Chuuk and Palau.

1224 - *Actinopyga* sp. * Holothuridae * Aspidochirotida * Papua New Guinea * Madang * Pig Island lagoon * 40 ft (12 m). This holothurian may be undescribed.

1225 - *Bohadschia argus* * Holothuridae * Aspidochirotida * Federated States of Micronesia * Chuuk Atoll * Lagoon reef * 30 ft (9 m). This is an easily recognized holothurian. It has a scale worm on it as seen in the photo.

1226 - *Bohadschia goeffreyi* * Holothuridae * Aspidochirotida * Federated States of Micronesia * Chuuk Atoll * algal flat * 100 ft (30 m).

1227 - *Bohadschia graeffei* * Holothuridae * Aspidochirotida * Federated States of Micronesia * Chuuk Atoll * lagoon * 33 ft (10 m).

1228 - *Bohadschia marmorata* * Holothuridae * Aspidochirotida * Federated States of Micronesia * Chuuk Atoll * 30 ft (9 m).

1229 - *Bohadschia* sp. * Holothuridae * Aspidochirotida * Federated States of Micronesia * Chuuk Atoll * 40 ft (12 m).

1230 - *Holothuria atra* * Holothuridae * Aspidochirotida * Hawaii * 40 ft (12 m). This holothurian is uniformly black or dark brown, soft and often covered with fine sand. It is found throughout most of the Indo-Pacific.

1231 - *Holothuria edulis* * Holothuridae * Aspidochirotida * Federated States of Micronesia * Chuuk Atoll * Tonoas Island * 40 ft (12 m). This species is easily recognized, being black dorsally and bright pink ventrally. It occurs from east Africa and Madagascar to the western Pacific Islands.

1232 - *Holothuria flavomaculata* * Holothuridae * Aspidochirotida * Federated States of Micronesia * Chuuk Atoll * Eten Island * 40 ft (12 m). The dark brown body with light brown papillae tips is distinctive in this species. It lives in lagoon reef areas and is known from the Red Sea to the western Pacific.

1233 - *Holothuria fuscopunctata* * Holothuridae * Aspidochirotida * Papua New Guinea * Port Moresby * Lion Island * 60 ft (18 m). This is a large species, found on open sandy bottoms.

1234 - *Holothuria hilla* * Holothuridae * Aspidochirotida * Marshall Islands * Enewetak Atoll * night * 10 ft (3 m). This holothurian is spotted usually with soft spires sticking out from the light areas. It hides during the day and extends itself out at night. In Hawaii this species often lives under rocks.

1226 - ***Bohadschia goeffreyi*** * Federated States of Micronesia

1227 - ***Bohadschia graeffei*** * Federated States of Micronesia

1228 - ***Bohadschia marmorata*** * Federated States of Micronesia

1229 - ***Bohadschia* sp.** * Federated States of Micronesia

1230 - ***Holothuria atra*** * Hawaii

1231 - ***Holothuria edulis*** * Federated States of Micronesia

1232 - ***Holothuria flavomaculata*** * Federated States of Micronesia

1233 - ***Holothuria fuscopunctata*** * Papua New Guinea

1234 - *Holothuria hilla* * Marshall Islands

1235 - *Holothuria leucospilota* * Hong Kong

1236 - *Holothuria nobilis* * Federated States of Micronesia

1237 - *Stichopus chloronotus* * Philippines

1235 - *Holothuria leucospilota* * Holothuridae * Aspidochirotida * Hong Kong * Cape d'Aguilar * sand and rock bottom * 3 ft (1 m). This is the most common holothurian in Hong Kong and is found throughout the tropical Indo-Pacific.

1236 - *Holothuria nobilis* * Holothuridae * Aspidochirotida * Federated States of Micronesia * Chuuk Atoll * Onang Island * 30 ft (9 m). This species is somewhat variable. In Hawaii it is reportedly black, rock hard and usually covered with sand.

1237 - *Stichopus chloronotus* * Stichopodidae * Aspidochirotida * Philippines * Cebu * Pescadero Island * 23 ft (7 m). This holothurian is easy to recognize being black with rows of pointed papillae along the corners of the body. It is common in shallow water areas from the Indian Ocean throughout the tropical Pacific to Hawaii.

1238 - *Stichopus horrens* * Stichopodidae * Aspidochirotida * Marshall Islands * Enewetak Atoll * quarry * night * 10 ft (3 m). The appearance of this species does not vary much. The milky color of the body contrasts with the spires with dark circles at their base and dark tips. This species is highly prized for bech-d'mer, dried holothurian body wall. The species is known from the Maldives, Indonesia, the Philippines, Australia to Hawaii.

1239 - *Stichopus noctivagus* * Stichopodidae * Aspidochirotida * Palau * Lighthouse Reef * 60 ft (18 m).

1240 - *Stichopus "variegatus"* * Stichopodidae * Aspidochirotida * Palau * Ngechesar * reef * 20 ft (6 m). This may well be a complex of species, as the individuals put into this species are quite variable in external appearance. The species complex is found throughout Indo-Pacific.

1241 - *Stichopus variegatus* * Stichopodidae * Aspidochirotida * Federated States of Micronesia * Chuuk Atoll * lagoon reef * 30 ft (9 m). This photo shows a large individual, over one half meter in length, identified as *S. variegatus*. This species is common in lagoonal areas of Micronesia.

1242 - *Stichopus* sp. * Stichopodidae * Aspidochirotida * Federated States of Micronesia * Chuuk Atoll * lagoon reef * 60 ft (18 m). This large holothurian may well be an undescribed species. It is found in a number of areas of Micronesia, and is common in the deeper portions of Chuuk lagoon. The fecal castings of the holothurian can be seen alongside its body.

1243 - *Tyrone* sp. * Stichopodidae * Aspidochirotida * Hong Kong * Breaker Reef * 60 ft (18 m). This ruby red sea cucumber may be an undescribed member of *Tyrone*, a poorly known genus.

1244 - *Thelenota ananas* * Stichopodidae * Aspidochirotida * Papua New Guinea * Madang * Rasch Passage * 100 ft (30 m). This is a large holothurian, reaching well over 2 feet in length. It is easily recognized by its abundant papillae which, although they appear thorn-like and sharp, are soft and flexible.

1245 - *Thelenota anax* * Stichopodidae * Aspidochirotida * Marshall Islands * Kwajalein Atoll * barrier reef * 20 ft (6 m). This is another large species which is common on reefs in Micronesia. Its shape is reminescent of a loaf of bread and it is often mottled in color. The photographed individual is sitting of a bed of the green alga *Tydemannia expeditionis*.

1246 - *Theletonota rubrolineata* * Stichopodidae * Aspidochirotida * Papua New Guinea * Port Moresby * barrier reef * 100 ft (30 m). This is a large species with fine red lines over the entire body and many spires along the corners of the body. It is found largely in deep reef environments down to at least 200 feet in Papua New Guinea and Micronesia, but the limits of its distribution are poorly known.

1247 - *Neothyonidium magnum* * Phyllophoridae * Dendrochirotida * Papua New Guinea * Port Moresby * 66 ft (20 m). This certainly is one of the more unusual sea cucumbers as it looks more like a sea anemone than anything else. The white body

1238 - *Stichopus horrens* * Marshall Islands

1239 - *Stichopus noctivagus* * Palau

1240 - *Stichopus "variegatus"* * Palau

1241 - *Stichopus variegatus* * Federated States of Micronesia

1242 - *Stichopus* sp. * Federated States of Micronesia

1243 - *Tyrone* sp. * Hong Kong

1244 - *Thelenota ananas* * Papua New Guinea

1245 - *Thelenota anax* * Marshall Islands

1246 - ***Theletonota rubrolineata*** * Papua New Guinea

1247 - ***Neothyonidium magnum*** * Papua New Guinea

1248 - ***Colochirus crassus*** * Hong Kong

1249 - ***Pseudocholchirus tricolor*** * Hong Kong

is buried in sand and only the branchial tree protrudes anemone-like above the bottom. The arms of the tree capture plankton and food particles and are inserted, one at a time, into the mouth, and the food items sucked off as the arm is pulled out of the closed mouth. This behavior must be seen to be appreciated.

1248 - *Colochirus crassus* * Cucumaridae * Dendrochirotida * Hong Kong * Breaker Reef * 50 ft (15 m). Unlike the previous species, this holothurian has the body exposed, attaching to rocks with its tube feet, and spreading its tentacles to capture particulate food which is conveyed to the mouth. The papillae visible on the body are not really sharp, but are firm, being neither soft nor flexible.

1249 - *Pseudocolchirus tricolor* * Cucumaridae * Dendrochirotida * Hong Kong * Breaker Reef * 60 ft (18 m). Commonly called the sea apple, this holothurian has a hard shell-like body. It is anchored to the bottom by its tube feet and extends the tentacles into the water to capture particulate food items. This is either the listed species or possibly *Pseudocolochirus axiologus*, but the exact identity is uncertain. Its tentacles are purple and white.

1250 - *Cucumaria miniata* * Cucumaridae * Dendrochirotida * Phlippines * Pamalican Island * 20 ft (6 m). This small yellow holothurian can occur in large numbers. It has been recorded from Indonesia, the Philippines, and Palau.

1251 - Genus species unknown * Family unknown * Dendrochirotida * Palau * Marine lake * 1 ft (0.5 m). These tiny brown and pink holothurians, less than an inch (2 cm) long, are known only from Palau where they occur in immense numbers in some marine lakes and other inshore environments. They occur on rocky bottoms and on other living organisms such as sponges and ascidians. The photograph shows a few hundred individuals on a sponge. They are probably represent an undescribed species.

1252 - *Euapta godeffroyi* * Synaptidae * Apodida * Papua New Guinea * Madang * Pig Island lagoon * night * 10 ft (3 m). The synaptid holothurians look like giant worms which are active at night. They are very soft and flexible, the body being greatly expanded with water. They might appear to be dangerous, but are not. Their sticky surface, however, makes handling them unpleasant to most people, so they are better left alone. This species is common throughout the Indo-west Pacific.

1253 - *Opheodesoma* sp. * Synaptidae * Apodida * Philippines * Cebu * Mactan Island * night * 6 ft (2 m). Synaptid holothurians crawl across the bottom with surprising rapidity . Their bodies, when inflated with water, are soft and acordian-like so they can move forward by rapid elongation and retraction of the body.

1254 - *Opheodosoma spectabilis* * Synaptidae * Apodida * Hawaii * Oahu * Coconut Island * 10 ft (3 m).

1255 - *Opheodesoma* sp. * Synaptidae * Apodida * Papua New Guinea * Dyaul Island * fringing reef * 10 ft (3 m). This photograph shows a close view of the feeding tentacles which are used to pick up organic matter from the bottom. This individual is crawling over a bed of the brown alga *Padina.*

1256 - *Synaptula* sp. * Synaptidae * Apodida * Papua New Guinea * Bagabag Island * 30 ft (9 m).

1257 - *Synaptula media* * Synaptidae * Apodida * Federated States of Micronesia * Chuuk Atoll * lagoon reef * 50 ft (15 m). This small synaptid reaches only a few inches in length, but occurs in large numbers on sponges. The photographed individuals were seen on a species of *Xestospongia*. Like other synaptids, it is sticky. It is known from Micronesia, New Caledonia and Papua New Guinea.

1250 - *Cucumaria miniata* * Philippines

1251 - Unidentified holothurian * Palau

1252 - *Euapta godeffroyi* * Papua New Guinea

1253 - *Opheodesoma* sp. * Philippines

1254 - *Opheodosoma spectabilis* * Hawaii

1255 - *Opheodesoma* sp. * Papua New Guinea

1256 - *Synaptula* sp. * Papua New Guinea

1257 - *Synaptula media* * Federated States of Micronesia

Ascidians

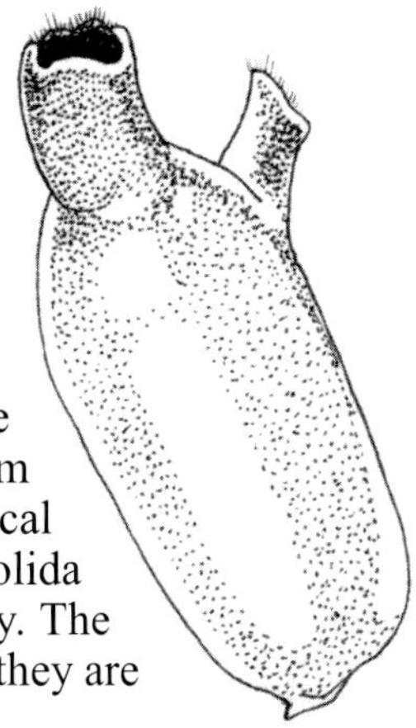

The Phylum Chordata includes not only vertebrates (which are beyond the limits of this book) but also invertebrates. Of the invertebrate chordates, the tunicates, comprising the subphylum Urochordata, are important inhabitants of the shallow water tropical Pacific. Several classes of tunicates (e.g. Pyrosoma, Salpa, Doliolida and Appendicularia) are entirely pelagic and will be covered briefly. The remaining class, Ascidiacea will be treated in most detail here since they are very common benthic animals throughout the tropical Pacific.

Class Ascidiacea - Ascidians or Sea Squirts. The ascidians are one of the those groups of invertebrates, seldom noticed or distinguished by casual divers and snorkelers, which are exceedingly interesting, diverse and important. As adults they live attached to the bottom with both solitary and colonial forms, many of which are brightly colored. They are the invertebrates most closely related to vertebrates, having common membership in the phylum Chordata, but are probably most often confused with sponges, a group near the opposite extreme of invertebrate phylogeny. Ascidians have been recognized as a group only since the end of the 19th century and in the tropical Pacific represent one of the least documented groups of marine macroorganisms. No one knows exactly how many species occur in the Indo-west Pacific region, but it numbers several hundred, if not a few thousand, species.

Ascidians take many different forms, but can be conveniently divided into solitary and colonial species. The solitary species are easiest to recognize and they are often relatively large and usually have two readily apparent siphons, even when the body is covered with other organisms growing on the surface of the ascidians. The solitary species range from the size of a grain of rice to the dimensions of a soccer ball. Solitary ascidians generally live as isolated individuals, but sometimes occur in such high densities that they resemble colonial species. The colonial species are made up of small individual units, called zooids, which can number many thousands in a large colony. They grow as sheets, stalked bouquets, large masses and lumps, and occur on many types of living and dead substrata including live coral, dead coral, rock, sponges, gorgonians, and other ascidians. Some hang, appearing to drip like candle wax, from whip coral or gorgonians. They occur in a bewildering variety of color and patterns, often with a great deal of intraspecific variation, producing some of the most spectacular visual treats to be found in the ocean.

Solitary and colonial species have a body wall, or mantle, with two openings to the outside, the oral and cloacal siphons. Long bundles of muscle fibers in the body wall can contract the body quickly, producing a fine stream of water out the siphons, thus the common name of "sea squirts". Within the body wall is the branchial sac, containing the gas exchange and food-gathering structures, plus the stomach, circulatory system, gonads and other internal organs. The colonial species often are inflated with water and if touched, the entire colony deflates, appearing to pull closer to the substratum.

Opposite- The orange bodies of these colonial ascidians, *Didemnum* sp. stand out on this reef wall in the Philippines. Ascidians are found in many reef environments, but their presence is often overlooked by divers in favor of the more familiar hard and soft corals.

Top- The yellow ascidian *Phallusia julienia*, with its large twin siphons, is an excellent example of a solitary ascidian (Philippines). Center- Each of these small colonial didemnid ascidians has a cloaca where excurrent water is vented and many tiny incurrent siphons, each one leading to an individual zooid (Indonesia). Bottom- This transparent colonial ascidian has multiple zooids radiating from the excurrent cloaca (Indonesia).

The siphons, particularly in the solitary species, are equipped with sphincters which can close the opening in an instant if the ascidian is disturbed or touched. This is their only possible response to potential danger, since they cannot flee and have no defensive structures like pincers or teeth. Some also react to the approach of a diver by siphon closure, either through detection of the pressure wave produced by a large object moving through the water or by changes in light. The sensory perception abilities of ascidians are not understood, although cells supposedly sensory in nature have been described from the epithelium of the body wall. It is uncertain what stimuli they can detect, however.

The bodies of ascidians are encased in a tunic, a somewhat flexible exoskeleton, composed largely of the polysaccharide tunicin (chemically similar to plant cellulose) in a protein matrix. The tunic anchors the animal to the substrate and maintains body shape while also playing a role in the removal and storage of wastes. Among the various species the tunic ranges from hard and rough to soft and slimy. In some the tunic is very clear, the zooids seemingly embedded in a gelatin-like substance, and such species can be extremely fragile. In quite a few ascidians, the tunic is covered with a variety of epibionts, including algae, sponges, bryozoans, and other ascidians, so that the sea squirt beneath is largely hidden.

Water enters the ascidian through the oral siphon and exits at the cloacal siphon. Within the body is a branchial sac, a perforated basket with slits called stigmata. The stigmata have cilia whose beating produces the water current through the branchial sac and ultimately the ascidian. Reportedly a 3 inch long ascidian can pass 3-4 quarts of water through its body per hour. Food particles are filtered from the water passing through the branchial sac by capture on very fine mucous webs which are moved by ciliated cells on the interior surface of the branchial sac. The fine mucous webs are rolled into a single rope-like strand and the mucous strand, with captured food particles, is ingested. Ascidians are not selective in what kind of particles are filtered from water. If excess sediment or unwanted particles are brought into the branchial sac, the ascidian contracts strongly, flushing the contents of the branchial sac back out the oral siphon, the ascidian equivalent of a cough! Ascidian blood plasma is colorless, although pigmented blood cells may be present. The circulatory system of ascidians is noted for the ability of the heart to reverse the direction of its pumping every few minutes.

Ascidians are often colorful, from a wide variety of pigments found in the blood cells, the tunic and elsewhere. Color can also come from calcareous spicules in the tunic and body wall, which occur among members of three diverse families (Didemnidae, Polycitoridae, Pyuridae). A number of ascidians incorporate sand grains into their tunics as they grow, making the animal more resistant to abrasion from sediment particles in the water, and these add color to the overall organism. Color is often not a useful character to aid in identifying ascidians, but can be used with caution to assist in assigning a specific name.

Ascidians are hermaphrodites, with independent male and female gonads in the same individual. All species shed sperm directly into the water. In some the eggs are released and externally fertilized, the fertile eggs eventually developing in open water into a tadpole-like larval form. This tadpole larvae is a free-swimming stage with a notochord and neural tube; these structures showing the common link with other chordates. In others, the eggs are fertilized inside the body and brooded until they are tadpole larvae, then released. Within a few hours of release, however, the larvae metamorphoses into a bottom-dwelling ascidian. While the tadpoles normally swim upward towards light, when ready to metamorphose they seek the bottom, attaching themselves with adhesive papillae and quickly developing into miniature ascidians. The metamorphosed ascidian quickly loses the notochord and neural tube, structures which identified it as a member of the Chordata.

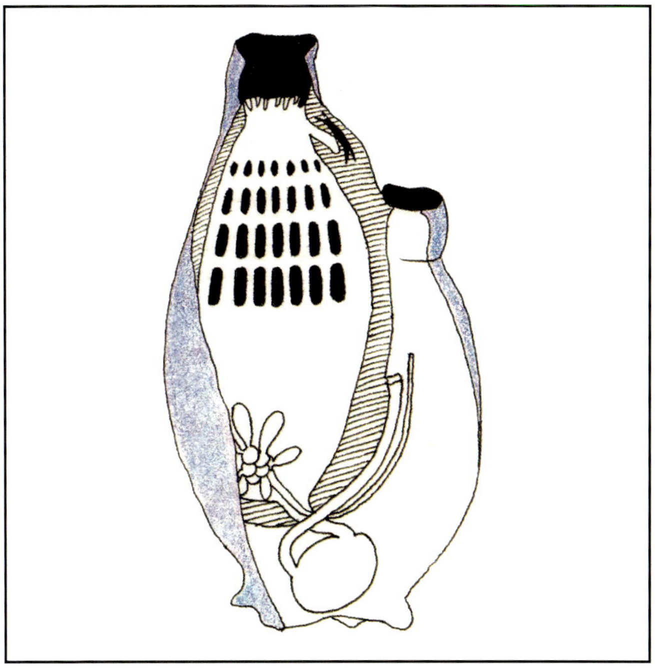

Internal structure of a solitary ascidian, with tunic and body wall (hatched), stigmata (dark marks) and internal organs.

Colonial ascidians, even those comprised of thousands of zooids covering several square feet, form from a single zooid by various methods. These include budding of individual zooids and budding (lobulation) of colonies. Colonial ascidians can grow by stolons, outgrowths of the mantle, which allow the colony to expand or to grow in one direction. Bouquet-like colonies, found in a number of genera, result when the tunic unites the colony only at the substratum. Others have the zooids embedded in a common tunic, and some of these have their cloacal apertures empty into a common cavity which then opens to the water in a common cloacal opening, with the channels in the tunic clearly visible. Among colonial species zooids are arranged in many different patterns, in rows, as rosettes, and swirls.

Top- This unidentified solitary ascidan is densely covered with a small species of sea cucumber, several types of algae and various small invertebrates (Palau). Bottom- This common unidentified ascidian from Palau has small tree-like hydroids and an orange sponge covering much of it.

Ascidians are generally found in all types of tropical marine habitats, but do not do well in soft substrata, or in areas with fluctuating or reduced salinity. In the tropical Pacific they are found from inshore areas on mangrove roots to the outer deep reefs and below. The settlement site of the larvae determine where the adult, if it survives, will spend its life. The geographic limits of distribution are not known for many species. Interestingly, there are certain ones which are cosmopolitan in the tropics. It is believed these species were transported worldwide as fouling organisms on the hulls of ships, a theory lent credence because the circumtropical species often seem to be found mostly in the vicinity of major harbors. Other ascidians are known only from one or a few locations, often the original site the ascidian was described from (type locality). Those species which do not adapt well for dispersal by ship's hull or on other drifting objects may have only limited means of dispersal, principally through their often short-lived planktonic larval stage.

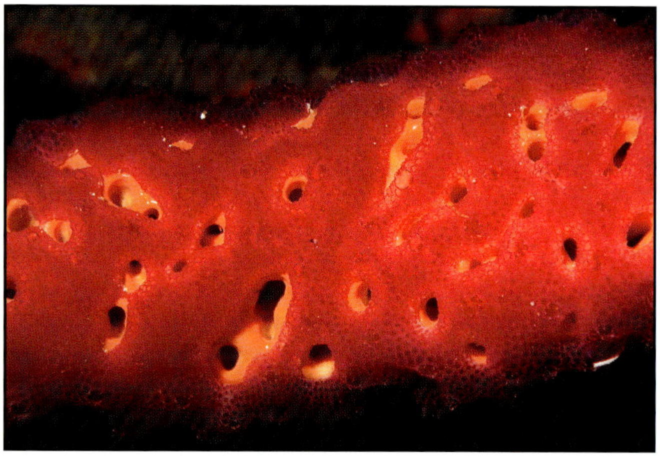

Above- Ascidians are found growing on a wide variety of other organisms. In some cases, such as this red ascidian growing on an orange sponge, they do not seem to harm their host.

Above top- Color variation in small *Clavelina* ascidians. Above- These beautiful bouquet-like *Oxycorynia fascicularis* appear to be growing on stalks. Each zooid has its own pair of siphons without a common cloaca for numerous zooids, as is found in many other colonial ascidians.

The colonial ascidians are fierce competitors for space, capable of rapid overgrowth of many other types of organisms, including stony corals, sponges, bryozoans and bivalve molluscs. This often results in the death of the overgrown organism. The opposite situation, colonial ascidians being overgrown to the point of being overwhelmed, occurs much less frequently. Solitary species, however, which are possibly long lived, are often densely covered with epibionts, including colonial ascidians. Some of these other organisms are commensals and parasites of ascidians, and include shrimps, copepods, amphipods, molluscs, and ciliate protozoans. Additional small mobile organisms can take up residence, so large solitary ascidians can become a discrete biological community, covered to the point that only the open siphons are exposed. Predators of ascidians include asteroids, polychaete worms, and some fishes and are, at best, poorly known.

Above- Some colonial ascidians occurs as a single stalk with head containing many zooids on it (Phillipines).

Ascidians are known to grow on man-made objects. Ascidians are resistant to many pollutants, so that polluted harbors are often an environment where these creatures do exceedingly well and they actually help in cleaning up polluted waters. They are capable of filtering bacteria from seawater and can concentrate and store heavy metals, such as vanadium, and hydrocarbons in their tunics. A number of cytotoxic compounds are also found in ascidians and the group has become a prime candidate for discovery of potential medicinal compounds from the sea.

Ascidians consist of three orders which are separated using morphological characters of the branchial sac. The Aplousobranchia have the simplest branchial sac and are all colonial species; they comprise three large families, Polyclinidae, Polycitoridae and Didemnidae. The Phlebobranchia have a more complicated branchial sac and there are several families which include both solitary and colonial forms. The Stolidobranchia have the most complicated branchial sac, often folded inside, and while both solitary and colonial, include many of the largest species in its three families, Styelidae, Pyuridae and Molgulidae.

Two genera of unicellular algae, the prochlorophyte *Prochloron* and the procyanophyte *Synechocystis*, are symbiotic in ascidians. Members of *Prochloron* are found on and near the ascidian's surface, where they can be easily dislodged by rubbing the colony, and within the cloacal chambers of some species, such as the very common *Didemnum molle*. Additionally, *Prochloron* and *Synechocystis* occur within the tissues and tunic of some didemnids, the latter alga tending to color colonies pink with their red pigments. Interestingly, the ascidians do not harbor the symbiotic dinoflagellate *Symbiodinium microadriaticum* (zooxanthellae), found in many other tropical invertebrates, including stony corals, sponges and cnidarians.

Class Thaliacea- This group includes the salps, the pyrosomes and the doliolids. They are pelagic urochordates, part of the "jelly plankton" and are occasionally seen around reefs. Some are very clear and transparent and most are generally small, on the order of two to rhree inches long. *Pyrosoma atlanticum*, a colonial species, can sometime grow so large a diver can enter the central cloacal cavity. *Pyrosoma* can be spectacularly bioluminescent. *Tethys* can also attain large sizes by forming long chains of medium-sized individuals.

Class Appendicularia- The appendicularians or larvaceans are a small group in which the adult retains throughout its life some larval characteristics, such as the "tail", which is lost in the metamorphosis of ascidians and salps. Larvaceans feed by straining food drawn into the "house" by their beating tail.

Above top- *Procholoron* algae inside a large *Didemnum molle* are visible as a deep green color in the area inside the cloaca. Above- These small *Lissoclinum bistriatum* are green from symbiotic algae present in their tissues.

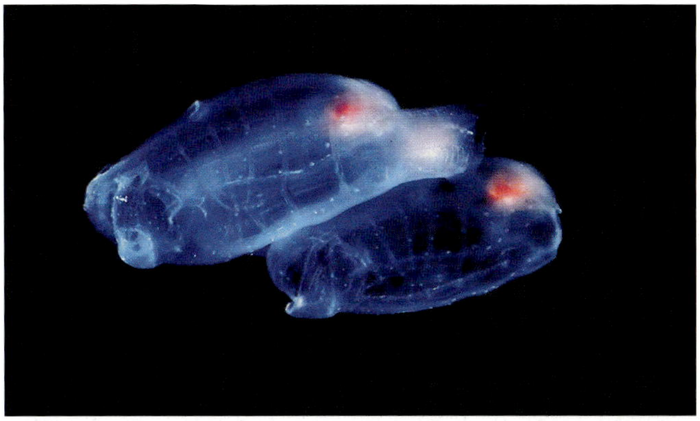

Above top- Two individuals of the salp *Tethys vagina* were photographed in open water. Above- *Doliola* sp. is similar in appearance to salps, but belongs in the closely related doliiolids.

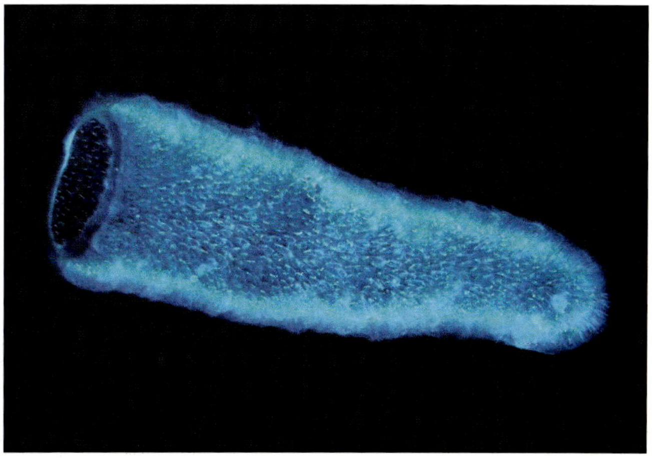

Above- The salp *Pyrosoma atlanticum* can reach lengths of more than 10 ft (3 m) and is found in open water.

1258 - *Aplidiopsis* sp. * Indonesia

1259 - *Pseudodistoma* sp. * Papua New Guinea

1260 - *Pseudodistoma* sp. * Federated States of Micronesia

1261 - *Pseudodistoma fragilis* * Papua New Guinea

1258 - *Aplidiopsis* sp. * Polyclinidae * Aplousobranchia * Indonesia * Manado * 23 ft (7 m). This beautiful ascidian, probably an undescribed species, has a circular system of zooids around a common cloaca. This is a series of colonies, the zooids embedded in a firm tunic, and growing among sponges, soft corals and crinoids. Crinoid arms lie above the ascidian and a contracted pink *Dendronephthya* soft coral is visible in a lower corner.

1259 - *Pseudodistoma* sp. * Polyclinidae * Aplousobranchia * Papua New Guinea * Madang * Rasch Passage * 60 ft (18 m). This is probably a new species. Ascidians of this form can be found in a wide variety of colors.

1260 - *Pseudodistoma* sp. * Polyclinidae * Aplousobranchia * Federated States of Micronesia * Chuuk Atoll * Pizion Reef * 115 ft (35 m) depth. This is an undescribed species which may be found only in Micronesia. It occurs only on the dropoff of the barrier reef, starting at depths of about 100 ft (30 m). There are many zooids on each stalk and many stalks make up a cluster of this beautiful ascidian.

1261 - *Pseudodistoma fragilis* * Polyclinidae * Aplousobranchia * Papua New Guinea * West New Britain * Kimbe Bay * 83 ft (25 m). This is a gelatinous translucent ascidian, and as its name implies, quite fragile. Each zooid has a single white line, possibly due to colored spicules, and a number of these radiate out from a cloaca where water is expelled. A few other colonies can be seen alongside and in the background of the photo. This species seems to be found only on deep reef dropoffs, living on overhanging portions of ledges and caves.

1262 - *Aplidium longithorax* * Polycitoridae * Aplousobranchia * Palau * Rock Islands * 40 ft (12 m). This attractive yellow species is fairly common in Palau where it lives in some of the inshore channels. The colony in the photograph has a black coral *Antipathes elegans* growing around it.

1263 - *Aplidium tabascum* * Polycitoridae * Aplousobranchia * Papua New Guinea * Madang Channel * 100 ft (30 m). This is a colorful species encrusting on exposed rocky areas of dropoffs.

1264 - *Aplidium* sp. * Polycitoridae * Aplousobranchia * Indonesia * Manado * Torowitan * 110 ft (33 m). This white ascidian has a subtle beauty, with a delicate pattern of grooves and zooids visible on its outer surface. It occurs along deep dropoffs on exposed rock, particularly on overhangs.

1265 - *Aplidium* sp. * Polycitoridae * Aplousobranchia * Papua New Guinea * New Ireland * Albatross Channel mouth * vertical wall * 100 ft (30 m). These fingery ascidians are easy to confuse with sponges, particularly at depth where light is low and the fine structure hard to distinguish. The fine pattern of zooids and excurrent cloaca can be seen on close examination of the photos. This species is undescribed and occurs only on deep reef vertical walls.

1266 - *Clavelina detorta* * Polycitoridae * Aplousobranchia * Philippines * Pamalican Island * 40 ft (12 m). The genus *Clavelina* has some of the most beautiful ascidians with transparent tunics and brightly colored structures internally. In this species the branchial basket is green with the gut gold in color. These occur in colonies of individual zooids, all connected by stolons, but they are actually separate individuals. This colony is growing on a red sponge. This species is widespread in the region.

1267 - *Clavelina flava* * Polycitoridae * Aplousobranchia * Indonesia * Biak Island * 33 ft (10 m). This small species shows the twin siphons of each zooid clearly. The small individual ascidians grow in groups of up to many thousands of zooids and can occur on any hard substrate of the reef.

1268 - *Clavelina meridionalis* * Polycitoridae * Aplousobranchia * Indonesia * Manado * 133 ft (40 m). This tiny deep reef ascidian is quite unusual in appearance, but the twin siphons on the outer end give it away instantly as an ascidian. We have seen this species only in Indonesia where it was found growing as widely scattered individuals on sloping rock bottom in deep water.

1269 - *Clavelina moluccensis* * Polycitoridae * Aplousobranchia * Philippines * Pamalican Island * 40 ft (12 m). This translucent species has colored markings near the siphons of each zooid, a feature which is found in many species of *Clavelina* and *Paraclavelina*. The branchial baskets of this cluster of ascidians are visible internally. A number of tiny hydroids with white polyps stick up among the zooids in the photo.

1270 - *Clavelina zobustra* * Polycitoridae * Aplousobranchia * Indonesia * Manado * reef overhang * 40 ft (12 m). This is quite a common species throughout the region. It does show some geographic variation in the markings around the siphons. It is found from about 3 to over 100 ft (30 m) depth.

1262 - *Aplidium longithorax* * Palau

1263 - *Aplidium tabascum* * Papua New Guinea

1264 - *Aplidium* sp. * Indonesia

1265 - *Aplidium* sp. * Papua New Guinea

1266 - *Clavelina detorta* * Philippines

1267 - *Clavelina flava?* * Indonesia

1268 - *Clavelina meridionalis* * Indonesia

1269 - *Clavelina moluccensis* * Philippines

1270 - *Clavelina zobustra* * Indonesia

1271 - *Clavelina* sp. * Papua New Guinea

1272 - *Clavelina* sp. * Federated States of Micronesia

1273 - *Distaplia regina* * Federated States of Micronesia

1271 - *Clavelina* sp. * Polycitoridae * Aplousobranchia * Papua New Guinea * West New Britain * 66 ft (20 m). This is quite a colorful species of *Clavelina* found along reef drop offs in Kimbe Bay.

1272 - *Clavelina* sp. * Polycitoridae * Aplousobranchia * Federated States of Micronesia * Chuuk Atoll * Nematon Bay * 33 ft (10 m). This is a picture of another attractive *Clavelina*. In this case the common attachment point of several zooids can be seen. There are many species of *Clavelina*, many of them unknown or poorly known.

1273 - *Distaplia regina* * Polycitoridae * Aplousobranchia * Palau * Malakal Harbor * 120 ft (36 m). This small hemispherical ascidian occurs on sediment bottoms at 110-130 feet in Palau where, because of the low light and usually poor visibility, it looks like a little lump of nothing on the bottom. While not particularly large, a number of much smaller red ascidians (somewhat covered with sediment) also occur on the bottom in the photo. We did not even know these were there, being so difficult to see at depth, but the photo reveals their presence.

1274 - *Eudistoma laysani* * Polycitoridae * Aplousobranchia * Philippines * Cebu * 0-3 ft (1 m).

1275 - *Eudistoma laysani* * Polycitoridae * Aplousobranchia * Palau * Ngercheu * 0-3 ft (1 m) (intertidal). This species is a common inhabitant of mangrove roots throughout much of the region. The color varies somewhat from area to area, but these are believed to all be the same species.

1276 - *Eudistoma reginum* * Polycitoridae * Aplousobranchia * Indonesia * Manado * Bunaken Island * caves * 60 ft (18 m). This purple ascidian has a heavy thick tunic in which the zooids are deeply embedded. This variety of this *Eudistoma* was common in the Bunaken Marine Park in North Sulawesi, Indonesia and illustrates the color variety that can occur in one species of ascidian. This is the same species as 1278, but a different color.

1277 - *Eudistoma reginum* * Polycitoridae * Aplousobranchia * Indonesia * Manado * 60 ft (18 m). The genus *Eudistoma* has many species, some of which (but by no means all) have their zooids arranged circularly around the common cloaca and embedded deeply in a thick, tough tunic. This chocolate brown ascidian looks good enough to eat! Also seen in the photograph is a white encrusting didemnid ascidian on the left and a distinctive yellow and purple veined *Polycarpa auriculata i*n the lower left.

1278 - *Eudistoma* sp. * Polycitoridae * Aplousobranchia * Indonesia * Manado * 16 ft (5 m). This greenish sheet ascidian illustrates another of the diverse forms of the genus *Eudistoma*, a thin sheet without a clear organization to the zooids and the cloacal apertures on the apex of the small raised papillae of the colony. Color can be quite variable in members of this genus.

1279 - *Eudistoma* sp. * Polycitoridae * Aplousobranchia * Indonesia * Manado * 140 ft (43 m). This colonial encrusting species has sand grains incorporated into its tunic, not only increasing the resistance to abrasion of the colony, but also effectively camouflaging it on a rocky bottom. The colony is thick, tough and very firmly attached to the bottom. The cloaca are visible, but if the colony is disturbed these close so tightly they essentially disappear. This colony cannot be ascribed to a particular species.

1280 - *Exostoma* sp. * Polycitoridae * Aplousobranchia * Federated States of Micronesia * Chuuk Atoll * lagoon fringing reef * 16 ft (5 m). This sort of ascidian is easy to confuse with sponges. They form sizable masses, the size of a fist or larger, and appear to have the ostia and oscules of a sponge. One quick way to tell if it is sponge or ascidian is to gently touch it. If it is an ascidian, it will usually contract and close the siphons. If it is sponge, it usually will not react.

1281 - *Exostoma* sp. * Polycitoridae * Aplousobranchia * Federated States of Micronesia * Chuuk Atoll * lagoon fringing reef * 16 ft (5 m). This species is fairly common on lagoon reefs in Chuuk.

1282 - *Sigillina* sp. * Polycitoridae * Aplousobranchia * Indonesia * Talisei * 60 ft (18 m). The genus is generally not found in tropical waters.

1283 - *Didemnum* sp.* Didemnidae * Aplousobranchia * Papua New Guinea * Manam Island * 50 ft (15 m). Many species of *Didemnum* grow as encrusting sheets over a surface. Once well established on that surface, they may grow outward. In this case, the ascidian is growing extensions which hang vertically in the water. If they eventually grow to the point they touch another hard surface, the ascidian will begin to grow out on that.

1274 - ***Distaplia regina*** * Palau

1275 -***Eudistoma laysani*** * Palau

1276 - ***Eudistoma reginum*** * Indonesia

1277 - ***Eudistoma reginum*** * Indonesia

1278 - ***Eudistoma* sp.** * Indonesia

1279 - ***Eudistoma* sp.** * Indonesia

1280 - ***Exostoma* sp.** * Federated States of Micronesia

1281 - ***Exostoma* sp.** * Federated States of Micronesia

1282 - ***Sigillina* sp.** * Indonesia

1283 - ***Didemnum* sp.*** Papua New Guinea

1284 - ***Didemnum molle*** * Federated States of Micronesia

1285 - ***Hypodistoma deeratum*** * Papua New Guinea

1284 - *Didemnum molle* * Didemnidae * Aplousobranchia * Federated States of Micronesia * Chuuk Atoll * lagoon reef *20 ft (6 m). *Didemnum molle* is common on reefs and rocky areas of the Indian and Pacific Oceans. This is undoubtedly the most abundant ascidian on nearly every Pacific coral reef and, along with *Polycarpa aurata*, the most often noticed. This is a colonial ascidian with a very flaccid body inside its surface "skin". The distinct white and green individuals are the same species. The green coloration comes from *Prochloron* algae in the tissues, while the variation in whiteness comes from the coloration of star-shaped spicules in the tunic of the ascidian. The tiny incurrent openings are visible on the surface as tiny pores. The large excurrent cloaca dominates the upper surface and the green color indicates plenty of *Prochloron* inside the ascidian.

1285 - *Hypodistoma deeratum* * Polycitoridae * Aplousobranchia * Papua New Guinea * West New Britain * Kimbe Bay * reef * 50 ft (15 m). This is a large fleshy ascidian common in Papua New Guinea. Usually they resemble the colony in the photograph, but we have also seen examples which had only a single conical mass with large excurrent cloaca, which were uniformly brown with no whitish mottling.

1286 - *Oxycorynia fascicularis* * Polycitoridae * Aplousobranchia * Papua New Guinea * Madang * Pig Island * 30 ft (9 m). This ascidian has a very clear stalk on which the zooids sit. This is one of the many forms that the ascidians take.

1287 - *Sigillina* sp. * Polycitoridae * Aplousobranchia * Papua New Guinea * Eastern Fields * 100 ft (30 m). This is another stalked species with a translucent head with the zooids. This species has only been found along deep reef dropoffs.

1288 - *Sigillina signifera* * Polycitoridae * Aplousobranchia * Federated States of Micronesia * Chuuk Atoll * Yanagi Island * 10 ft (3 m). This species is common in Micronesia where it occurs as thick gelatinous masses on shallow reefs, occasionally covering large areas of bottom. The intensity of the dark blue-green color of the zooids varies from area to area. The species may well be capable of killing some corals and taking over space from them.

1289 - *Didemnum gutatum* * Didemnidae * Aplousobranchia * Papua New Guinea * Eastern Fields * 66 ft (20 m). This species occurs as a rubbery sheet which covers over rock and corals. It is evidently capable of overgrowing a number of species of living corals, such as the *Porites nigrescens* shown in the photograph, killing the coral by cutting off light and water circulation. From the basal mass of the ascidian, it grows up the branches of the coral, eventually covering the entire branch. Some branches in the photograph still have their tips uncovered, but it is only a matter of time until those too are smothered. This is an excellent example of the continuous competition for space on a healthy reef.

1290 - *Didemnum molle* * Didemnidae * Aplousobranchia * Philippines * Pamalican Island *20 ft (6 m). *Didemnum molle* in some areas grows to the size of a fist, still flaccid, with their deep green *Prochloron* algae easily visible in the tissues of the ascidian.

1291 - *Didemnum mosleyi* * Didemnidae * Aplousobranchia * Federated States of Micronesia * Chuuk Atoll * Basis Island patch reef * 13 ft (4 m). This didemnid has tiny colonies, with a few dozen zooids clumped together. The colonies, though, can occur in huge numbers so that areas many square feet across can be dominated by these tiny ascidians. There is believed to be considerable color variation in *D. mosleyi*, from near white to orange and yellow.

1292 - *Didemnum* sp. * Didemnidae * Aplousobranchia * Philippines * Pamalican Island * 50 ft (15 m). This is possibly one of the color variations of *Didemnum mosleyi*, but its identity is not certain.

1293 - *Didemnum psammathodes* * Didemnidae * Aplousobranchia * Papua New Guinea * Port Moresby * Pt. Osbourne * 33 ft (10 m). This gray, nondescript ascidian occurs in silty environments where it overgrows any hard objects available. It can be quite common in formerly healthy reef areas that are being smothered by fine sediment, as the dead coral and gorgonians make excellent sites for growth of this sheet ascidian.

1294 - *Didemnum rubium* * Didemnidae * Aplousobranchia * Palau * Airai Channel * 33 ft (10 m). This species occurs as encrusting sheets made

1286 - *Oxycorynia fascicularis* * Papua New Guinea

1287 - *Sigillina* sp. * Papua New Guinea

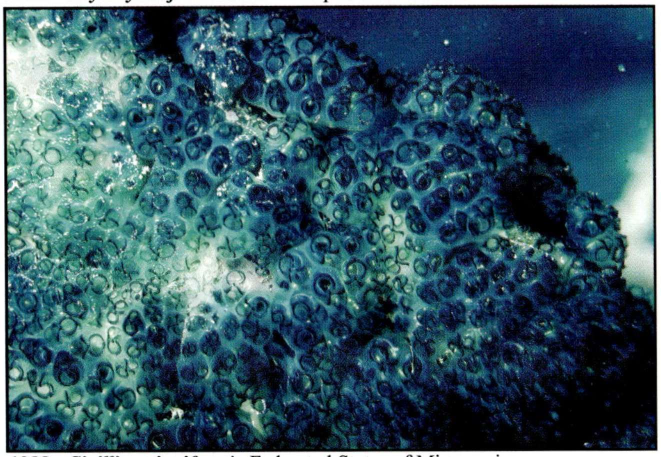
1288 - *Sigillina signifera* * Federated States of Micronesia

1289 - *Didemnum gutatum* * Papua New Guinea

1290 - *Didemnum molle* * Philippines

1291 - *Didemnum mosleyi* * Federated States of Micronesia

1292 - *Didemnum* sp. * Philippines

1293 - *Didemnum psammathodes* * Papua New Guinea

1294 - ***Didemnum rubium*** * Palau

1295 - ***Didemnum* sp.** * Papua New Guinea

1296 - ***Didemnum* sp.** * Papua New Guinea

1297 - ***Didemnum* sp.** * Philippines

up of mammilate groups of zooids clustered around a single cloaca. It is usually inflated with water, resulting in the appearance shown in the photograph, but it can also retract to form a thin featureless sheet which makes it hard to recognize as the same ascidian. This species is commonly found on the sides of clear water channels in Micronesia.

1295 - *Didemnum* sp. * Didemnidae * Aplousobranchia * Papua New Guinea * Dyaul Island * 50 ft (15 m). This white didemnid is growing alongside a dark sponge on a rock ledge. It is not easy to distinguish the ascidian from a sponge underwater.

1296 - *Didemnum* sp. * Didemnidae * Aplousobranchia * Papua New Guinea * Duke of York Islands * Ulu Reef * 33 ft (10 m). This white didemnid is shown among a wide variety of other benthic invertebrates. *Dendronephthya* soft corals occur on either side, with ascidians, including a *Polycarpa aurata*, occupying much of the remainder of the photo.

1297 - *Didemnum* sp. * Didemnidae * Aplousobranchia * Philippines * Pamalican Island *20 ft (6 m). This attractive ascidian always has the pale areas over the excurrent channels. It is known from Micronesia and the Philippines. Like many other didemnids, it is probably undescribed. The family Didemnidae has relatively few morphological characters which are useful for distinguishing species.

1298 - *Didemnum* sp. * Didemnidae * Aplousobranchia * Federated States of Micronesia * Chuuk Atoll * lagoon reef * 33 ft (10 m). Another of the "drippy" encrusting didemnids, this species is common in Micronesia. Notice how similar this species is to the previous species, although they are believed to represent separate species!

1299 - *Didemnum* sp. * Didemnidae * Aplousobranchia * Federated States of Micronesia * Chuuk Atoll * lagoon reef * 66 ft (20 m). This white didemnid, although similar in appearance to previous white didemnids, is believed to be a different species. It is found in Chuuk in some caves in lagoon reefs and channels, plus on some of the shipwrecks in the lagoon.

1300 - *Didemnum* sp. * Didemnidae * Aplousobranchia * Indonesia * Talisei * fringing reef * 133 ft (40 m). This is an encrusting species, its thin tunic closely covering the area it is overgrowing so that every detail on the structure underneath is visible. The illustrated specimen is growing on dead coral. It is known from Palau and Indonesia, similar to *D. viride.*

1301 - *Didemnum* sp. * Didemnidae * Aplousobranchia * Federated States of Micronesia * Chuuk Atoll * Ozen Island * 50 ft (15 m). This is another new species of *Didemnum* found in Micronesia which grows as a thin encrusting sheet over rock and coral. The green color almost certainly comes from symbiotic algae.

1302 - *Diplosoma similis* * Didemnidae * Aplousobranchia * Indonesia * Manado * 27 ft (8 m). This greenish ascidian is usually found growing around the branches of dead *Acropora* coral. It is found in tropical areas of the Pacific and Indian Oceans.

1303 - *Diplosoma virens* * Didemnidae * Aplousobranchia * Palau * Marine River * 1 ft (0.5 m). This lovely green species is full of symbiotic algae. The species is known from the Indian Ocean and much of the tropical Pacific.

1304 - *Diplosoma* sp. * Didemnidae * Aplousobranchia * Papua New Guinea * Eastern Fields * vertical wall * 66 ft (20 m). This clear ascidian is undescribed. It is known only from sheer reef walls at Eastern Fields, an atoll in the Coral Sea. In different areas of the outer wall of Eastern Fields, with identical habitat, the ascidians were either fairly common or it simply did not occur. This may be a result of the limited dispersal abilities. Many ascidians can only spread widely during their swimming larval stage.

1305 - *Leptoclinides reticulatus* * Didemnidae * Aplousobranchia * Papua New Guinea * Duke of York Islands * Mioko Reef * 40 ft (12 m). This is an encrusting species which has faint orange rings around the incurrent siphons and fine white rings around the cloacal openings.

1306 - *Leptoclinides* sp. * Didemnidae * Aplousobranchia * Indonesia * Sulawesi * Ruang Island * reef overhang * 83 ft (25 m). This is another encrusting species, typical of the genus, which forms a thin well-attached sheet over rock. There are a great many color variations of these ascidians, some, representing separate species others, intraspecific variation.

1307 - *Leptoclinides* sp. * Didemnidae * Aplousobranchia * Federated States of Micronesia * Chuuk Atoll * Nematon Bay * Polle Reef * 33 ft (10 m). This is another thin sheet with variable coloration.

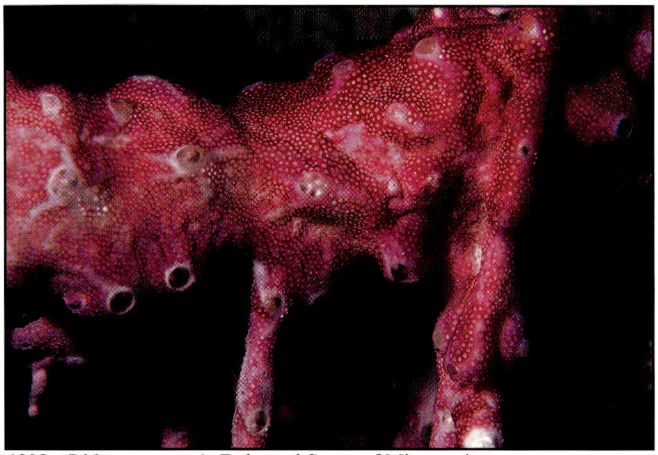
1298 - **Didemnum sp.** * Federated States of Micronesia

1299 - **Didemnum sp.** * Federated States of Micronesia

1300 - **Didemnum sp.** * Indonesia

1301 - **Didemnum sp.** * Federated States of Micronesia

1302 - **Diplosoma similis** * Indonesia

1303 - **Diplosoma virens** * Palau

1304 - **Diplosoma sp.** * Papua New Guinea

1305 - **Leptoclinides reticulatus** * Papua New Guinea

1306 - ***Leptoclinides* sp.** * Indonesia

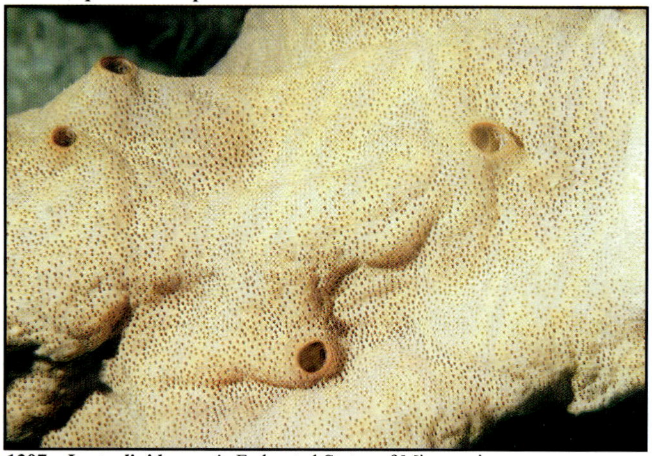

1307 - ***Leptoclinides* sp.** * Federated States of Micronesia

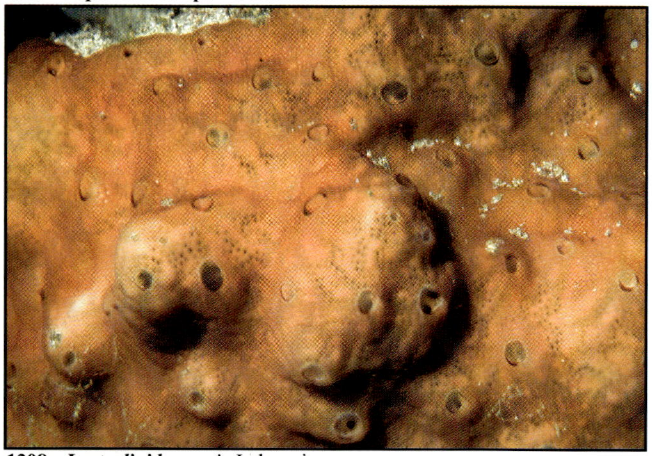

1308 - ***Leptoclinides* sp.** * Indonesia

1309 - ***Lissoclinum bistratum*** * Palau

1308 - *Leptoclinides* sp. * Didemnidae * Aplousobranchia * Indonesia * Manado * 125 ft (38 m). This is an encrusting didemnid which occurs on dead corals and rocky substratum in deep reef environments. Like all didemnids it has tiny zooids. The cloacal opening of each zooid empties into a system of channels, which combine to form the large cloacal openings seen here. This colony is inflated. The genus *Leptoclinides* possesses spicules.

1309 - *Lissoclinum bistratum* * Didemnidae * Aplousobranchia * Palau * inshore reef * 6 ft (2 m). This small ascidian looks more like an alga than an animal, but again the symbiotic algae are responsible for the color of the ascidian. This species encrusts on rock and other hard substrates. The shape of the individual colonies can vary, from the rounded ones seen in the photograph to flattened, button-like colonies.

1310 - *Lissoclinum japonicum* * Didemnidae * Aplousobranchia * Papua New Guinea * Eastern Fields * 60 ft (18 m). This bronze colored ascidian is not common, and can be confused with large *Didemnum molle*, which it resembles except for color which may not be apparent at depth.

1311 - *Lissoclinum patella* * Didemnidae * Aplousobranchia * Papua New Guinea * Madang * Pig Island reef * 10 ft (3 m). This is an ascidian, that forms large flattened masses with distinct valleys and depressions on its upper surface. The species varies from near white to dark green depending on the density of symbiotic algae, often with differences in color between valleys and ridges on the surface. The photographed individual is living among live coral, a common locality for this species which likes level reef bottoms with abundant light. The species is well known and is widely distributed, including the Indian Ocean to the central tropical Pacific.

1312 - *Lissoclinum* sp. * Didemnidae * Aplousobranchia * Federated States of Micronesia * Chuuk Atoll * Nematon Bay * patch reef * 33 ft (10 m). This is another new species, with an attractive mottled pattern. It is known from Chuuk at depths of six to seventy feet on lagoon reefs.

1313 - *Lissoclinum* sp. * Didemnidae * Aplousobranchia * Papua New Guinea * Eastern Fields * 215 ft (65 m). This is another new species known only from deep water at Eastern Fields, an isolated Coral Sea atoll about 100 miles from Port Moresby. It was found on a vertical wall growing on a finely branched gorgonian.

1314 - *Trididemnum cyclops* * Didemnidae * Aplousobranchia * Papua New Guinea * Kranket Island * lagoon shore * 3 ft (1 m). This is another one of the dark green ascidians, again due to symbiotic algae. This species strongly resembles *Diplosoma virens*, but is different upon close examination.

1315 - *Diazona* sp. * Diazonidae * Phlebobranchia * Indonesia * Manado * 40 ft (12 m). This translucent species has a beautiful circle of white spots around each siphon. The translucent and gelatinous ascidians have the same general appearance, but close examination will show that they all have significant differences in color and morphology.

1316 - *Diazona* sp. * Diazonidae * Phlebobranchia * Indonesia * Manado * 66 ft (20 m). This delicate ascidian, possibly an undescribed species, occurs as clusters of zooids with a common base, often near the previous species which it superficially resembles. The tunic is translucent and the white lines on each zooid stand out. Close comparison of this and the previous species of *Diazona* show numerous differences.

1317 - *Diazona* sp. * Diazonidae * Phlebobranchia * Philippines * Cebu * Mactan Island * cave * 80 ft (24 m). This member of *Diazona* was found in a dark reef cave in the Philippines.

1318 - *Rhopalaea* sp. * Diazonidae * Phlebobranchia * Philippines * Cebu * Mactan Island * 50 ft (15 m). This *Rhopalaea* is very similar to the previous species, but has gold rims around the openings of the siphons. It may represent another species.

1319 - *Rhopalaea* sp. * Diazonidae * Phlebobranchia * Palau * Ngerkuul Pass * 16 ft (5 m). This species of *Rhopalaea* is common in the rock channels around Palau, even in very shallow water.

1320 - *Ascidia* sp. * Ascidiidae * Phlebobranchia * Indonesia * Manado * 100 ft (30 m). This large solitary ascidian appears similar to *Rhopalaea*, but is in a different family. Identifications of most ascidians rely on internal structures of the zooid and unless a species is otherwise distinctive and well known, it is risky to base ascidian identifications on photographs.

1321 - *Diazona chinensis* * Diazonidae * Phlebobranchia * Indonesia * Manado * 43 m. This diazonid has the zooids deeply embedded in its thick, firm tunic. The tunic can be colorless with the zooids having a slight yel-

1310 - *Lissoclinum japonicum* * Papua New Guinea

1311 - *Lissoclinum patella* * Papua New Guinea

1312 - *Lissoclinum* sp. * Federated States of Micronesia

1313 - *Lissoclinum* sp. * Papua New Guinea

1314 - *Trididemnum cyclops* * Papua New Guinea

1315 - *Diazona* sp. * Indonesia

1316 - *Diazona* sp. * Indonesia

1317 - *Diazona* sp. * Philippines

1318 - ***Rhopalaea* sp.** * Philippines

1319 - ***Rhopalaea* sp.** * Palau

1320 - ***Ascidia* sp.** * Indonesia

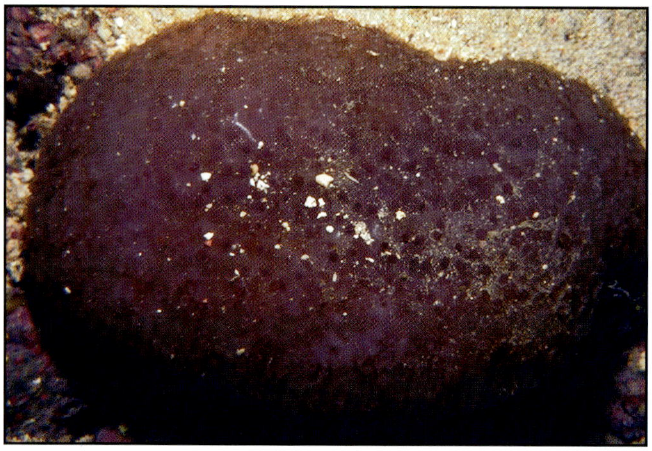
1321 - ***Diazona chinensis*** * Indonesia

lowish ring around the incurrent opening. It is found in deep reef dropoff areas below 100 ft (30 m).

1322 - *Diazona* sp. * Diazonidae * Phlebobranchia * Indonesia * Manado * 110 ft (33 m). This *Diazona* has the zooids deeply embedded in a thick, translucent tunic. It superficially resembles a coral, except it is soft, but firm.

1323 - *Rhopalaea crassa* * Diazonidae * Phlebobranchia * Papua New Guinea * Bagabag Island * fringing reef * 66 ft (20 m). The clear and colored tunics of *Rhopalaea* are graced with delicate colored lines which differ between species. These are among the most lovely of ascidians. This species is found on reef dropoffs at 12 to 100 ft (30 m) depths.

1324 - *Rhopalaea* sp. * Diazonidae * Phlebobranchia * Indonesia * Manado * 33 ft (10 m). This small blue species is close to *R. crassa*, but differs in several regards. The genus is known through the tropical Indo-Pacific, with probably several undescribed species similar to *R. crassa*.

1325 - *Ascidia* sp. * Ascidiidae * Phlebobranchia * Papua New Guinea * New Britain * Kimbe Bay * Restorf Island * 33 ft (10 m). It is common for large solitary ascidians to grow in crevices with only their siphons protruding, as is seen in this photograph. Different species have different color markings around the siphons.

1326 - *Ascidia* sp. * Ascidiidae * Phlebobranchia * Philippines * Santa Rosa * 66 ft (20 m). This solitary *Ascidia* shows another variation in siphon color. Again the ascidian is growing in a crevice with the siphons exposed.

1327 - *Ascidia* sp. * Ascidiidae * Phlebobranchia * Marshall Islands * Enewetak Atoll * Medren pinnacle * 10 ft (3 m). This translucent species usually grows beneath dead coral plates or other dark areas with good water circulation.

1328 - *Phallusia arabica* * Ascidiidae * Phlebobranchia * Indonesia * Bunaken Island * 66 ft (20 m). This solitary ascidian is known from the Red Sea, Madagascar, Sri Lanka, to the Philippines and northeastern Australia. This individual was on a reef face in the Bunaken Marine Park in North Sulawesi, Indonesia and is embedded in a crevice among the sand-lined tubes of an unidentified worm. Unlike its bright yellow relative *P. julinea, P. arabica* does not advertise its presence.

1329 - *Phallusia julinea* * Ascidiidae * Phlebobranchia * Marshall Islands * Enewetak Atoll * patch reef *20 ft (6 m). This is among the brightest colored of the Ascidiidae. It is reported to be the most common big ascidian in New Caledonia, reaching lengths of one foot. The yellow color comes from the pigmented blood cells. The tunic is clear and is never covered with epibionts. Other members of this family in the Ascidia are often nearly colorless and transparent, such as *A. munda.*

1330 - *Plurella* sp. * Plurellidae * Phlebobranchia * Papua New Guinea * Manam Island * black sand slope * 66 ft (20 m). Members of *Plurella* often occur nearly buried in sand, where only the siphons are exposed. In such a situation, if the colony is touched, all the siphons close immediately making it appear that the sand reacts when touched. This appears to be the most common species of *Plurella*, at least in Papua New Guinea, with the yellow siphon rims distinctive.

1331 - *Plurella* sp. * Plurellidae * Phlebobranchia * Philippines * Cebu * Mactan Island * 66 ft (20 m). This is a second species of *Plurella*, with a different shape and color from the previous species. The taxonomy of the genus is poorly known, so it is uncertain whether these species are undescribed and what their geographic ranges might be.

1332 - *Perophora namei* * Perophoridae * Phlebobranchia * Indonesia * Biak Island * 100 ft (30 m). This lovely ascidian is one of a group which have the tiny zooids on a stem that grows out from the reef face. Needless to say, such ascidians are not immediately obvious and only with careful searching in the right area can they be found.

1333 - *Botryllus* sp. * Styelidae * Stolidobranchia * Papua New Guinea * Port Moresby * Basilisk Passage * 100 ft (30 m). This undescribed species of *Botryllus* occurs in Papua New Guinea and perhaps Indonesia. It varies somewhat in color, but is always encrusting as is shown in the photograph.

1334 - *Botryllus* sp. * Styelidae * Stolidobranchia * Indonesia * Manado * 66 ft (20 m). *Botryllus* contains some of the most colorful of ascidians. Color within the species of the genus is also quite variable. This colony from

1322 - ***Diazona* sp.** * Indonesia

1323 - ***Rhopalaea crassa*** * Papua New Guinea

1324 - ***Rhopalaea* sp.** * Indonesia

1325 - ***Ascidia* sp.** * Papua New Guinea

1326 - ***Ascidia* sp.?** * Philippines

1327 - ***Ascidia* sp.** * Marshall Islands

1328 - ***Phallusia arabica*** * Indonesia

1329 - ***Phallusia julinea*** * Marshall Islands

1330 - ***Plurella* sp.** * Papua New Guinea

1331 - ***Plurella* sp.** * Philippines

1332 - ***Perophora namei*** * Indonesia

1333 - ***Botryllus* sp.** * Papua New Guinea

Indonesia is certainly distinctive, but can also be a bright yellow. A tiny nudibranch can be seen on the lower right of the ascidian, a fringe of white outlining its mantle. It is known from Indonesia and Port Moresby, Papua New Guinea.

1335 - *Botryllus* sp. * Styelidae * Stolidobranchia * Papua New Guinea * Duke of York Islands * Makada Reef * 50 ft (15 m). This is a beautiful example of delicate color combinations found in *Botryllus*.

1336 - *Botryllus* sp. * Styelidae * Stolidobranchia * Indonesia * Manado * 66 ft (20 m). The color combinations found in *Botryllus* are hard to imagine, and certainly serve to make the colonies stand out on the reef.

1337 - *Botryllus* sp. * Styelidae * Stolidobranchia * Philippines * Batangas * Pulang Buli * 33 ft (10 m). This picture shows another color form of *Botryllus*, although the few examples shown here hardly begin to document the geographic variation in the genus.

1338 - *Eusynstyela latericius* * Styelidae * Stolidobranchia * Indonesia * Manado * 40 ft (12 m). This ascidian encrusts tightly to rocky substrates, under coral heads and overhangs, and each zooid is encased in a leather-like tunic. In this individual growing on a dead branching coral, the zooids are only connected by a thin layer of tunic, and the siphons of each zooid are visible at the ends of the elongate zooid.

1339 - *Eusynstyela* or *Symplegma* sp. * Styelidae * Stolidobranchia * Palau * Babulukes * 66 ft (20 m). This colonial ascidian forms a tough sheet with each zooid closely locked into the overall structure. Notice the fine red ring which passes through both the siphons of each zooid. A *Phyllidia* nudibranch grazes the surface of the colony.

1340 - *Eusynstyela* sp. * Styelidae * Stolidobranchia * Palau * marine lake * 6 ft (2 m). This group of bright red ascidians is actually a colony, each zooid connected by at least a fine stolon to its neighbors. A yellow sponge occurs to one side in the photograph.

1341 - *Polycarpa argentata* * Styelidae * Stolidobranchia * Papua New Guinea * Madang * dock pilings * 3 ft (1 m). This solitary ascidian was growing on a dock in Madang. There is an orange *Botryllus* sp. growing over much the surface of the piling. These are good examples of the ability of ascidians to colonize (foul) man-made surfaces.

1342 - *Polycarpa aurata* * Styelidae * Stolidobranchia * Palau * lagoon reef * 16 ft (5 m). This ascidian is probably the most distinctive species found in the entire Pacific region. The yellow or orange tunic with purple veining certainly stands out. This species is very common, being found in nearly all reef environments where there is a hard surface to which it can attach.

1343 - *Polycarpa captiosa* * Styelidae * Stolidobranchia * Palau * Jellyfish Lake * 10 ft (3 m). This ascidian is known from Indonesia, Singapore, New Caledonia, Australia and Palau. In Palau it is common in marine lakes.

1344 - *Polycarpa contecta* * Styelidae * Stolidobranchia * Palau * Airai Dredge * 6 ft (2 m). This species forms dense groups of solitary individuals and is a common fouling organism in Micronesia.

1345 - *Polycarpa cryptocarpa* * Styelidae * Stolidobranchia * Palau * marine lake * 23 ft (7 m). The white speckled siphons are distinct and attractive in this species. This helps to identify the species as its tunic is often densely covered with other organisms. A number of *Didemnum molle* occur above and near the *P. cryptocarpa* and a second small green ascidian can also been seen in the photograph.

1346 - *Polycarpa papillata* * Styelidae * Stolidobranchia * Papua New Guinea * Madang * Rasch Passage * 50 ft (15 m). These small *Polycarpa* seem almost naked with a minimum of epibionts growing on them.

1347 - *Polycarpa* sp. * Styelidae * Stolidobranchia * Federated States of Micronesia * Chuuk Atoll * Dublon Island * 133 ft (40 m). This large dark ascidian has a black tunic which is usually heavily overgrown with other organisms. The orange ring around the siphons is the only thing which makes the ascidian stand out. This species is found exposed on deep reef dropoffs and in deep lagoon bottom areas.

1348 - *Polycarpa* sp. * Styelidae * Stolidobranchia * Papua New Guinea * Kavieng * Albatross Channel mouth * vertical wall * 33 ft (10 m). This is a beautiful solitary ascidian known only from Papua New Guinea at 30 to

1334 - ***Botryllus* sp.** * Indonesia

1335 - ***Botryllus* sp.** * Papua New Guinea

1336 - ***Botryllus* sp.** * Indonesia

1337 - ***Botryllus* sp.** * Philippines

1338 - ***Eusynstyela latericius?*** * Indonesia

1339 - ***Eusynstyela* or *Symplegma* sp.** * Palau

1340 - ***Eusynstyela* sp.** * Palau

1341 - ***Polycarpa argentata*** * Papua New Guinea

1342 - *Polycarpa aurata* * Palau

1343 - *Polycarpa captiosa* * Palau

1344 - *Polycarpa contecta* * Palau

1345 - *Polycarpa cryptocarpa* * Palau

90 foot depths. The white and gold colors of the siphons, contrasted with the dark tunic, make them stand out along a reef wall.

1349 - *Polycarpa* sp. * Styelidae * Stolidobranchia * Papua New Guinea * Kavieng * Albatross Channel * 66 ft (20 m). This large ascidian can attain the size of a coconut. It is largely buried in the sediment on the bottom of Albatross Channel near Kavieng. If the siphons contract, the ascidian looks exactly like an algae covered rock.

1350 - *Polycarpa* sp. *Styelidae * Stolidobranchia * Indonesia * Manado * 23 ft (7 m). Due to taxonomic problems within *Polycarpa*, it is not possible to assign a specific name to this species.

1351 - *Polycarpa* sp. * Styelidae * Stolidobranchia * Papua New Guinea * Eastern Fields * 10 ft (3 m). Because of the taxonomic confusion within *Polycarpa* it is impossible to put an accurate scientific name on this ascidian which occurred on the reef top at isolated Eastern Fields in the Coral Sea.

1352 - *Symplegma viride* * Styelidae * Stolidobranchia * Federated States of Micronesia * Chuuk Atoll * Basis Patch Reef * 26 ft (8 m). Many of the smaller ascidians are not immediately noticed by the average diver, but there is near endless variation and remarkable coloration found in the group. They are also a biological frontier as very little is known of the biology and interactions of the many species.

1353 - *Herdmania momus* * Pyuridae * Stolidobranchia * Indonesia * Manado * 33 ft (10 m). This species is found in all warm seas and it is a common fouling organism. Within weeks of setting up sea water aquariums in Chuuk, we had *Herdmania momus* growing all over the tanks.

1354 - *Pyura* sp. * Pyuridae * Stolidobranchia * Papua New Guinea * Madang * offshore reef * 33 ft (10 m). This ascidian was quite common on the shallower portions of a small reef called Planet Rock several miles offshore of Madang. The epibionts growing on the tunic make the ascidian blend in quite well with the organisms growing on the rocky bottom and it is the siphons which easily reveal the presence of the ascidians.

1346 - *Polycarpa papillata* * Papua New Guinea

1347 - *Polycarpa* sp. * Federated States of Micronesia

1348 - *Polycarpa* sp. * Papua New Guinea

1349 - *Polycarpa* sp. * Papua New Guinea

1350 - *Polycarpa* sp. * Indonesia

1351 - *Polycarpa* sp. * Papua New Guinea

1352 - *Symplegma viride* * Federated States of Micronesia

1353 - *Herdmania momus* * Indonesia

1354 - *Pyura* sp. * Papua New Guinea

SUGGESTIONS FOR FURTHER READING

Introduction.

Coral Reefs:

Allen, Gerald R. and Roger Steene. 1994. Indo-Pacific Coral Reef Field Guide, Tropical Reef Research, Singapore, 378 pp.

Ryan, Paddy. 1994. Snorkleller's Guide to the Coral Reef. University of Hawaii Press, Honolulu, 184 pp.

Steene, R. 1990. Coral Reefs Natures Richest Realm, Mallard Press, New York, 336 pp.

Micronesia:

Devaney, D.M., E.S. Reese, B.L. Burch and P. Helfrich. 1987, The Natural History of Enewetak Atoll. U.S. Dept. of Energy (DOE/EV/0073-T1), 2 vols.

Faulkner, D. 1974, This Living Reef, Quadrangle/The New York Times Book Co., New York, 183 pp.

Myers, R.F. 1989. Micronesian Reef Fishes. Coral Graphics, Guam, 299 pp.

While dealing with the fishes of the Micronesia, this book has an excellent introduction to the area.

New Caledonia:

Laboute, P., M. Fuega and R. Garndperrin. 1991. Le Plus Deau Lagon du Monde, Editions Alizes, Noumea, New Caledonia, 272 pp.

Laboute, P. and Y. Magnier. 1979. Underwater guide to New Caledonia, les Editions du Pacifique, 160 pp.

The Great Barrier Reef

Bennett, Isobel. 1988. The Great Barrier Reef, Lansdowne Press, Sydney, 184 pp.

Mather, P. and I. Bennett. 1993. Coral Reef Handbook, A Guide to the Geology, Flora and Fauna of the Great Barrier Reef. Surrey Beatty and Sons, Norton, NSW, Australia, 4th ed. 264 pp.

Reader's Digest. 1983. Reader's Digest Book of the Great Barrier Reef, Reader's Digest, Sydney, 384 pp.

Hawaiian Islands

Devaney, Dennis M. and Lucius G. Eldredge (eds). Reef and Shore Fauna of Hawaii, Section 1. Protozoa through Ctenophora; Section 2. Platyhelminthes through Phoronida; Section 3. Sipncula through Annelida; Section 4. Mollusca. Bishop Museum Special Publication 64 (1-4).

Fielding, Ann and Ed Robinson. 1987. An Underwater Guide to Hawai'i, University of Hawaii Press, Honolulu, 156 pp.

Hobson, Edmund and E.H. Chave. 1990. Hawaiian Reef Animals, University of Hawaii Press, Honolulu, 137 pp.

Hong Kong and South China Sea:

Morton, Brian and John Morton. 1983. The Seashore Ecology of Hong Kong, Hong Kong University Press, 350 pp.

Phylum Porifera

Hooper, J.N.A. and F. Wiedenmayer. 1994. Zoological Catalogue of Australia Volume 12 Porifera, CSIRO Melbourne, 624 pp.

Phylum Cnidaria and Ctenophora

Faulkner, D. and R. Chesher. 1979. Living Corals, Clarkson N. Potter, New York, 311 pp.

Fautin, Daphne G. and Gerald R. Allen. 1992. Field Guide to Anemonefishes and their Host Sea Anemones. Western Australian Museum Press, 160 pp.

Veron, J.E.N. 1986. Corals of Australia and the Indo-Pacific, Angus and Robertson Publ., 644 pp. (reissued in 1992 by University of Hawaii Press).

Worms

Stephen, A.D. and S.J. Edmonds. 1972. The phyla Sipuncula and Echiura. British Museum Natural History, London.

Fauchaud, K. 1977. The polychaete worms. Natural History Museum of Los Angles County, Science Series 28: 188 pp.

Molluscs

Bertsch, H. and S. Johnson. 1981. Hawaiian nudibranchs, Oriental Publishing Co, Hawaii, 112 pp.

Burgess, C.M. 1970. The Living Cowries, A.S. Barnes and Co., Cranbury, New Jersey, 389 pp.

Hinton, Alan. 1972. Shells of New Guinea and the Central Indo-Pacific. The Jacaranda Press, Milton, Qld. Australia, 94 pp.

Pechar, Chris, Chris Prior and Brian Parkinson. no date. Mitre Shells from the Pacific and Indian Oceans. Robert Brown and Associates, Australia, no pagination.

Radwin, G.E. and A. D'Attilio. 1976. Murex Shells of the World. Stanford University Press, 284 pp.

Willan, Richard C. and Neville Coleman. 1984. Nudibranchs of Australasia. Australasian Marine Photographic Index, Sydney, Australia

Arthropods- Crustaceans

Debelius, Helmut, 1983. Armoured Knights of the Sea. Kernen Verlag, Essen, 120 pp.

Lophophorate Phyla.

Emig, C.C. 1982. The biology of Phoronida. Adv. Mar. Biol. 19:1-89.

Soule, D.F. and J.D. Soule. 1976. The status of faunistic information on tropical reef bryozoans. Micronesica 12: 157-164.

Echinoderms

Guille, Alain, Pierre LaBoute and Jean-Louis Menou. 1986. Guide des etoiles de mer, oursins et autres echinodermes du lagon de Nouvelle-Caledonie, Editions de l'ORSTOM, Paris, 238 pp.

Chordates- Ascidians

Monniot, C., F. Monniot and P. Laboute. 1991. Coral Reef Ascidians of New Caledonia, Editions de l'ORSTOM, Paris, 248 pp.

This book is an excellent source for information about the biology of coral reef ascidians.

INDEX - Genus and species

Charles Arneson is Executive Director of the Coral Reef Research Foundation. He had his first experiences with tropical marine life while assisting scientists at the University of Puerto Rico's marine laboratory during the summers between college semesters. He obtained a Master's degree from the University of Puerto Rico in 1976 and made his first visit to the tropical Pacific at Enewetak in 1979. He subsequently was employed at the Scripps Institution of Oceanography where he gained a reputation as an outstanding naturalist. More recently, he operated live abroard dive boats in the western Pacific. He specializes in cnidarian biology of sea anemones and jellyfishes and marine bioluminescence and has authored numerous scientific papers in those fields. His underwater photographs have been published in many books and magazines.

Patrick Colin is President of the Coral Reef Research Foundation and has been in love with the tropical ocean since seeing, at age 12, coral reef fishes in aquariums in his native midwest. He received a Ph.D. from the University of Miami in 1973 while working on the biology of western Atlantic reef fishes. He has since lived in Puerto Rico, the Marshall Islands, Papua New Guinea, the Federated States of Micronesia and more recently Palau. His work has taken him to tropical reef areas around the world and encompasses a wide range of topics, from the taxonomy and reproductive biology of reef fishes to the relationships between sediments and organisms. His 20 years of experience in the tropical Pacific has resulted in an intmate aquaintance with and admiration for the creatures of this region. He has written over 60 scientific papers, plus two books, on tropical marine life.